Exterior Ballistics with Applications

Third Edition

EXTERIOR BALLISTICS WITH APPLICATIONS

Skydiving, Parachute Fall, Flying Fragments

THIRD EDITION

GJERGJ KLIMI

This book was printed in the United States of America.

The front cover displays a collage of the photos of the sculptures of J. Paco (1914-1991), which are in the private collection of Paco's family, and George & Dorina Klimi's.

Exterior Ballistics with Applications: Skydiving, Parachute Fall, Flying Fragments, Third Edition, is updated in September 2011.

To order additional copies of this book, contact:
Xlibris Corporation
1-888-795-4274
www.Xlibris.com
Orders@Xlibris.com
46607

CONTENTS

To my adorable wife Dorina,
and my sons Ivi and Erio

To my mother Violeta.

To my lovely friends:
Tom and Ellen Moore and their son Douglas,
and William and Brigid Sullivan

In memory of
my father Vlash,
and my father-in-law, the great sculptor Janaq Paco.

A WORD FROM THE AUTHOR

What's New

Together with corrections and alterations, the 2011 improved and updated edition of *Exterior Ballistics with Applications* contains new original methods and techniques for the solution of exterior ballistics problems for small arms as well as new applications and exercises.

Thus, we have introduced a new original method to determine the function of resistance of a bullet using the existing range tables or the data (velocity as a function of range) from the shooting tests.

Another innovative method that allows us to find the ballistics elements of the trajectory of bullet flight analytically using an algebraic equation of the trajectory of flight is also presented in the book (co-author James A. Boatright).

The new methods add to the already existing exterior ballistics methods of study new techniques to solve the problems of exterior ballistics without measuring experimentally the form coefficient and the ballistics coefficient, BC.

Acknowledgments

My gratitude and sincere thanks go to Prof. James Lewis (Marquette University) for the invaluable technical review, comments, suggestions, corrections and encouragement to update the Exterior Ballistics with Applications.

Prof. Lewis thoughtful feedbacks and his writings presented in the book were a great contribution to improve the book, and I again express my high esteem.

I would like to thank James A. Boatright (retired aerospace engineer) for his co-authorship work in the article *A Contemporary Method in Exterior Ballistics* presented in the book.

George Klimi
New York, September 2011

Prof. James Lewis writes:
"I've come late to your text, *Exterior Ballistics with Applications,* so perhaps the errata suggested have been previously catalogued".
"At the end of these suggested errata I express comments which may be of help in future editions, and I express a conundrum which I desire you to address".

PREFACE

*E*xterior Ballistics with Applications: Skydiving, Parachute Fall, and Flying Fragments is an exterior ballistics book, but it can be considered as an interdisciplinary applied mathematics and physics book for the vast mathematics and physics techniques and concepts employed. It is a wonderful source for applications in calculus, differential equations, numerical methods, physics and PC programming as well.

Exterior Ballistics with Applications is addressed to readers interested in exterior ballistics such as the military students and scholars, army personnel, and ballistics experts, as well as amateurs and professionals interested in problems of flying projectiles or projectile fragments, skydiving and paratroops, sport shooting, hunting, etc

It might be appealing to a broader audience: mathematics or physics students, and faculties; graduate students and scholars engaged in the field of wound ballistics, engineering, and military applications.

Exterior Ballistics with Applications is a unique book in the literature of exterior ballistics for the innovative approach to exterior ballistics problems, for the simplicity of mathematics and physics techniques and models employed, and for the huge amount of problems of exterior ballistics that are approached and solved analytically and numerically, with the use of PC programs as well.

The book methodically presents the problem of motion of projectiles in presence of air resistance, employing a variety of ballistics methods, as well as mathematical and physics models to analyze and solve the differential equations of projectile flight with satisfying practical accuracy.

The book uses a modern and simple analytical **Siacci's function** of air resistance that allows us to update and simplify the analytical methods for solving the two main problems of exterior ballistics:

- For a given projectile launched with a known speed, determine the launching angle needed to hit a target located at a known range, as well as all the other elements of the projectile trajectory.
- Determine the range and all the other trajectory elements for a given projectile launched with a given angle at a given muzzle velocity.

A characteristic of the book is the use of different approaches to solve the differential equations of projectile motion—among them the Siacci method and the numerical integration.

The results obtained through integration of equations of projectile flight are mostly analytical formulas that describe the projectile trajectory and are necessary to make the exterior ballistics a comprehensible science.

Employing the analytical formulas, we are able to obtain very accurate results for the elements of projectile trajectory that are close to the results given in the range tables of different artillery or infantry firearms.

The Siacci method we employ to study the projectile motion is based entirely on the integration of differential equations of projectile flight, avoiding the use of different tabulated functions (Siacci's, Ingall's, I. D. Lowry's, or G-function tables) that most of the modern ballistics literature still employ.

Though we use the Siacci solution method, there is no need for the tabulated Siacci functions that are still used in most of contemporary literature. But for the ballisticians who are familiar and prefer to use the tabulated functions to solve the ballistics problems, we give the tabulated data of the modern Siacci analytical functions in use in this book.

Differential Equations of Projectile Flight are numerically integrated using some original PC programs that can be easily modified to be used in similar or other new scenarios. There are around 19 QBasic PC programs that accompany the book and give the reader the possibility to solve a great variety of exterior ballistics problems.

Each section of the book is illustrated with solved examples that demonstrate the use of formulas or the solution methods obtained theoretically to find the elements of the projectile trajectories.

Most of the exercises present practical ballistics problems related to different artillery and infantry firearms, ammunitions, and other projectiles.

For computation, at least a graphing calculator (TI-83 Plus) is needed and some knowledge in numerical methods and programming to enjoy the automatic PC solutions and the methods used in preparation of the PC programs.

The solutions of the ballistics problems are compared most of the time with outcomes given by different authors or with the results obtained by numeric integration of differential equations of projectile flight using PC programs.

Actually there are around 140 solved ballistics problems throughout the book.

The book includes two interesting topics as well that can be considered as applications of exterior ballistics:

- Skydiving and parachute falling related with the trajectory of a parachutist launched from a horizontally flying airplane with undeployed parachute in different meteorological conditions and in presence of air resistance and wind.
- The ballistics of fragments is an important element of **terminal ballistics** necessary to study the effectiveness of fragmentation ammunitions on the personnel, as well as other problems related to forensic sciences.

The approach to the skydiving and parachute fall in different meteorological conditions presents an in-depth study that better reflects the practice of skydiving and is far from the simplified approach contained in physics textbooks.

Introduced as well is a PC program to find the coordinates of the skydiver at any moment during the flight and the landing coordinates of the parachutist as well.

The reader can "jump" to the ninth chapter (skydiving), ignoring the study of preceding chapters, though some knowledge they contain are essential for understanding the fall of the parachutist.

Interesting results for the motion of the projectile fragments, launched from the detonation of antipersonnel ammunitions, are obtained as applications of the ballistics methods already developed for regular symmetrical projectiles.

Gjergj (George) Klimi, PhD
Pace University and City Tech, New York
Email: gklimi@pace.edu; Iven24@aol.com
November 2007

ACKNOWLEDGMENTS

I would like to thank my colleagues and friends for their invaluable assistance to the preparation of the PC programs:

- Col. Genc Kokoshi (Military Academy of Tirana), Guido LoVechio, PhD, Carlo Ferrario, PhD, and Dr. Marco Merli (University of Ferrara, Italy)

I acknowledge with profound gratitude Edoardo Mori, Esq., for the generous permission to adapt some of the tables presented in his wonderful book *Balistica Teorica e Practica* shown at htpp://earmi.it/ballistics.

A special thanks to my professors and colleagues:

- Aleko Minga, PhD, Mina Naqo, PhD, and Floran Vila, PhD, at Tirana University for their encouragement and suggestions in writing the book.

A special depth of gratitude to the following for their efforts and generous support that made possible for me in 2000 to come to the wonderful and welcoming City of New York and be a teacher in Mathematics Department at Pace University:

- My best friends Brigid and William Sullivan, Thomas and Ellen Moore, and Blanche Fierstein.
- My colleagues Prof. Marilyn Jaffe-Ruiz, John Sharkey, PhD, and Emeritus Faculty Irwin Kabus. (Pace University).

I would like to thank the professionals at Xlibris: Lynn Moore, Michael Bonghanoy, Donald Villamero, Pierre Pobre, Sherwin Soy, Manolito Bastasa, Floramie Tuastomban, Ethan Maghuyop, Riki Sayon (2011 edition) for the help they gave in publishing the book.

INTRODUCTION

Exterior Ballistics—A Modern Approach

1. The Object of Exterior Ballistics

P rojectile motion is the motion of an object, usually a regular symmetrical body (artillery shell, rocket, bullet, etc.), thrown into the air at an angle above or below the horizontal line. The path followed by the projectile during its flight in air is the ballistics trajectory of the projectile.

The study of projectile motion and its ballistics trajectories is the object of exterior ballistics, which aims to give physical and mathematical models for the solution of problems related with the motion of projectiles.

The final scope of the projectile study is the compilation of accurate range tables. For that reason, the exterior ballistics elaborates special methods to solve a set of differential equations of the projectile motion that are the result of Newton's second law of dynamics and to obtain data necessary to create range tables.

The solution of the differential equations of the projectile motion considering all factors that influence the flight of a projectile is a very complicated problem.

In general, the differential equations of flight describe the motion of a projectile launched in air with a known initial velocity $\vec{v}_0$ considering the gravity, the air resistance and all the other relevant factors related with the ballistics characteristics of the given projectile, as well as the characteristics of the atmosphere (for example, the shape and the diameter of a projectile, the temperature of air, the wind, the air density and the speed of sound, etc.).

The solutions of the differential equations of a projectile considering almost all "significant factors" are not analytical functions. That is the reason we use the numerical integration of the equations of flight, a procedure that accumulates numerical errors as the solution procedure evolves. Though the use of computers has simplified the integration of the differential equations of projectile flight, it is almost impossible to have absolutely exact results because it is not possible to know in advance accurately all the significant factors that influence at any moment the projectile flight.

Nevertheless, exterior ballistics provides special methods to analyze and solve, with satisfying practical accuracy, the problems related with the flight of projectiles in air.

2.   A Modern Approach

Precisely, the book *Exterior Ballistics with Applications* is a modern approach to introduce the basics of exterior ballistics and its methods starting from the simple ideal model of the projectile motion to the automatic integration of the differential equations of projectile flight using the PC programs.

The Ideal Model

One of the most familiar illustrations of physics and mathematical concepts and methods in physics and mathematics textbooks is related with the ideal model of projectile motion assuming a constant gravity and the absence of air resistance.

In general, the results obtained for the ideal flight of the projectile are far from the outcomes observed in practice for the motion of bodies thrown in air with relatively high speeds (over 30–50 m/s).

The ideal model approach gives approximate results for the motion of projectiles thrown in the air with relatively low speeds, 30–50 m/s (98.4–164 ft./s.).

The ideal model of the projectile flight is treated briefly in the first chapter of the book for the sake of tradition and to reach some valuable conclusions that are used later for the projectile flight in the presence of drag.

The Equations of Projectile Motion

The second chapter deals with the fundamental concepts and the basic equations of the exterior ballistics, which are indispensable to understand the methods of exterior ballistics and the brilliance of the founders of the modern exterior ballistics: Mayevski, Zaboudski, Siacci, Garnier (the Gâvre Commission), Ingalls, etc.

The study of motion of a projectile in air, as a first approach, is modeled by **a system of differential equations** that is the result of Newton's second law of mechanics, which considers only the gravity and the drag force exerted on the (point-mass) projectile flying in **standard atmosphere** but temporarily neglects all other factors that influence the projectile flight.

In this chapter, a modern, simple analytical **Siacci function of resistance** is presented, which allows us to solve throughout the rest of the book a large variety of ballistics problems relatively easy in an analytical way, avoiding the traditional use of the enormous amount of tabulated data of different drag functions that are still employed nowadays.

Elaborated as well are the methods we use to determine the Siacci "ballistics coefficients" and "form coefficients" of projectiles that are necessary to estimate the projectile trajectory elements, such as the launching angle, the horizontal range, the time of flight, the maximum altitude, and so on.

Shown are some original ways to estimate the Siacci ballistics coefficient and the form coefficient of a projectile when the Ingalls or Mayevski ballistics coefficient of that projectile are known.

Approximate Solutions

In the third and fourth chapters, we study the motion of projectiles launched in air respectively with relatively high speeds (greater than 256 m/s) and low speeds (less than 256 m/s).

We show some different approximate solution methods of differential equations that describe the projectile flight in standard atmosphere considering the projectile a point of mass.

The solutions of the systems of differential equations of the projectile flight, obtained using the model of a point-mass projectile, are analytical formulas that enable us to get very approximate results for the elements of trajectories of projectiles flying in standard atmosphere for some particular cases, such as for the flight of projectiles for relatively narrow launching angles or for uphill and downhill shooting and when the time of flight is relatively short.

The theoretical outcomes are close enough to practical results observed in infantry and artillery fire and other related fields, such as in the flight of bullets, artillery shells, fragments of antipersonnel ammunitions, skydiving, paratroops, etc.

The approximate solution methods of ballistics problems developed here are simple mathematical models and interesting applications of different techniques of calculus and differential equations.

Siacci's Method

In chapter 5, introduced is the outstanding Siacci method that has been successfully employed since the 1880s to solve complicated ballistics problems.

Using the Siacci concept of the "pseudo-speed" of the projectile, based on the "correction coefficient" and the modern Siacci function of resistance that we have shown in preceding chapters, we obtain original analytical solutions of the equations of flights that represent a set of formulas. Employing those formulas, we are able to find the main elements of the trajectory of the projectiles without employing the tabulated Siacci functions that are indispensable in the traditional Siacci method.

As a summery, chapters 3 to 5 enables us to use simplified models and different theoretical methods to solve analytically and with great practical accuracy many ballistics problem for any projectile flying in standard atmosphere.

Theory of Corrections

In chapter 6, we consider the deviations of the atmospheric air conditions from the standard atmosphere, the influence of the range wind and the crosswind, as well as the deviation of the projectile characteristics from the nominal data to "correct" the projectile trajectory elements already obtained for that projectile in standard atmosphere.

Note that for the unguided projectiles, it is not possible to make corrections in their trajectory as they fly since the path of projectile is already determined by the launching initial conditions.

We use the Siacci correction method to find the elements of the trajectory in nonstandard atmosphere, to solve the main problems of Exterior Ballistics, and to create the range tables using a theoretical and experimental approach.

The study of a projectile as a rotating body that is related with the six degrees of freedom differential equations is beyond the scope of this book. The six degrees of freedom projectile models are useful for in-depth studies by ballistics experts. According to R. L. McCoy, those models are not needed to solve the traditional exterior ballistics problems.[1]

Anyway, assuming the projectile is point-mass and considering the practical results obtained during firearms tests—we make some necessary "corrections" in the coefficient of projectile form—and then solving the differential equations of the point-mass projectile, we are able to give accurate data for use in the practice of shooting of infantry and artillery firearms, in whatever atmospheric conditions and fire situations (uphill, downhill, etc.).

---

[1] McCoy, Robert L. *Modern Exterior Ballistics*, p. 212. Schiffer Publishing Ltd., 1999.

The Range Tables

The seventh chapter deals with the main task of the exterior ballistics: the compilation of range tables for a firearm and the corresponding projectile(s).

Range tables for a particular firearm allow the personnel to set the launching (or projection) angle to fire and hit a given target with an acceptable accuracy. They are very useful to solve problems encountered in the practice of shooting.

Shown are two theoretical-practical models to construct the range tables:

- The Siacci method that considers the ballistics coefficient a function of launching angle
- The method of numerical integration of differential equations of flight that considers the ballistics coefficient a function of the projectile speed

The methods are illustrated with examples using different firearm projectiles.

Projectile Trajectory PC Programs

In chapter 8, we present again the effects of meteorological factors in the flight of projectiles. Here we study the variation of air density and pressure with altitude over the ground, as well as the effect of humidity of air in the flight of projectile that is reflected in the so-called virtual temperature.

The Siacci function of resistance and the density function are modified to take into account the "virtual temperature" of atmospheric air and the effect of the speed of sound.

Using the modified Siacci function and the density function, we set up the differential equations of projectile flight necessary to solve the ballistics problems using numerical integration and PC programs.

The solution of the differential equations of projectile flight is presented in this chapter with the introduction of some original PC programs that consider the main characteristics of the projectile and the atmospheric factors, including the wind.

The use of PC programs is a great technique that allows as to automaticaslly calculate the elements of the trajectory in real time and with great accuracy.

Skydiving and Parachute Fall

In chapter 9, is included the study of skydiving related with the trajectory of a parachutist launched from a horizontally flying airplane with undeployed parachute.

There are considered three typical problems of skydiving and parachute fall:

- The fall of a skydiver from the flying airplane until the moment the skydiver opens the parachute
- The flight of the skydiver/parachutist with parachute from the moment he deploys the parachute till he lands on the ground
- The flight of the parachutist jumping from a hovering helicopter or a tower

The concept of "terminal speed" of skydiving (parachuting) is further elaborated to reflect the practice of skydiving that is somehow ignored in physics.

The original approach and the use of the PC programs make it possible to find with good accuracy the coordinates of the skydiver at any moment during the flight, considering the influence of wind and other atmospheric factors in the trajectory of flight, and the landing of the parachutist.

Ballistics of Projectile Fragments

In the last chapter (chapter 10), we apply some of the results of the third and fourth chapters to study the problems related with the ballistics of fragments (irregular or spherical) of antipersonnel ammunitions and antiaircraft fragmentation projectiles (mines, grenades, artillery fragmentation ammunitions, etc.).

The ballistics of fragments is an important element of terminal ballistics necessary to study the effectiveness of fragmentation ammunitions on the personnel, as well as the problems of terminal ballistics, or other problems related with forensic sciences.

Illustration Examples

Each chapter is illustrated with solved ballistics problems that serve as models to demonstrate the way to use the theoretical results obtained integrating the differential equations of projectile motion and the use of other ballistics methods.

To show the validity and accuracy of the different methods used in different problems of exterior ballistics, the problem solutions obtained in the examples are compared most of the time with outcomes given by different authors or with the outcomes that can be found in range tables of corresponding projectiles we have considered for problem illustrations.

Sometimes a problem is solved using two different methods. The results obtained using the Siacci methods, or other approximate methods, are sometimes compared with the results obtained using a PC program.

Actually there are around 140 solved ballistics problems throughout the book.

Almost all examples represent ballistics problems related with firearm projectiles.

1

The Motion of Projectile in Free Space

Introduction

One of the most familiar and well-known problems used to illustrate some concepts and methods in physics and mathematics is related to the motion of a projectile in the so-called "ideal" atmosphere where the resistance of air is ignored and the only force acting on the projectile during the flight is the gravitational force.

The projectile motion in ideal atmosphere is restricted to objects thrown near the earth's surface with relatively low speeds. The gravity is considered to be constant since the distance traveled and the maximum height of the projectiles above the earth are short compared to the earth's radius. Such model is ideal and not really observed in practice.

In general, the results obtained using the ideal model are far from the outcomes observed in practice for the motion of bodies thrown in the air with relatively high speeds (over 30–50 m/s). However, the model of the ideal motion of projectiles can be employed for motions of projectiles that are observed in practice when they are thrown in the air with relatively low speeds (some 10 m/s) or for projectiles traveling in great altitudes where the density of air is low and air resistance can be neglected.

The ideal model of the projectile flight is treated briefly in this chapter of the book for the sake of tradition and to reach some valuable conclusions that are used later for projectile flight in presence of drag.

1.1 The Equations of Projectile Motion in Free Space

To study the motion of a projectile in ideal atmosphere, we consider a projectile with mass m thrown with initial velocity $\vec{v}_0$ at an angle α_0 above the horizon in a space where the only force acting on the projectile is the gravitational attraction.

The launching angle α_0 represents the angle between the initial velocity $\vec{v}_0$ and the positive direction of x-axis (figure 1).

We assume the following:

- The earth's surface is considered plane.
- The projectile is a particle with constant mass m.
- The force acting on projectile at any moment after it is launched in the air is the gravitational force $\vec{F}=m\vec{g}$.
- The gravitational acceleration $\vec{g}$ is constant, perpendicular to the ground. Its value is g =9.80665m/s (g = 32.17405ft/s).

Under such conditions, we say that a projectile moves in free space.

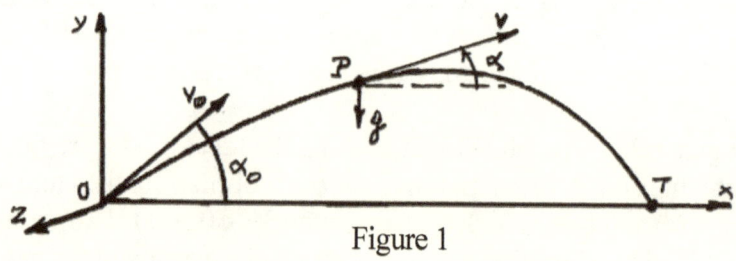

Figure 1

To analyze the motion of the projectile, we draw a three-dimensional rectangular coordinate system such that the x-axis is along the horizon while y-axis is drawn perpendicular to x-axis and is in the same plane with the initial velocity $\vec{v}_0$ and x-axis.

Since the components of the initial velocity $\vec{v}_0$ along x-, y-, and z-axes are respectively

$$v_{0x}=v_0\cos\alpha_0, \qquad v_{0y}=v_0\sin\alpha_0, \qquad v_{0z}=0, \qquad (1.1.1)$$

the initial velocity of projectile at the time of launch ($t = 0$) is

$$\vec{v}_0=v_0\cos\alpha_0\vec{i}+v_0\sin\alpha_0\vec{j}, \qquad (1.1.2)$$

where v_0 is the initial speed, $\vec{i}$ and $\vec{j}$ are the unit vectors along x- and y-axes respectively.

The study of motion of projectiles in free space is governed by the Newton's second law of mechanics,

$$\vec{F}=m\frac{d\vec{v}}{dt}, \qquad (1.1.3)$$

where $\vec{v}=d\vec{r}/dt$ is the velocity of the projectile at a time t, while $\vec{r}=\vec{r}(t)$ is its position vector.

Substituting the gravitational force $\vec{F}=m\vec{g}$, which acts on the projectile, in the second law of Newton (1.1.3), we obtain the differential equation of motion

$$m\frac{d\vec{v}}{dt}=m\vec{g},$$

which can be written in the form

$$\frac{d\vec{v}}{dt}=\vec{g}. \qquad (1.1.4)$$

1.2 The Projectile Velocity

Integrating the differential equation of the projectile motion (1.1.4) for the initial conditions ($t=0$, $\vec{v}=\vec{v}_0$), we obtain the velocity of projectile at any time t during the flight

$$\vec{v}=\vec{v}_0+\vec{g}t. \qquad (1.2.1)$$

Since the components of the gravitational acceleration $\vec{g}$ along x-, y-, and z-axes are 0, -g, and 0 respectively, we can write,

$$\vec{g}=-g\vec{j} \ .$$

(1.2.2)

Substituting (1.1.2) and (1.2.1) in (1.2.2) for the velocity of projectile, we obtain

$$\vec{v}=(v_0\cos\alpha_0)\vec{i}+(v_0\sin\alpha_0-gt)\vec{j} \ .$$

(1.2.3)

Hence, for the components of the projectile velocity at any time t during the flight, we obtain the parametric equations

$$v_x=v_0\cos\alpha_0, \quad v_y=v_0\sin\alpha_0-gt \ , \quad v_z=0 \ .$$

(1.2.4)

From (1.2.4), for the speed v of the projectile at a time t, we can write

$$\vec{v}^2=[(v_0\cos\alpha_0)\vec{i}+(v_0\sin\alpha_0-gt)\vec{j}]^2=v_0^2-2g(v_0\sin\alpha_0 t-\tfrac{1}{2}gt^2)$$

Thus, for the projectile speed, we have

$$v^2=v_0^2-2g(v_0\sin\alpha_0 t-\tfrac{1}{2}gt^2) \ .$$

(1.2.5)

The direction of velocity at any time t is determined by the initial angle α_0 the velocity vector $\vec{v}$ makes with x-axis. For the tangent of the angle α that the projectile velocity forms with x-axis, we can write

$$\tan\alpha=\frac{v_y}{v_x} \ .$$

Substituting v_y and v_x from (1.2.4) in the above formula, we find that

$$\tan\alpha=\frac{v_0\sin\alpha_0-gt}{v_0\cos\alpha_0} \ .$$

(1.2.6)

From (1.2.4), we see that the horizontal component of the velocity $v_x = v_0 \cos\alpha_0$ remains constant during the projectile flight while the vertical component $v_y = v_0 \sin\alpha_0 - gt$ changes linearly with time t.

1.3 The Parametric Equations of Projectile Motion

To find the equation of the projectile trajectory, we substitute the velocity $v = d\vec{r}/dt$ in (1.2.1). We can write

$$\frac{d\vec{r}}{dt} = \vec{v}_0 + \vec{g}t \quad .$$

$$(1.3.1)$$

Integrating with respect to t, on condition that at time $t = 0$, the projectile is at the origin of the coordinates (position vector $\vec{r}_0 = 0$), we obtain the vector equation of the projectile motion

$$\vec{r} = \vec{v}_0 t + \frac{1}{2}\vec{g}t^2 \quad .$$

$$(1.3.2)$$

The components of the position vector $\vec{r}$ of the given projectile along x-, y-, and z-axes at any time t are respectively

$$x = (v_0 \cos\alpha_0)t \quad , \quad y = (v_0 \sin\alpha_0)t - \frac{1}{2}gt^2 \quad , \quad z = 0 \quad .$$

$$(1.3.3)$$

Equations (1.3.3) are the parametric equations of projectile motion in free space.

Remark. If the initial position of the projectile, thrown with velocity $\vec{v}_0$ at an angle α_0 with the horizon, will be at an altitude h above the horizon ($y_0 = h$), i.e., at $t = 0$ the position vector is $\vec{r}_0 = y_0\vec{j}$, then integrating (1.3.1), we obtain

$$\vec{r} = \vec{r}_0 + \vec{v}_0 t + \frac{1}{2}\vec{g}t^2 \quad .$$

$$(1.3.4)$$

Hence, for the parametric equations of the projectile flight, we have

$$x=(v_0\cos\alpha_0)t \quad y=y_0+(v_0\sin\alpha_0)t-\frac{1}{2}gt^2 \quad z=0 \tag{1.3.5}$$

Parametric equations (1.3.3) or (1.3.5) determine the coordinates of a projectile in terms of the initial speed and the launching angle at any time t during the flight, i.e., they describe the projectile trajectory.

The trajectory of a projectile in motion is a geometric characteristic of the flight.

Note that at any time t during the flight

- the horizontal component of the projectile velocity $\vec{v}$, given in (1.2.4), during the time of flight remains constant and equal to $v_x=v_0\cos\alpha_0$;
- the component of the velocity along z-axis at any time is zero, $v_z=0$; and
- the component z of the position vector $\vec{r}$ along z-axis remains zero, $z=0$.

Since the component of the position vector along z-axis at any time t is $z=0$, we conclude that

- the motion of projectile is plane. The projectile trajectory is on the launching plane (xy-plane).

We also note that

- the mass of projectile is not present either in the projectile velocity formulas or in the formulas that describe the position of the projectile during the flight.

This means that the mass m of the projectile does not have an effect on the projectile motion in ideal atmosphere. In other words, the velocity and the position of the projectile in motion in free space—

hence, the projectile trajectory—does not depend on the projectile mass.

The trajectory and the velocity of a projectile in motion are determined only by the initial position and initial velocity (launching speed and launching angle).

1.4 The Equation of the Projectile Trajectory

The parametric equations of the projectile trajectory determine at any time the position of the projectile in free space. For practical applications, it is convenient to express the projectile path through the coordinates of the projectile.

Solving for t the first equation of (1.3.3) and then replacing the expression obtained for t in the second equation of (1.3.3), we obtain the well-known "equation of projectile trajectory in free space"

$$y=(\tan\alpha_0)x-(\frac{g}{2v_0^2\cos^2\alpha_0})x^2 . \tag{1.4.1}$$

Equation (1.4.1) shows that the trajectory of a projectile in free space is a parabola.

Substituting the second equation of (1.3.3) in equation (1.2.5), we can express the speed of the projectile through its y-coordinate.

$$v^2=v_0^2-2gy , \tag{1.4.2}$$

where y can be computed employing (1.3.3) or (1.4.1).

Differentiating the equation of the projectile trajectory (1.4.1) with respect to x and taking into consideration that $dy/dx = \tan\alpha$, we obtain another relation for the angle α the projectile velocity makes with x axis

$$\tan\alpha=\tan\alpha_0-(\frac{g}{v_0^2\cos^2\alpha_0})x . \tag{1.4.3}$$

1.5 The Impulse of the Launching Force

At the launching point, the "thrusting mechanism" gives the projectile a momentum and energy needed to overcome the work of gravitational force $\vec{F}=-mg\vec{j}$.

From the formulas of the velocity vector (1.2.1) and position vector (1.3.2), we conclude that

- Projectiles of different masses with the same launching velocity $\vec{v}_0$ and initial position vector $\vec{r}_0$ have identical trajectories.

On the other hand, the momentum of a projectile,

$$\vec{p}=m\vec{v}_0 ,$$
(1.5.1)

launched at the same point with the same initial velocity depends on the projectile mass.

Thus, the impulse of force $\vec{f}$ needed to be applied on a projectile by the launching mechanism during the launch time τ is

$$\int_0^\tau \vec{f}dt=m\vec{v}_0 .$$
(1.5.2)

The average thrust-force,

$$\vec{f}_{ave.}=\frac{m\vec{v}_0}{\tau} ,$$
(1.5.3)

depends on the projectile mass m. That means that to give to two projectiles of different masses the same initial velocity $\vec{v}_0$, the launching mechanism needs to apply different average thrust forces; the average thrust force is greater for the projectile with greater mass m.

1.6 The Energy and Momentum of a Projectile

While the trajectories and kinematics properties of projectiles thrown with the same initial velocity in free space do not depend in their masses, the dynamic properties (momentum, kinetic energy) of projectiles that have identical trajectories but different masses are different.

The work $W = \int_{\vec{r}_0}^{\vec{r}} \vec{f} \cdot d\vec{r}$ done by a given force $\vec{f}$ acting on a particle of mass m is equal to the change in kinetic energy ($E_k = mv^2/2$)

$$\int_0^{\vec{r}} \vec{f} d\vec{r} = \frac{mv^2}{2} - \frac{mv_0^2}{2} . \qquad (1.6.1)$$

Substituting in (1.6.1) the gravitational force $\vec{f} = -mg\vec{j}$ acting on the projectile, we find that

$$-mg\vec{j} \cdot \vec{r} = \frac{mv^2}{2} - \frac{mv_0^2}{2} .$$

Since

$$\vec{r} = x\vec{i} + y\vec{j} + 0\vec{k} ,$$

from the above equation, we find that the change in kinetic energy of the projectile is equal to the change in potential energy

$$\frac{mv^2}{2} - \frac{mv_0^2}{2} = -mgy . \qquad (1.6.2)$$

Hence, for the kinetic energy of projectile at any point of the trajectory, we can write

$$E_k = \frac{mv_0^2}{2} - mgy . \qquad (1.6.3)$$

From (1.6.2) we get the same result for the velocity of projectile at altitude y

$$v^2 = v_0^2 - 2gy .$$

$$(1.6.4)$$

The last two formulas show that the speed and the kinetic energy of a projectile decreases with the height y of projectile above the horizon.

At the point of impact $(x_T, 0)$, the coordinate $y = 0$. Substituting $y = 0$ in (1.6.3) and (1.6.4), we find that the speed and kinetic energy of a projectile are respectively equal to their initial values

$$v = v_0, \qquad E_k = \frac{mv_0^2}{2} .$$

$$(1.6.5)$$

We conclude that during the projectile motion in free space,

- the kinetic energy and the speed of projectile changes with height y above the horizon according to the relation $E_k = mv_0^2/2 - mgy$;
- the projectile in free space transmits to the point of impact on the horizon the energy given to it by the launching mechanism; and
- the projectile in free flight is a "mechanisms" that transmits energy from the launching point to the target (point of impact).

1.7 The Elements of Projectile Trajectory

The equation of trajectory (1.4.1), or its parametric equations (1.3.3), show that the coordinates of a given projectile in flight depend on the initial velocity $\vec{v}_0$ (the initial speed v_0 and the initial angle (launching angle) α_0) and the time of flight t.

In practice, we are able to measure or determine the initial speed of the projectile and the initial angle of flight. Thus the position of the projectile for a given initial velocity depends only on the time of flight t.

Knowing the initial velocity $\vec{v}_0$, we are able to determine the location (x, y) of the projectile at any time t, i.e., to solve the so-called "straight problem of the projectile motion."

Now we consider the four parametric equations of projectile motion,

$$v_x = v_0 \cos\alpha_0, \quad v_y = v_0 \sin\alpha_0 - gt ; \tag{1.7.1}$$

$$x = (v_0 \cos\alpha_0)t, \quad y = (v_0 \sin\alpha_0)t - \frac{1}{2}gt^2 ; \tag{1.7.2}$$

and

$$v^2 = v_0^2 - 2g[(v_0 \sin\alpha_0)t - \frac{1}{2}gt^2], \quad \tan\alpha = \frac{v_0 \sin\alpha_0 - gt}{v_0 \cos\alpha_0}, \tag{1.7.3}$$

as well as all other equations that are obtained using the above equations, i.e.,

$$y = (\tan\alpha_0)x - (\frac{g}{2v_0^2 \cos^2\alpha_0})x^2, \tag{1.7.4}$$

$$v^2 = v_0^2 - 2gy, \tag{1.7.5}$$

$$\tan\alpha = \tan\alpha_0 - (\frac{g}{v_0^2 \cos^2\alpha_0})x. \tag{1.7.6}$$

When we fire a projectile, we aim to hit a target located at a known point (x, y) that can be in the same level as, above, or below the firing point. In many ballistics problems, we are interested to solve the following problem:

For a given location of the target (x, y) and a known initial speed of the projectile v_0, we need to find that launching angle α_0 for which the projectile will hit the given target, the time of flight t to the target, as well as the velocity at the target.

Now, let's find some important characteristics of projectile trajectory in free space.

For that we consider the equation of the trajectory (1.7.4), i.e.,

$$y=(\tan\alpha_0)x-(\frac{g}{2v_0^2\cos^2\alpha_0})x^2 .$$

Horizontal Range

At the point of impact that is at the same level as the gun the y-coordinate is zero. Substituting $y=0$ in the trajectory equation, we have

$$(\tan\alpha_0)x-(\frac{g}{2v_0^2\cos^2\alpha_0})x^2=0 .$$

Hence, we find

$$x=0 \quad \text{and} \quad x_T=\frac{v_0^2}{g}\sin2\alpha_0 . \qquad (1.7.7)$$

The first solution corresponds to the launching point while the second one corresponds to the impact point of projectile.

Launching Angle

From (1.7.7) for the launching angle needed to hit the target located at the horizontal range x_T, we obtain

$$\alpha_0=\frac{1}{2}\sin^{-1}(\frac{g\cdot x_T}{v_0^2}) . \qquad (1.7.8)$$

Impact Speed

Substituting in (1.7.5), $y=0$ we find that the speed of the projectile at the point of impact x_T is the same as the speed v_0 of projectile at the launching point,

$$v_T^2 = v_0^2 - 2gy = v_0^2 - 2g(0) = v_0^2 . \tag{1.7.9}$$

Impact Angle

Substituting the abscissa x_T of impact point into (6), we have

$$\tan\alpha_T = \tan\alpha_0 - (\frac{g}{v_0^2 \cos^2 \alpha_0})x_T = \tan\alpha_0 - (\frac{g}{v_0^2 \cos^2 \alpha_0})(\frac{v_0^2}{g}\sin2\alpha_0) .$$

Hence, since $\sin2\alpha_0 = 2\sin\alpha_0 \cos\alpha_0$, we find that the impact angle in absolute value is equal to the launching angle,

$$\tan \alpha_T = -\tan \alpha_0, \text{ or } \alpha_T = -\alpha_0 . \tag{1.7.10}$$

Time of Flight

Substituting $x_T = v_0^2 \sin2\alpha_0 / g$ in the first equation of (1.7.2), $x_T = (v_0\cos\alpha_0)t_T$, we find that the time of flight to the target is

$$t_T = \frac{2v_0 \sin\alpha_0}{g} \tag{1.7.11}$$

Trajectory Vertex

At the vertex of trajectory, the angle of flight is zero, $\alpha_m = 0$. Substituting $\alpha_m = 0$ in (1.7.6), we have

$$\tan\alpha_0 - (\frac{g}{v_0^2 \cos^2 \alpha_0})x_m = 0$$

Solving for x_m the above equation, we find the abscissa of the vertex,

$$x_m = \frac{v_0^2}{2g}\sin2\alpha_0 , \tag{1.7.12}$$

or, considering (1.7.7),

$$x_m = x_T/2 .$$

Substituting the above value in the equation of trajectory (1.7.4), we obtain the ordinate of the vertex that is the maximum altitude of the trajectory

$$y_m = (\tan \alpha_0)x_m - (\frac{g}{2v_0{}^2 \cos^2 \alpha_0})x_m^2$$

$$= (\tan \alpha_0)(\frac{v_0^2}{2g}\sin 2\alpha_0) - (\frac{g}{2v_0^2 \cos^2 \alpha_0})(\frac{v_0^2}{2g}\sin 2\alpha_0)^2 = \frac{v_0^2 \sin^2 (\alpha_0)}{2g} \quad \cdot (1.7.13)$$

Thus, the trajectory vertex is located at the point with coordinates x_m and y_m determined respectively by (1.7.12) and (1.7.13).

Time of Flight to the Vertex Point

Employing the first equation of (1.7.2) and substituting in it the abscissa of the vertex, we find the time of flight to the vertex

$$t_m = \frac{v_0 \sin \alpha_0}{g} , \quad\quad\quad (1.7.14)$$

or, considering (1.7.11), we find that

$$t_m = t_T/2 .$$

Maximum Range

For a given projectile (given launching speed), the range (1.7.7),

$$x_T = \frac{v_0^2}{g}\sin 2\alpha_0$$

depends on the launching angle α_0.

The maximum value of sine function is 1. So the range is maximum when $\sin2\alpha_0=1$, i.e., when the launching angle $\alpha_0=45°$. The value of the maximum range for $\alpha_0=45°$ is

$$x_{max} = \frac{v_0^2}{g}.$$ (1.7.15)

Again on the Maximum Altitude of the Parabolic Trajectory

Employing (1.7.13) and (1.7.14) we obtain another formula to calculate the maximum altitude y_m of the parabolic trajectory of projectile:

$$y_m = \frac{gt_T^2}{8}$$ (1.7.16)

where $t_T = \frac{2v_0\sin\alpha_0}{g}$ is the time of flight to the point of impact with coordinates $(x_T, 0)$.

Another formula for the maximum altitude y_m of the parabolic trajectory that can be easily obtained using (1.7.7) and (1.7.13), is

$$y_m = \frac{1}{4}x_T \tan\alpha_0.$$ (1.7.17)

Notes

- The formulas (1.7.13), (1.7.16) and (1.7.17) can be used as well to estimate the maximum altitude of the trajectory of the projectile that is launched in the presence of drag force. The time of flight t_T to the point of impact $(x_T, 0)$ can be estimated using (1.7.11) or it can be obtained from the firing tables of the firearm. The maximum altitude that we obtain using the above mentioned formulas are always smaller than the real maximum altitude.

In other words, the calculated time of flight t_T to the impact point as well as the maximum altitude of the parabolic trajectory y_m can be used as approximate values respectively for the time of flight to the point of impact and maximum altitude of the projectile flying in presence of drag. (See table 1 of example 1).

A better approximation for the maximum altitude of the projectile flying in presence of resistance of air is given by the formulas (see: Cranz, C and Becker, K, *Exterior Ballistics*, p.97, London 1921. Okunev, B. N, *Fundamentals of Ballistics*, p.85-86, 1943):

$$y_m = \frac{1}{8} x_T (\tan \alpha_0 + \tan|\alpha_T|), \qquad (1.7.18)$$

or

$$y_m = \frac{1}{4} x_T \sqrt{\tan \alpha_0 \cdot \tan|\alpha_T|}, \qquad (1.7.19)$$

The x-coordinate x_m of the trajectory vertex of the projectile launched in presence of resistance of air can be estimated using the Valier's formula:

$$x_m = x_T (0.5000 + 0.0001 \cdot v_0), \qquad (1.7.20)$$

or Hamilton's formula:

$$x_m = \frac{x_T}{1 + \tan \alpha_0 / \tan|\alpha_T|}, \qquad (1.7.21)$$

where the launching angle α_0 and the impact angle α_T can be obtained from the firing tables.

Again on the parametric equations of flight in the absence of drag

Employing the second formula of (1.7.2) and (1.7.11) we obtain another expression for the y-coordinate of the projectile in flight:

$$y = \frac{gt(t_T - t)}{2} \qquad (1.7.22)$$

that corresponds to

$$x = v_0 \cos\alpha_0 t . \qquad (1.7.23)$$

The Average Altitude of Flight

The average altitude of the parabolic trajectory of flight $\bar{y}$ can be obtained by the formulas:

$$\bar{y} = \frac{1}{t_T}\int_0^{t_T} y \, dt , \qquad \bar{y} = \frac{1}{x_T}\int_0^{x_T} y \, dx \qquad (1.7.24)$$

Substituting (1.7.22) in the first formula of (1.7.24) we have:

$$\bar{y} = \frac{1}{t_T}\int_0^{t_T}\frac{gt(t_T - t)}{2} \, dt = \frac{gt_T^2}{12} = \frac{2}{3}\frac{gt_T^2}{8} \qquad (1.7.25)$$

Considering (1.7.16), from equation (1.7.25), we find that the average altitude of flight is

$$\bar{y} = \frac{2}{3}y_m . \qquad (1.7.26)$$

The last formula is used to estimate the average density of atmospheric air in artillery fire (see chapter 5).

We obtain the same formula using the second relation of (1.7.24).

Example 1. A projectile is fired from a 60 mm trench mortar with a speed of 89 m/s under an angle of $45°$. Determine the elements of the trajectory of flight.

Solution

Using the formulas obtained in this section we find

Horizontal Range

$$x_T = \frac{v_0^2}{g}\sin 2\alpha_0 = \frac{(89)^2}{9.80665}\sin(2\cdot 45°) = 807.72m$$

Impact Speed

$$v_T = v_0 = 89m/s$$

Impact Angle

$$\alpha_T = \alpha_0 = 45°$$

Time of Flight

$$t_T = \frac{2v_0\sin\alpha_0}{g} = \frac{2(89)\sin(45°)}{9.80665} = 12.83sec$$

Coordinates of Trajectory Vertex

$$x_m = x_T/2 = 807.72/2 = 402.36m$$

$$y_m = \frac{v_0^2\sin^2(\alpha_0)}{2g} = \frac{(89)^2\sin^2(45°)}{2(9.80665)} = 202m$$

Time of Flight to the Vertex Point

$$t_m = t_T/2 = 12.83/2 = 6.42sec.$$

Table 1 compares some elements of the trajectory obtained above for the free space with the elements of trajectory of the same projectile in presence of resistance of air (see "example 1, of 4.6").

Table 1.

Atmosphere	Departure Angle	Terminal Angle	Range	Time	Maximum Altitude
Ideal	45 degree	- 45.00 degree	807.72	12.84 sec.	202
Drag	45 degree	- 48.17 degree	713.80	12.40 sec.	189

From the table, we see that the results obtained for the elements of the trajectory in free space are far from the results observed in practice.

Example 2. A projectile of a 122mm Russian canon, model 1960, is launched in air with a speed $v_0 = 885m/s$ and angle $\alpha_0 = 7.533°$. Find the maximum altitude of flight (y_m) and the corresponding x-coordinate (x_m) if the projectile range is $x_T = 11200m$, the angle of impact is $\alpha_T = 14°$ and the time of flight is $t_T = 20s$.

Solution

Using the approximate formulas (1.7.17), (1.7.18) and (1.7.19) we obtain respectively:

$$y_m = \frac{1}{4}x_T \tan\alpha_0 = \frac{1}{4}(11200)\cdot\tan(7.533°) = 370m,$$

$$y_m = \frac{1}{8}x_T(\tan\alpha_0 + \tan|\alpha_T|) = \frac{1}{8}(11200)\cdot(\tan(7.533°)+\tan|-14°|) = 534m,$$

$$y_m = \frac{1}{4}x_T\sqrt{\tan\alpha_0 \cdot \tan|\alpha_T|} = \frac{1}{4}(11200)\sqrt{\tan(7.533°)\cdot\tan|-14°|} = 508m.$$

Using the formulas (1.7.20) and (1.7.21) we find the x-coordinate of trajectory vertex:

$$x_m = x_T(0.5 + 0.0001v_0) = 11200\cdot(0.5 + (0.0001)\cdot(885)) = 6591m,$$

and

$$x_m = \frac{x_T}{1+\tan\alpha_0/\tan|\alpha_T|} = \frac{11200}{1+\tan(7.533°)/\tan|-14°|} = 7311m.$$

Note:
- The exact altitude given in the range table of the given projectile is y_m = 490m. In this case the better approximation is the value obtained by formula (1.7.19).
- The exact corresponding x-coordinate is x_m = 6398m.

1.8 The Angle of Projection. The Slant Distance

We will show a general formula to determine the angle of projection A for any trajectory in free space, considering that the target is located at a point with coordinates (x, y), at a distance d (slant range) from the origin of coordinates (figure. 3). The projection angle A is the angle between the line of departure and the line of sight that joins the the launching point (0, 0) and impact point (x, y).

We denote E the angle that the line of sight forms with x-axis; $-90° < E < 90°$. The angle E is called "angle of sight."

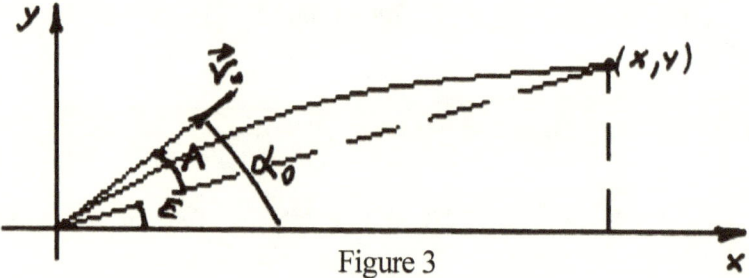

Figure 3

We can easily see that the launching angle is

$$\alpha_0 = E + A.$$

Consider the trajectory equation (1.4.1),

$$y = (\tan\alpha_0)x - (\frac{g}{2v_0^2 \cos^2 \alpha_0})x^2.$$

Substituting in the equation of trajectory the coordinates of the target,

$$x=d\cos E, \quad y=d\sin E \tag{1.8.1}$$

and

$$\alpha_0 = E + A,$$

we obtain

$$\sin E = \cos E \cdot \tan(E+A) - \frac{g\cos^2 E}{2v_0^2 \cos^2(E+A)} \cdot d \tag{1.8.2}$$

Solving the last equation for d, we have

$$d = \frac{2v_0^2}{g\cos^2 E} \cdot \cos(E+A)\cdot\sin A. \tag{1.8.3}$$

The last equation is an equation that relates the distance to the target d located at the point (x, y) and the projection angle A for a given initial speed v_0.

In practice, we are interested to determine the projection angle A for a given location of a target with respect to the position of the firing mechanism. Transforming the equation (1.8.3), we obtain a formula that allows us to find the projection angle A as a function of the angle of sight E, initial speed v_0, and the slant distance d

$$\sin(2A+E) = \frac{gd}{v_0^2}\cdot\cos^2 E + \sin E. \tag{1.8.4}$$

Now we apply formula (1.8.4) when the target is on the horizon at the distance d from the launching point. In this case, the angle of sight is $E=0$ and the projection angle is $A=\alpha_0$. Substituting $E=0$ and $A=\alpha_0$ in (1.8.4), we find

$$\sin 2\alpha_0 = \frac{gd}{v_0^2}. \tag{1.8.5}$$

Substituting (1.8.5) in (1.8.4), we obtain the following formula:

$$\sin(2A+E)=\sin2\alpha_0\cdot\cos^2 E+\sin E ,\qquad(1.8.6)$$

to compute the angle of projection A. Formula (1.8.6) relates the projection angle A that corresponds to a slant distance d of the target with the launching angle α_0 that corresponds to the zero range d of the target.

We can give to formula (1.8.6) a simple form when the angle of sight is narrow. Employing the trigonometric identity for the sine of the sum of two angles, we can write

$$\sin2A\cos E+\cos2A\sin E=\sin2\alpha_0\cos^2 E+\sin E .\qquad(1.8.7)$$

Now, we assume that projection angle is narrow, such that we can consider

$$\cos2A\approx1 ,\qquad \cos\alpha_0\approx1 .\qquad(1.8.8)$$

Considering the first equation (1.8.8), from (1.8.7), we can write

$$\sin2A=\sin2\alpha_0\cos E\qquad(1.8.9)$$

or

$$2\sin A\cos A=2\sin\alpha_0\cos\alpha_0\cos E$$

From the above equation (since $\cos\alpha_0\approx1$, $\cos A\approx1$), we obtain an approximate formula to estimate the angle of projection:

$$\sin A=\sin\alpha_0\cos E .\qquad(1.8.10)$$

Remarks

- Formula (1.8.6) can be used for the infantry firearms when the projectiles fly in presence of drag (see "example 3 section 3.5").

- In the formula (1.8.6), the value of slant range d is not present. The value of d is used to calculate the angle of departure α_0 when the target is on the horizon at a zero range d that is equal to the slant range (figure 3).
- If we consider a projectile launched in air in presence of air drag, then the formulas (1.8.6) or (1.8.10) can be employed for the projectile flying in air on condition that the angle of sight α_0 is calculated for the horizontal range d, but in presence of drag.

Example 1. Find the projection angle needed to hit a target located at a point on a hill that is in a slant distance $d=46m$. The angle of sight is $E=20°$. The initial speed is $v_0=25m/s$.

Solution

Substituting in (1.8.4), we have

$$\sin(2A+E)=\frac{gd}{v_0^2}\cos^2 E+\sin E=\frac{9.80665\cdot(46)}{(25)^2}\cos^2(20°)+\sin(20°)=0.979359 \quad .$$

Hence,

$$(2A+E)=\sin^{-1}(0.979359)=78.34°$$

and

$$A=(78.34°-E)/2=(78.34°-20°)/2=29.17° \quad .$$

The launching angle is

$$\alpha_0=(E+A)=(20°+29.17°)=49.17° \quad .$$

In the same way, we find the other value of the launching angle

$$(2A+E)=\sin^{-1}(0.979359)=(180°-78.34°)=101.66°$$

that is not acceptable.

Example 2. A projectile launched with speed $v_0=20m/s$ under an angle $\alpha_0=15°$ hits the ground at a horizontal distance of $x_T=20.39m$ from the launching point.

Find the projection angle needed to hit a target located in a slant distance $d=20.39m$ if the angle of sight is $E=12°$.

Solution

Substituting in formula (1.8.6),

$$\sin(2A+E)=\sin2\alpha_0\cdot\cos^2 E+\sin E$$,

we have

$$\sin(2A+12°)=\sin2(15°)\cdot\cos^2(12°)+\sin(12°)=0.483452$$.

Hence, we find the projection angle

$$A=\sin^{-1}[(0.483452)-12°)/2]=8.46555°$$

and the launching angle

$$\alpha_0=(E+A)=(8.46555°+20°)=20.46555°$$.

2

The Equations of Projectile Motion
In Presence of Drag

Introduction

The results obtained in chapter 1 using the ideal model of the projectile are far from the outcomes observed in practice for the motions of projectiles thrown in the air with relatively high speeds, greater than 30–50 m/s.

The study of projectile motion in air considering the influence of all factors (gravitational force, drag force, projectile spin, earth revolution, wind, change of temperature, density of air, etc.) is a very complex problem. For that reason, the exterior ballistics employs simplified mathematical models by ignoring some factors that have insignificant influence in the accuracy of the practical results and temporarily neglecting some other factors whose influence can be reflected later in the obtained results as "corrections" (see chapter 6).

Thus, for relatively small ranges of fire, the influence of the spherical shape of the earth and its rotation in projectile flight can be neglected. The earth surface can be considered plane and earth itself not a rotating body. Those assumptions allow us to consider the gravity constant, to neglect the Coriolis effect, and to obtain relatively easy the solutions of the problems of projectile flight.

On the other hand, for example, the influence of wind or the influence of the variation of air temperature, air density and atmospheric pressure in the projectile flight are significant and cannot be neglected.

To simplify the study of the projectile flight in air, we consider the motion of a projectile near the earth surface in presence of gravity and air resistance assuming that there are no deviations in temperature, density, or other significant factors from their standard values and in the absence of wind.

Since the distance that the projectile travels and the maximum height of the trajectory above the ground are short compared to the earth's radius, the ground can be assumed flat and the gravity constant.

The projectile motion in presence of gravity and air resistance (drag force) can be modeled by a system of differential equations. Solving the differential equations of projectile flight, we find the projectile trajectory and all the other characteristics of flight.

Then, using appropriate methods, we make "corrections" in the results already obtain from the differential equations to reflect the results of firing tests, the influence of atmospheric factor, or other ballistics characteristics in the trajectory of projectile flight, etc.

The projectile trajectory has the following characteristic elements:

the launching angle α_0, the initial speed v_0, the ballistics coefficient c, the horizontal range x_T, the terminal speed v_T, the angle of impact on the target α_T, the time of flight to the target t_T, and the maximum trajectory altitude y_{max} (coordinates x_m, y_m of the trajectory vertex). Using the methods of exterior ballistics, we can determine all elements of a trajectory if there are known three of them, mainly three out of four ballistics characteristics: α_0, v_0, c, and x_T.

The main problems exterior ballistics deals with are the following:

- For a given ballistics coefficient c and a launching velocity $\vec{v}_0$ (initial speed v_0, launching angle α_0) of a projectile find the range of fire x_T and all other elements of the trajectory.
- For a given initial speed v_0 of a projectile, range of fire x_T and the ballistics coefficient c find the launching angle α_0 and all the other trajectory elements.
- For a given launching angle α_0, initial speed v_0 and range x_T, find the ballistics coefficient c and all the other elements of projectile trajectory.
- For a given launching angle α_0, range x_T and ballistics coefficient c, find the initial speed v_0 and the rest of elements of the projectile trajectory.

In the following chapters we approach the solution of those problems employing a diversity of ballistics methods.

The actual chapter is basic to lay the foundations of the methods of solutions of the problems of exterior ballistics.

2.1 The Resistance of Air. The Drag Coefficient and the Form Factor

A projectile flying in air experiences the resistance of air, called drag force. The drag force is a resisting force that air exerts on a projectile due to the interaction of air particles and the projectile. The magnitude of the aerodynamic drag force, also called the frontal resistance, is

$$D = A\frac{\rho v^2}{2}C(\frac{\rho v d}{\eta}, \frac{v}{a}), \tag{2.1.1}$$

where A is the cross-sectional area of the projectile, ρ is the air density, v is the speed of projectile, η is the viscosity of air, and a is the speed of sound in air. The density of air is a function of altitude of the projectile above the sea level ($\rho = \rho(y)$).

The quantity $C(\frac{\rho v d}{\eta}, \frac{v}{a})$ is called "drag coefficient," $\frac{\rho v d}{\eta}$ is the Reynolds number R_e, while v/a is the Mach number M, i.e.,

$$R_e = \frac{\rho v d}{\eta}, \qquad M = \frac{v}{a}. \tag{2.1.2}$$

The Reynolds number is a characteristic of kinematic viscosity of air. The viscosity plays a significant role in the frontal resistance for relatively low speeds of projectiles.

The influence of the Reynolds number R_e on the drag coefficient $C(\frac{Vd}{\eta}, \frac{V}{a})$ for artillery projectiles and other army ammunitions can be neglected.[2] Thus the magnitude of the drag force for a given projectile can be written as

$$D = A\frac{\rho v^2}{2} C(\frac{v}{a}) \tag{2.1.3}$$

or

$$D = A\frac{\rho v^2}{2} C(M). \tag{2.1.4}$$

From equation (2.1.3), we have

$$C(\frac{v}{a}) = \frac{(D/A)}{(\rho v^2/2)}.$$

The above equation shows that the drag coefficient is the ratio of the total drag acting on the unit of the cross-sectional area of the projectile and the dynamic pressure $\rho v^2/2$.

---

[2]   E. J. McShane, J. L. Kelley. F. V. Reno, "Exterior Ballistics", p. 166. The University of Denver Press, 1953.

The magnitude of drag (2.1.3) or (2.1.4) includes the drag coefficient $C(v/a)$, which is a characteristic of a given projectile and depends on the Mach number $M=(v/a)$.

In general, the drag coefficient $C(v/a)$ is different for different projectiles. The drag coefficient can be determined experimentally for each projectile, for example, using wind tunnels, Doppler radars, ballistics chronographs, etc.

However, in practice, instead of determining the drag coefficient $C(v/a)$ for each projectile, exterior ballistics determines experimentally a drag coefficient for a standard projectile that is considered as representative of a group of projectile shapes. We denote $C_D(v/a)$ the drag coefficient of the standard projectile.

The drag coefficient $C_D(v/a)$ is usually given in the form of a table of values of drag coefficient, called "the table of drag coefficient" (see for example the drag coefficient of Mayevski-Zabudski's function given in table 1 below).

The drag coefficient $C(v/a)$ of a nonstandard projectile with respect to a particular standard projectile is determined by the equation

$$C(\frac{v}{a})=iC_D(\frac{v}{a}),\qquad\qquad(2.1.5)$$

where i is a dimensionless number called "form factor" or "form coefficient" of the projectile.

The form factor i is estimated experimentally using firing tests, wind tunnels wind tunnels, Doppler radars, ballistics chronographs, etc.

In exterior ballistics of small arms the form factor of a given projectile is assumed to be constant, with an average value that gives an optimal approximation to the trajectory elements. Though the form factor i is defined as a number, in fact, as (2.1.5) shows, it is a variable quantity that depends on the Mach number $M=(v/a)$.

Substituting (2.1.5) in (2.1.3), we find that the magnitude of the drag force D of a nonstandard projectile can be written through the drag coefficient $C_D(v/a)$ of the standard projectile:

$$D=A\frac{\rho v^2}{2}iC_D(\frac{v}{a})$$. (2.1.6)

The projectile acceleration caused by drag force is

$$a_D=\frac{D}{m}=A\frac{\rho v^2}{2m}iC_D(\frac{v}{a})$$. (2.1.7)

The form coefficient can be seen as a quantity that is used to compare projectiles drag. A standard projectile has a form coefficient equal to 1. A nonstandard projectile whose form coefficient relative to a given standard projectile is less than 1, as (2.1.7) shows, experiences less drag in flight than the standard one does.

The form coefficient depends as well on the particular standard drag coefficient $C_D(v/a)$ we use in equation (2.1.5).

Indeed, for a given projectile in flight, the drag coefficient that corresponds to a given Mach number $M=(v/a)$ is $C(v/a)$. Now we consider two different standard projectiles whose drag coefficients, are respectively $C_{D1}(v/a)$ and $C_{D2}(v/a)$. Using (2.1.6), for the magnitude of drag force the air exerts on the given nonstandard projectile, we can write

$$D=A\frac{\rho v^2}{2}i_1C_{D1}(\frac{v}{a}) , \quad D=A\frac{\rho v^2}{2}i_2C_{D2}(\frac{v}{a})$$. (2.1.8)

Since the magnitude of drag force D and the projectile characteristics are the same in both formulas of (2.1.8), we find that

$$\frac{i_1}{i_2}=\frac{C_{D2}(v/a)}{C_{D1}(v/a)})$$. (2.1.9)

Since $C_{D1}\neq C_{D2}$ from the last equation, we find that $i_1\neq i_2$. That means that the form coefficient of the same projectile is relative and depends on the drag coefficient $C_D(v/a)$ of the standard projectile in use. In other words, the form coefficient of a projectile has more than one value that depends on the standard projectile we use in (2.1.5).

The Mayevsk-Zabudski's drag coefficient for a standard projectile, for the speed of sound $a_0 = a_{0N} = 340.83 m/s$, is given in the following table.[3]

Table 1. Mayevski -Zaboudski's drag coefficient
$C(v/a)$ for a standard projectile

$M = v/a$	0.1	0.2	0.3	0.4	0.5	0.6	0.7	0.8
$C_M(v/a)$	0.228	0.228	0.228	0.228	0.228	0.228	0.228	0.259
$M = v/a$	0.9	1.0	1.1	1.2	1.3	1.4	1.5	1.6
$C_M(v/a)$	0.313	0.430	0.573	0.626	0.643	0.643	0.643	0.643
$M = v/a$	1.7	1.8	1.9	2	2.1	2.2	2.3	2.4
$C_M(v/a)$	0.633	0.623	0.613	0.603	0.594	0.586	0.578	0.571
$M = v/a$	2.5	2.6	2.7	2.8	2.9			
$C_M(v/a)$	0.559	0.549	0.540	0.531	0.521			

Example 1. The drag coefficient of the Mayevski standard projectile for Mach number $M=v/a=2$ is $C_M(v/a)=0.603$. The drag coefficient of a projectile of caliber $d=7.62mm$ and mass $m=7.9g$, measured when its speed is $v=683.66$ (i.e. when $M=v/a=2$), has the value $C(v/a)=C(2)=0.3777$. Find the Mayevski form coefficient i for the given bullet, the drag force D (the atmospheric air applies on the projectile), as well as the drag acceleration $a_D = D/m$. The density of air is $\rho = 1.210 kg/m^3$.

Solution

Using (2.1.5), we find the form coefficient

$$i = \frac{C(v/a)}{C_D(v/a)} = \frac{C(2)}{C_D(2)} = \frac{0.378}{0.603} = 0.627.$$

---

[3]   Mori, Edoardo. *Balistica teorica e pratica*.
http://www.earmi.it/balistica/formi.htm [Web site]

The cross-section area of the bullet is $A=\pi \cdot d^2/4$. Substituting in (2.1.3), we find that the magnitude of the drag force is

$$D = A\frac{\rho v^2}{2}C(\frac{v}{a}) = \frac{\pi \rho d^2 v^2}{8}C(\frac{v}{a}) = \frac{\pi(1.21)\cdot(0.00762)^2(683.66)^2}{8}\cdot(0.3777) = 4.8706N$$

The magnitude of acceleration due to drag is

$$a_D = D/m = (4.8706)/(0.0079) = 616.53m/s^2.$$

Note: The magnitude of the bullet acceleration is about sixty-three times more than the gravity acceleration. The direction of acceleration is opposite to the projectile velocity.

2.2 The Differential Equations of Projectile Motion. The Drag Function and the Density Function

Now, as a first approach, we consider the motion of a projectile near the ground in presence of air resistance (drag force).

The projectile motion with air resistance can be modeled by a system of differential equations that is the result of Newton's second law of mechanics, $\vec{F}=m\vec{a}$, that considers only the gravity and the drag force that acts on the projectile but neglects all other factors that influence the projectile flight.

Solving that system of differential equations, we are able to obtain very approximate results for the elements of trajectories of motion of projectiles in real conditions.

To study the flight of a projectile, we will consider a three-dimensional Cartesian coordinate system. The projectile is launched at the origin of coordinates with an initial velocity $\vec{v}_0$ that makes an angle α_0 with the x-axis. The direction of the y-axis is such that the initial velocity $\vec{v}_0$ is on the xy-plane (figure 4).

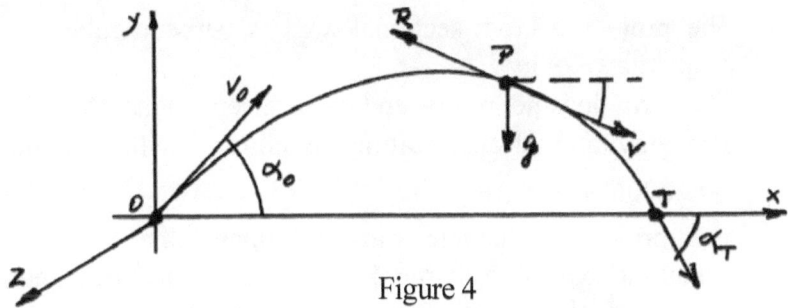

Figure 4

The xy-plane is called the launching plane of the projectile. The z-axis is perpendicular to the launching plane.

In exterior ballistics, in use are different systems of measurement units. We use the International System of Units (SI) unless otherwise stated.

Differential Equation of a Projectile in Flight
We assume the following:

- The projectile is a particle with constant mass m located at the projectile center of masses.
- The projectile does not rotate about its axis of symmetry; the projectile axis, at any moment in flight, has the direction of the projectile instantaneous velocity $\vec{v}$. In other words, the projectile is considered a point-of mass without spinning motion, with axis of symmetry along the tangent of the trajectory of flight.
- The forces acting on the projectile at any moment in flight are the gravitational force $\vec{F}_G = m\vec{g}$ and the resistance of air (drag force), opposite to the projectile velocity

$$\vec{D} = -A\frac{\rho v^2}{2}C(\frac{v}{a})\frac{\vec{v}}{v}, \quad \text{or} \quad \vec{D} = -A\frac{\rho v^2}{2}iC_D(\frac{v}{a})\frac{\vec{v}}{v},$$

where $(\vec{v}/v)$ is the unit vector in the direction of the projectile velocity $\vec{v}$.

- The projectile cross-sectional area is perpendicular to the projectile velocity $\vec{v}$.
- The earth does not rotate, and its surface is plane.
- The gravitational acceleration $\vec{g}$ is constant with magnitude $g=9.80665 m/s^2$.
- The projectile characteristics: the mass, the caliber, the launching explosive charge (black powder), etc., are standard, identical to the manufacturing characteristics.
- The launching barrel is brand-new.

Newton's second law for a given projectile in flight at a time t with respect to the three-dimensional Cartesian coordinate system of figure 1 is written

$$m\frac{d^2\vec{r}}{dt^2}=m\vec{g}-A\frac{\rho v^2}{2}iC_D(\frac{v}{a})\frac{\vec{v}}{v},\qquad (2.2.1)$$

where $\vec{r}=\vec{r}(t)$ is the position vector of the projectile at a time t,

$$\vec{r}(t)=x(t)\vec{i}+y(t)\vec{j}+z(t)\vec{k}.$$

Since the gravitational force and the drag force at any time t during the projectile flight are in the launching plane (xy-plane), and no other force acts on the direction of the z-axis, according to Newton's second law, we conclude the following:

- The trajectory of projectile flight (in absence of projectile rotation and wind) is a plane curve in the plane of flight (xy-plane). The z-coordinate of the projectile is zero at any instant during the projectile flight.

Note: From the assumption that the projectile does not rotate about its axis, it follows that the gyroscopic force directed along the z-axis is absent.

Dividing both sides of equation (2.2.1) with the projectile mass m, we obtain a second order differential equation in vector form

$$\frac{d\vec{v}}{dt}=\vec{g}-\frac{1}{m}A\frac{\rho v^2}{2}iC_D(\frac{v}{a})\frac{\vec{v}}{v}. \qquad (2.2.2)$$

Drag Acceleration. Function of Air Resistance

The magnitude of the projectile acceleration $(a_D=D/m)$ due to the drag is

$$a_D=A\frac{\rho v^2}{2m}iC_D(\frac{v}{a_{0N}}). \qquad (2.2.3)$$

Substituting in (2.2.3) the cross-sectional area of the projectile, $A=\pi{\cdot}d^2/4$, where d is the diameter of the cross-sectional area that is equal to the firearm caliber, we can write the projectile acceleration (2.2.3) as

$$a_D=i\frac{\pi{\cdot}d^2\rho}{8m}v^2C_D(\frac{v}{a}), \quad \text{or} \quad a_D=\frac{\pi{\cdot}d^2\rho}{8m}v^2C(\frac{v}{a}). \qquad (2.2.4)$$

As formula (2.2.2) shows the acceleration of the projectile $(d\vec{v}/dt)$ is the difference of the gravitational acceleration $\vec{g}$ (perpendicular to x-axis) and the acceleration due the drag force that is opposite to the projectile instantaneous velocity $\vec{v}$. The density of air ρ is a function of altitude y.

The diameter (caliber) and the mass of a standard projectile are measured respectively in millimeter and gram while the projectile speed is in meter per seconds.[4]

The density of air is measured in kilogram per cubic meter. Multiplying and dividing the right side of (2.2.4) by the density $\rho_{0N}=1.205 kg/m^3$ of air in standard atmosphere (see the definition of standard atmosphere at the end of the section), for the acceleration of the projectile, we obtain

---

[4]   In the U.S.A., the caliber is measured in inch, while the projectile mass is in pounds.

$$a_D = 0.473202 i \frac{d^2}{m} \cdot \frac{\rho}{\rho_{0N}} v^2 C_D(\frac{v}{a})$$

or

$$a_D = 0.473202 \frac{d^2}{m} \cdot \frac{\rho}{\rho_{0N}} v^2 C(\frac{v}{a}) \, .$$

Since the unit of acceleration in SI is meter per square second while the quotient (d^2/m) is expressed in $mm^2/gram$ to make the unit of the right side of the above equation m/s^2, we multiply the above equation by the converting factor $10^{-6}/10^{-3} = 10^{-3}$. So the drag acceleration of a nonstandard projectile in SI is

$$a_D = (4.73202 \cdot 10^{-4}) \cdot i \frac{d^2}{m} 1000 \frac{\rho}{\rho_{0N}} v^2 C_D(\frac{v}{a})$$

or

$$a_D = (4.73202 \cdot 10^{-4}) \cdot \frac{d^2}{m} 1000 \frac{\rho}{\rho_{0N}} v^2 C(\frac{v}{a}) \, . \qquad (2.2.5)$$

The exterior ballistics, for a projectile, instead of the drag coefficient employs the function $f_D(v)$ or $f(v)$ defined by the equation

$$f_D(v) = 4.732 \cdot 10^{-4} \cdot v^2 C_D(v/a)$$

or

$$f(v) = 4.732 \cdot 10^{-4} \cdot v^2 C(v/a) \, , \qquad (2.2.6)$$

respectively, for a standard projectile and a nonstandard one.

The function $f_D(v)$ is called "function of air resistance" of the standard projectile, or simply the "drag function," while $f(v)$ is "the function of resistance" of a nonstandard projectile. In fact, the function of resistance $f_D(v)$, or $f(v)$, is a function of the Mach number.[5] The correct notation would be $f_D(v/a)$. Because of the historical tradition, we maintain for the moment the notation $f_D(v)$, keeping in mind that actually it implies a function of the Mach number.

Using the first equation of (2.2.6), for the acceleration (2.2.5) of a nonstandard projectile, we find

$$a_D = \frac{id^2}{m} 1000 \frac{\rho}{\rho_{0N}} f_D(v), \text{ or } a_D = \frac{d^2}{m} 1000 \frac{\rho}{\rho_{0N}} f(v). \quad (2.2.7)$$

We denote

$$h(y) = \frac{\rho}{\rho_{0N}} \qquad (2.2.8)$$

the "function of air density" and

$$c = \frac{id^2}{m} 1000, \qquad (2.2.9)$$

the "ballistics coefficient."[6] Then the projectile acceleration (2.2.7) can be written

$$a_D = c \cdot h(y) f_D(v). \qquad (2.2.10)$$

The acceleration (2.2.3) can be written in the following form

---

[5] A variety of functions of resistance for some standard projectiles were obtained from the firing tests in the nineteenth and twentieth centuries. The well-known functions of resistance: (1) Mayevski-Zabudski's function, (2) Siacci's function, (3) Gavre's function, (4) Ingalls's function, etc. revolutionized the methods of exterior ballistics, transforming it into a modern science. Due to the function of resistance, it was made possible the solution of the differential equations of projectile flight and compilation of ballistics tables.

[6] The ballistics coefficient in practice in U.S. Army at sea level is $C = m/(i \cdot d^2)$, in pound per square inch.

$$\frac{dv_D}{dt}=c \cdot h(y) f_D(v)$$

(2.2.11)

Taking into consideration the projectile acceleration, the vector differential equation (2.2.2) takes the form

$$\frac{d\vec{v}}{dt}=\vec{g}-c \cdot h(y) f_D(v)\frac{\vec{v}}{v}$$

(2.2.12)

Substituting (2.2.6) in (2.2.12), we obtain another form of the differential equation of projectile flight

$$\frac{d\vec{v}}{dt}=\vec{g}-4.732 \cdot 10^{-4} c \cdot h(y) v^2 C_D(v/a)\frac{\vec{v}}{v}$$

(2.2.13)

The Density Function

The density function $h(y)$, for altitudes till around ten thousand meters above the sea level (see chapter 8), can be determined using the following formula:[7]

$$h(y)=(\frac{289.08-0.006328y}{289.08})^{4.4}$$

(2.2.14)

Another formula that can be used for the density function is

$$h(y)=(20,000-y)/(20,000+y)$$

(2.2.15)

The density function can be estimated as well employing the formula[8]

$$h(y)=e^{-0.0001036465y}$$

(2.2.16)

The density functions (2.2.14), (2.2.15), (2.2.16) are valid when the air density at sea level has the value $\rho_{0N}=1.205 kg/m^3$.

---

[7] Shapiro, J. M. *Vneshnaja Balistika*, page 54. Oborongiz 50'.

[8] Herrmann, E. E. *Exterior Ballistics*, p. 35. U.S. Naval Institute, The College Press, 1935.

When the density of air at sea level (or the firing site) is different from the standard value, $\rho_0 \neq 1.205 kg/m^3$, the density function (2.2.14) is written

$$h(y)=(\frac{\tau_0-0.006328y}{\tau_0})^{4.4},\qquad(2.2.17)$$

where τ_0 is the virtual temperature at the firing site (see chapter 8).

The density function is determined by the value of the air density measured at sea level (firing site). When the projectile is fired above sea level, using the density measured at the firing point, we find the density at sea level, and then we can use the above formulas to determine the density function $h_0(y)$.

Once More on the Form Factor

From both equations (2.2.7), we find that the form factor of a projectile can be determined also by the formula

$$f(v)=i \cdot f_D(v) \qquad (2.2.18)$$

that relates the function of resistance of a nonstandard projectile with the function of resistance of the projectile accepted as standard.

Important Note

- When solving ballistics problems, we should be careful when we need to use the data and formulae given by different authors, ballistics textbooks, or manuals. There is confusion not only due to units in use—SI system of units, English system of measurements, or other old systems of measurement units—but also due to the function of resistance in use and, related with that, the form factors.

In the following sections are given some models to adapt the form factor or the ballistics coefficient for use with the function of resistance we prefer, the Siacci function.

The Traditional Standard Atmosphere (TSA)

Exterior ballistics assumes some standard conditions for the atmospheric air to simplify the solution of the equation of projectile motion (2.2.2).

We assume that the projectile motion is in the traditional standard atmosphere (TSA):

- At sea level ($y=0$), the air density is $\rho_{0N}=1.205 kg/m^3$, the temperature of air (virtual temperature) is $\tau_{0N}=289.08K$, the atmospheric pressure is $p_{0N}=750$ mmHg, and the speed of sound is $a_{0N}=340.83$.
- The temperature, the pressure, and the density of air vary only with the altitude y above sea level.
- The projectile flight is in the absence of wind.

The equation (2.2.13) for the projectile flying in standard atmosphere is

$$\frac{d\vec{v}}{dt}=\vec{g}-4.732{\cdot}10^{-4}c{\cdot}h(y)v^2 C_D(v/a_{0N})\frac{\vec{v}}{v}, \qquad (2.2.19)$$

where

$$f_D(v)=4.732{\cdot}10^{-4}{\cdot}v^2 C_D(v/a_{0N}) \qquad (2.2.20)$$

is the function of resistance for the standard atmosphere.

Note that from now on, we assume that the projectile flies in the standard atmosphere unless otherwise stated.

Example 1. Calculate the ballistics coefficient of the projectile of a trench mortar 82 mm. Projectile mass is $m=3.153$ kg. The form factor of the projectile $i=0.60$.

Solution

Substituting in (2.2.9) $m=3.153 kg$, caliber $d=82mm=0.082m$, and the form factor $i=0.60$, we obtain:

$$c=\frac{id^2}{m}1000=\frac{(0.6)(0.082)^2}{3.153}\cdot1000=1.28m^2/kg$$

Example 2. Calculate the ballistics coefficient of the bullet of a Russian rifle of caliber $d=7.62$ mm, bullet mass $m=7.9$ g, form factor $i=0.56$.

Solution

Substituting in (2.2.9) the caliber $d=7.62$ mm $=0.00762$ m, mass $m=7.9$ g $=0.0079$ kg, and $i=0.56$, we find that the ballistics coefficient of the given projectile is

$$c=\frac{id^2}{m}1000=\frac{(0.56)(0.00762)^2}{(0.0079)}1000=4.116m^2/kg.$$

2.3 The Function of Resistance and the Drag Coefficient

Siacci's Function of Resistance

There are different types of function of resistance $f_D(v)$ that correspond to different standard projectiles used in different countries or firearm companies. Usually the standard functions of resistance are given analytically by a formula or in tableau form.

By the end of nineteenth century, the Italian mathematician, Col. Francisco Siacci introduced the function of resistance,[9] known as "Siacci's function,"

$$f_S(v)=0.2002\cdot v-48.05+[(0.1648v-47.95)^2+9.6]^{1/2}+\frac{0.0442\cdot v(v-300)}{371+(v/200)^{10}}, \quad (2.3.1)$$

where

$$0<v\le1200m/s.$$

---

9   Mori, Edoardo. *Balistica teorica e pratica.* *http://www.earmi.it/balistica.* [Web site]

The Siacci function was the result of the experimental data obtained by Mayevski's firing tests for a particular standard projectile.

Since the formula is complicated, Siacci created tables of values of function (2.3.1).

Approximate Siacci Function

Hereafter it is given a modern Siacci function of resistance $f_D(v)$, denoted $K_D(v)$, for projectiles (bullets, artillery shells, missiles, grenades, etc.),[10]

$$K_D(v) = \begin{cases} 1.212 \cdot 10^{-4} v^2 & for \quad v \leq 256 m/s \\ (v-240)/3 & for \quad v > 256 m/s \end{cases} \tag{2.3.2}$$

which especially for speeds till 750–800 m/s presents a good approximation to the Siacci function of air resistance.

We will use the approximate Siacci function of resistance (2.3.2) throughout the material. As we will see, the results obtained using (2.3.2) instead of the original Siacci function of resistance (2.3.1) are accurate.

Modified Approximate Siacci Function

For speed of projectile greater than 750–800 till 1,200 m/s, the function of resistance (2.3.2) will be quite close to the Siacci function (2.3.1) if we introduce the correction factor 1.060922 (see example 1); so we obtain a modified Siacci function of resistance

$$K_D(v) = 1.060922(v-240)/3 \tag{2.3.3}$$

for speeds in the interval (800–1,200 m/s).

---

[10]   Mucinov S.S., Shevcenko N.A. Zadacnik po Osnovami Strelbi is Strelkovova Oruzhie. Moskva 1969;
Gubinim, S. G., Gorovim, S. A. Ballistics. Handbook, *http://www.ssga.ru/AllMetodMaterial/metod_mat_for_ioot/metodichki/ballistica/index.htm*. [Web site]

Thus, the approximate Siacci function of resistance for projectile speed till 1,200 m/s can be written as

$$K_D(v)=\begin{cases} 1.212 \cdot 10^{-4} v^2 & v \leq 256 \\ (v-240)/3 & 256 < v \leq 800 \\ 1.060922(v-240)/3 & 800 < v \leq 1200 \end{cases} \qquad (2.3.4)$$

Using the fictive ballistics coefficient (see example 1),

$$c^* = 1.060922 \cdot c \qquad (2.3.5)$$

we can always use the Siacci approximate function (2.3.2) for speeds till 1,200 m/s to study the trajectory of a projectile. The differential equation of projectile flight (2.2.12) for the Siacci function (2.3.2), has the form

$$\frac{d\vec{v}}{dt} = \vec{g} - c \cdot h(y) K_D(v) \frac{\vec{v}}{v}. \qquad (2.3.6)$$

Siacci's and Mayevski's Form Factors

Siacci's function of resistance was result of the experiments performed by Mayevski, Bashford, Krupp company. Summarizing those results, Siacci introduced his function of resistance (2.3.1). The average converting factor from Mayevski's function of resistance, $f_M(v)$, to Siacci's function of resistance $f_S(v)$ is $(0.896)^{-1}$. Using (2.2.18), we can write[11]

$$f_S(v) = (0.896)^{-1} \cdot f_M(v). \qquad (2.3.7)$$

Considering standard Mayevski's projectile, we can interpret the converting factor $(0.896)^{-1}$ as the form coefficient of Siacci's projectile relative to Mayevski's standard projectile. Thus, the Mayevski form coefficient of the Siacci standard projectile is

---

[11]   Cline, Donna. *Trajectories, Trajectories*. 1999–2004
   www.Angelfire.com/poetry/u31240468/Aeroballistivcs.pdf;
   Shapiro, J. M. *Vneshnaja Balistika*, p.48 Oborongiz 50'.

$$i_m = (0.896)^{-1}.$$

$$(2.3.8)$$

For the respective drag coefficients, Siacci's drag coefficient $C_S(v/a)$ and Mayevski's drag coefficient $C_M(v/a)$, we can write

$$C_S(\frac{v}{a}) = (0.896)^{-1} C_M(\frac{v}{a})$$

$$(2.3.9)$$

Let's assume that the Siacci form coefficient of a nonstandard projectile is i_S and let i_M be the Mayevski form coefficient of the same projectile. For the function of resistance of the given nonstandard projectile, we can write

$$f(v) = i_S \cdot f_S(v)$$

$$(2.3.10)$$

and

$$f(v) = i_M \cdot f_M(v).$$

$$(2.3.11)$$

From (2.3.10) and (2.3.11), we have

$$i_S \cdot f_S(v) = i_M \cdot f_M(v).$$

$$(2.3.12)$$

From (2.3.7) and (2.3.12), we find that the Siacci form coefficient of a projectile is lower than the Mayevski form factor of the same projectile

$$i_S = (0.896) \cdot i_M$$

$$(2.3.13)$$

Formula (2.3.13) is valid as well for the Mayevski- Ingalls function of resistance (see section 2.5).

Comment

- The coefficient of form $i_m = (0.896)^{-1}$ is an average multiplication factor used to match the values of Mayevski's function of resistance with Siacci's function values.

Note. Another modified Siacci function that is close to the values of the Siacci function of resistance (2.3.1), for speeds in the interval (256–1,200 m/s), is

$$K_D(v)=\begin{cases} 1.212 \cdot 10^{-4} v^2 & for \quad v \le 256m/s \\ 1.0472357(v-240)/3 & for \quad v>256m/s \end{cases}. \quad (2.3.14)$$

The above modified Siacci function is obtained using the same procedure we use in example 1. The ballistics coefficient for the modified Siacci function (2.3.14) is

$$c^{**}=1.0472357 \cdot c. \quad (2.3.15)$$

We repeat again that we will use the modified Siacci function (2.3.2) throughout the material.

Example 1. (The Modified Siacci Function). Graph the original Siacci function (2.3.1) and the approximate Siacci function of resistance (2.3.2).

Adjust (correct) the approximate Siacci function (2.3.2) to obtain a better approximation for the data of the projectile speeds in the interval $(750m/s-1200m/s)$.

Solution

From both graphs presented in figure 5 and figure 6, we see that the values of the original Siacci function (the graph on top) are bigger than the corresponding values of the approximate Siacci function (the graph below).

Using a TI-83 Plus calculator, we find that the average value of the original Siacci function of resistance and the approximate Siacci function in the interval $(750m/s-1200m/s)$ are respectively

$$\overline{K}_D(v)=\frac{1}{1200-750}\int_{750}^{1200}[(v-256)/3]dx=\frac{110250}{(450)}=245$$

and

$$\bar{f}_s(v)=\frac{1}{1200-750}\int_{750}^{1200}f(v)dx=\frac{116966.65}{(450)}=259.9259$$.

Dividing the above equations, we obtain

$$\bar{f}(v)=1.060922\bar{K}_D(v)$$.

Thus, as a reasonable correction factor for projectile speeds over $800m/s$, we can assume $k=1.060922$. The approximate Siacci function for speeds greater than $800m/s$ is

$$K_D(v)=1.060922(v-240)/3,$$

where $800m/s<v\le 1200m/s$.

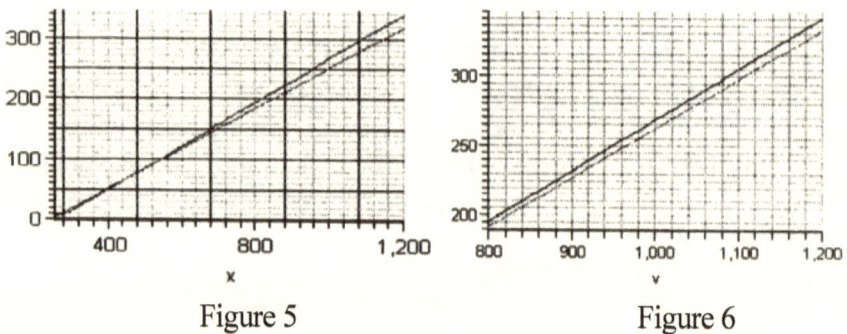

Figure 5   Figure 6

The modified "approximate" Siacci function is displayed in the third column of the table below.

Table 1. Values of Siacci's functions

Projectile Speed	Approximate Siacci Function (2.3.2)	Original Siacci Function (2.3.1)	Modified Siacci Function (2.3.4)
300	*20.000	15.448	**21.218**
400	*53.333	51.533	**56.583**
500	*86.667	87.085	**91.947**
600	*120.00	123.23	**127.31**
700	153.33	**159.62**	*162.67
750	170.00	**177.84**	*180.36
800	186.67	**196.07**	*198.04

900	220.00	232.55	*233.40
1,000	253.33	269.04	*268.77
1,100	286.67	305.54	*304.13
1,200	320.00	342.03	*339.50

Comment. From table 1, we see that the corrected function of resistance (2.3.4) gives approximate data (which are better than those obtained using the noncorrected function of resistance) till around 700–750 m/s.

Anyway, we can use the nonmodified approximate Siacci function of resistance (2.3.2) where, for speeds greater than 750 m/s, we use (consider) the fictive ballistics coefficient

$c* = 1.060922 \cdot c$

We will denote the "fictive" ballistics coefficient with c and use always the approximate Siacci function of resistance (2.3.2). The results we obtain using (2.3.2), as are demonstrated throughout the book, are very accurate.

Example 2. Find the magnitude of drag acceleration a_D of the projectile of example 1 of 2.2 at a time when the mine speed is $v = 150 m/s$. Consider the density function $h(y) \approx 1$.

Solution
Since the projectile speed $v = 150 m/s \leq 256 m/s$, we use the first equation of (2.3.2). Substituting $v = 150 m/s$ in the first equation of (2.3.2), we find

$$K_D(150) = 1.212 \cdot 10^{-4} v^2 = 1.212 \cdot 10^{-4} \cdot (150)^2 = 2.727$$

Substituting in (2.2.10), $c = 1.28$, $K_D(150) = 2.727$ and $h(y) \approx 1$ (calculated in example 1 of 2.2), we find the projectile acceleration

$$a_D = c \cdot h(y) K_D(v) \approx (1.28)(1)(2.727) \approx 3.49 m/s^2$$

Example 3. Find the magnitude of the drag acceleration a_D of the projectile of example 2 of 2.2 at the time when the speed of the projectile is $v=700m/s$. Consider the density function $h(y)\approx 1$.

Solution

Since the projectile speed $v=700m/s>256m/s$, using (2.3.2), we can write

$$K_D(700)=(v-240)/3=(700-240)/3=153.3.$$

Substituting in (2.3.15) $c=1.28$ (calculated in example 2), $K_D(V)=153.3$, and $h(y)\approx 1$, we find the acceleration of the projectile as

$$a_D=c\cdot h(y)K(v)\approx(4.116)(1)(153.3)\approx 631.12m/s^2.$$

Example 4. Evaluate the value of the density function $h(y)$ and the density of air at 1,000 m above sea level. Assume standard atmosphere.

Solution

(a) Substituting $y=1000$ in (2.2.14), we find that the density of air 1,000 m above sea level is

$$h(y)=[(289.08-0.006328y)/289.08]^{4.4}=[(289.08-0.006328(1,000))/289.08]^{4.4}=0.907.$$

Since $h(y)=\rho/\rho_{0N}$, we find that the density of air is

$$\rho=\rho_{0N}h(y)=(1.205)(0.907)=1.093kg/m^3.$$

(b) We get approximately the same results using (2.2.15). Indeed,:

$$h(y)=(20,000-y)/(20,000+y)=(20,000-(1,000))/(20,000+(1,000))=0.905$$

and

$$\rho=\rho_{0N}h(y)=(1.205)(0.905)=1.091kg/m^3.$$

(c) Quite the same result is obtained using (2.2.16):

$$h(y)=e^{-0.0001003937y}=e^{-0.0001003937\cdot(1000)}=0.9045$$

and

$$\rho=\rho_{0N}h(y)=(1.205)(0.905)=1.091kg/m^3 .$$

Example 5. Find the magnitude of the drag force for the projectile of example 1 (section 2.2) when the altitude of flight above sea level is $y=1000m$ and the speed of projectile is 150 m/s. Mass of projectile 3.153kg.

Solution
     The acceleration of projectile is

$$a_D=c\cdot h(y)K_D(v)=(1.28)\cdot(0.907)\cdot(2.727)=3.1659 .$$

The magnitude of drag force in Newton is

$$D=ma_D=(3.153)(3.1659)=9.9822N .$$

Example 6. Use equation (2.3.9), $C_S(v/a)=(0.896)^{-1}C_M(v/a)$, to estimate the Siacci drag coefficients for the following Mach numbers: $M=v/a=1$ and $M=v/a=2$.

Solution
     From table 1of section 2.2, for the Mayevski drag coefficient, we have

$$C_M(1)=0.430 , \text{ and } C_M(2)=0.603 .$$

Substituting in (2.3.9), we find the Siacci drag coefficients

$$C_S(1)=(0.896)^{-1}C_M(\frac{v}{a})=(0.896)^{-1}(0.430)=0.4799$$

and

$$C_S(2)=(0.896)^{-1}C_M(\frac{v}{a})=(0.896)^{-1}(0.603)=0.6899$$
.

Example 7. Find the drag coefficient $C_D(v/a)$ of the Siacci's "fictive" standard projectile that corresponds to the Siacci's function of resistance (2.3.2). Then calculate the values of the drag coefficient when:

The Mach number is $M = v / a = 2$.
The Mach number is $M = v / a = 0.6$.

At the sea level (firing site), consider the speed of sound $a = 340.83m / s$ (standard atmosphere).

Solution
Substituting the first equation of (2.3.2) on the left side of equation (2.2.6), i.e. in

$$f_D(v) = 4.732 \cdot 10^{-4} v^2 \cdot C_D(v/a),$$

we can write:

$$1.212 \cdot 10^{-4} v^2 = 4.732 \cdot 10^{-4} v^2 \cdot C_D(v/a).$$

Hence, when the projectile speed and the Mach number are respectively $v \le 256m / s$ and $M \le 0.751$, we find that

$$C_D(v/a) = 0.256. \tag{1}$$

The equation (1) shows that the drag coefficient is constant for Mach number $M \le 0.751$.
Substituting the second equation of (2.3.2) on the left side of equation (2.2.6), i.e. in

$$f_D(v) = 4.732 \cdot 10^{-4} v^2 \cdot C_D(v/a),$$

we obtain:

$$(v - 240) / 3 = 4.732 \cdot 10^{-4} v^2 \cdot C_D(v/a).$$

Hence, for the drag coefficient, when the projectile speed and the Mach number are respectively $v > 256m/s$ and $M > 0.751$, we have:

$$C_D(v/a) = \frac{v - 240}{1.4196 \cdot 10^{-3} \cdot v^2}. \qquad (2)$$

Considering (1) and (2), we find that the drag coefficient of the Siacci's "fictive" standard projectile that corresponds to Siacci's function of resistance (2.3.2) is

$$C_D(v/a) = \begin{cases} 0.256 & for \quad v/a \leq 0.751 \\ (v - 240)/(1.4196 \cdot 10^{-3} v^2) & for \quad v/a > 0.751 \end{cases}$$
$$(3)$$

,

When the speed of sound is $a = 340.83m/s$, to the Mach number $M = v/a = 0.6$

corresponds the projectile speed: $v = 204.50m/s$. Substituting in (1) we find:

$$C_D(v/a) = C_D(0.6) = 0.256.$$

When speed of sound is $a = 340.83m/s$, to the Mach number $v/a = 2$ corresponds

the projectile speed: $v = 681.66m/s$. Substituting in (2): we find that the drag coefficient is

$$C_D(2) = \frac{(681.66) - 240}{1.4196 \cdot 10^{-3} \cdot (681.66)^2} = 0.670.$$

Note: In the figure below it is represented the graph of the Siacci's drag coefficient $C_D(v/a)$ given by equation (3).

Graph of the Drag coefficient $C_D(v/a)$ of Siacci's "Fictive" Standard Projectile

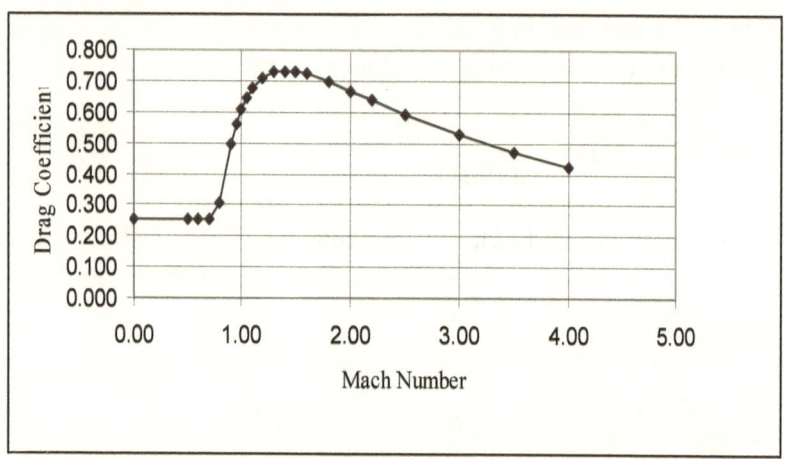

For comparison, hereafter is shown the graph of the drag coefficient that corresponds to the original Siacci's function of resistance (2.3.1). It is obtained in a similar way as the above graph, by substituting (2.3.1) in the left side of (2.2.6).

The graph is constructed by Prof. James Lewis, Marquette University.

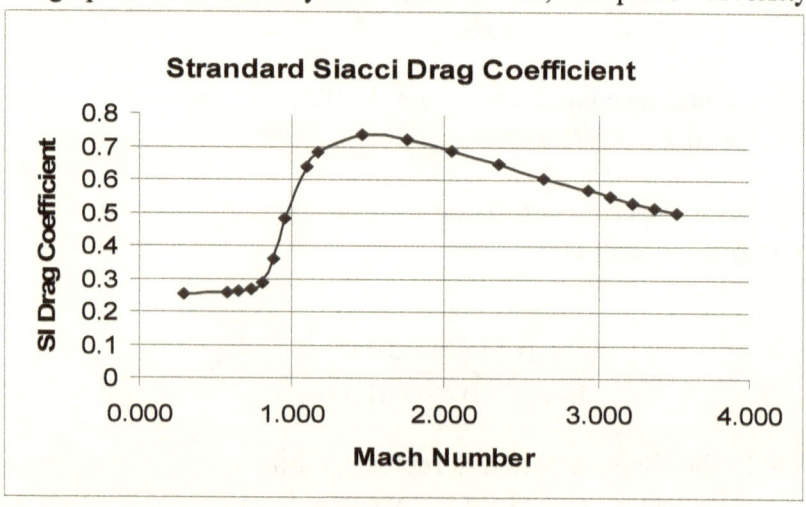

2.4 The Ingalls-Siacci Function in English Units

Col. Ingalls converted the Siacci function of resistance (2.3.1)

$$f(v)=0.2002 \cdot v-48.05+[(0.1648v-47.95)^2+9.6]^{1/2}+\frac{0.0442 \cdot v(v-300)}{371+(v/200)^{10}}, \qquad (2.4.1)$$

into English units. The Ingalls function of resistance is

$$I(v)=0.284746 \cdot v-224.221+[(0.234396v-223.754)^2+209.043]^{1/2}+\frac{0.019161v(v-984.261)}{371+(v/656.174)^{10}}$$

.

$$(2.4.2)$$

Instead of (2.4.2), we use the approximate Siacci function of resistance (2.3.2) converted in English units. For the acceleration (2.3.6) in m/s^2 of a given projectile, when the speed is greater than 256 m/s, we have

$$a_D = 1000 \cdot \frac{id^2}{m} h(y) \frac{v-240}{3}. \qquad (2.4.3)$$

Expressing the speed v in ft/s and the ballistics coefficient $c=id^2/m$ in $inch^2/pound$ and then introducing the multiplication factor 1.4223 to transform the right side of (2.4.3) into units of acceleration ft/s^2, we obtain the drag acceleration a_D expressed in English units $ft./s^2$

$$a_D = 1.4223 \cdot \frac{id^2}{m} h(y) \frac{v-787.4016}{3}, \qquad (2.4.4)$$

where d is the projectile diameter in inches (in), m is the projectile mass in pounds (lb), v is the projectile speed in feet per second ft/s. The formula (2.4.4) is valid when the projectile speed is greater than 840 ft/s.

In a similar way, when projectile speed is less than 256 m/s, the acceleration in English units is

$$a_D = 1.4223 \cdot \frac{id^2}{m} \cdot (1.1260 \cdot 10^{-5} v^2). \qquad (2.4.5)$$

Thus, the approximate Siacci function in English units is

$$K_I(v) = \begin{cases} 1.126 \cdot 10^{-5} v^2 & for \quad v \le 840\,ft./s \\ (v - 787.402)/3 & for \quad v > 840\,ft./s \end{cases}$$

(2.4.6)

The acceleration of the projectile for speed v from 0 till 3,937 ft/s can be written

$$a_D = 1.4223 \cdot \frac{id^2}{m} \cdot h(y) \cdot K_I(v)$$

(2.4.7)

The ballistics coefficient in the U.S. Army is defined by

$$C = \frac{m}{id^2}$$

(2.4.8)

From (2.4.7) and (2.4.8), we have

$$a_D = \frac{1.4223}{C} \cdot h(y) \frac{v - 787.402}{3}$$

(2.4.9)

We denote

$$c = \frac{1.4223}{C}$$

(2.4.10)

where c is in m^2/kg and C in $lb./in.^2$

Formula (2.4.10) can be used to convert the Ingalls ballistics coefficient into the Siacci ballistics coefficient and vice versa. It is obvious that the Ingalls form factor and the Siacci form factor are identical since they are related to the same Siacci standard projectile. In general, for projectile speed from 0 to 3,600 ft/s, the equation of projectile (2.3.6) in English units can be written as

$$\frac{d\vec{v}}{dt} = \vec{g} - c \cdot h(y) K_I(v) \frac{\vec{v}}{v}$$

(2.4.11)

where c is given by (2.4.10), while $K_I(v)$ by (2.4.6).

Standard Atmosphere in English Units

The characteristics of the atmospheric air are the Army Standard Metro[12] characteristics:

- The density of air is $\rho_{0N}=1.205kg/m^3=0.0752257lb/in^3$.
- Temperature of air is $T_{0N}=15°C=59°F$.
- Atmospheric pressure at sea level is $H_{0N}=750mm\ Hg.=29.52756in.Hg.$
- The relative air humidity is 50% and corresponds to the partial pressure of water vapor $e_{0N}=6.35mm\ Hg.=25in.$
- The speed of sound $a_{0N}=340.83m/s=1118.21ft./s$. Note that in USA army the standard temperature of the thrusting charge of the projectile is 21 degree Celsius.

The acceleration of gravity is $g=9.80665m/s=32.17405ft./s$. The density function that is present in (2.4.9) for the altitude in ft can be found using[13]

$$h(y)=\frac{\rho(y)}{\rho_{0N}}=e^{-0.00003159145y},\qquad(2.4.12)$$

where ρ_{0N} is the density of air at sea level, $\rho(y)$ is the density at altitude y above sea level.

Example 1. The Ingall's ballistics coefficient of a projectile is $C=0.454lb/in^2$. Find the Siacci ballistics coefficient.
Solution

Substituting in (2.4.10), $C=0.454lb/in^2$, we find that the Siacci ballistics coefficient is

---

[12] Relative humidity for the Army Standard Metro is 70%.
[13] Herrmann, Ernest. *Exterior Ballistics*, p. 35. U.S. Naval Institute, The College Press, 1935.

$$c = \frac{1.4223}{C} = \frac{1.4223}{0.454} = 3.133 m^2 / kg .$$

2.5 Mayevski-Zaboudski's and Mayevski-Ingalls's Functions

In the United States, the functions of resistance in use are the Mayevski-Zaboudski function (converted into English units by Col. Ingalls) and the Gâvre function. We call Mayevski-Zaboudski's function just Mayevski's function. The Mayevski function has the form

$$f_M(v) = A_m \cdot v^n .$$

(2.5.1)

Ingalls-Mayevski's function of resistance, referred as Ingalls's drag function is

$$f_I(v) = A \cdot v^n .$$

(2.5.2)

The values of the parameters A and A_m and the corresponding values of the exponent n are given in table 1.[14]

Table 1. Ingalls's and Mayevski-Zaboudski's Constants

Ingalls's Parameters & Exponents English Units			Mayevski-Zaboudski's Parameters & Exponents SI Units		
Projectile Speed, *fps*	Parameter A	Exponent n	Projectile Speed, *m/s*	Parameter A_m	Exponent n
3,600–2,600	4.0649013×10^{-3}	1.55	1,100–800	5.7080241×10^{-3}	1.55
2,600–1,800	1.2479581×10^{-3}	1.70	800–550	2.09422×10^{-3}	1.70
1,800–1,370	1.3160125×10^{-4}	2	550–419	3.154139×10^{-4}	2
1,370–1,230	9.5697369×10^{-8}	3	419–375	7.528305×10^{-7}	3
1,230–970	$6.3367999 \times 10^{-14}$	5	375–295	5.370842×10^{-12}	5
970–790	5.9352634×10^{-8}	3	295–240	4.670367×10^{-7}	3
790–0.00	4.6761669×10^{-5}	2	240–0	1.120888×10^{-4}	2

----

[14] Mori, Edoardo. Balistica Teorica e Pratica. http://www.earmi.it/balistica/formi.htm. [Web site]

Note: The parameter A_m in SI units is obtained considering the fact that Mayevski's ballistics coefficient is $c_m(v)=id^2\rho_0/w$, where the projectile diameter d is in millimeters, the weight w is in gram force. The air density is considered $\rho_0=1.225$ and not $\rho_0=1.205$.

The projectile acceleration related to the Mayevski function in meter per square second is

$$a_D=\frac{id^2}{m}1000A_m v^n,$$

(2.5.3)

where the projectile caliber d is in meters and the mass m is in kilograms. The form factor related to Mayevski's standard projectile is $i=1$.

The projectile acceleration in feet per seconds associated with the Mayevski-Ingalls function (2.5.2) is

$$a_D=\frac{\cdot 1}{m/(i\cdot d^2)}Av^n,$$

(2.5.4)

where the projectile caliber d is in inches, the mass m is in pounds. The standard projectile form factors is $i=i_{ing}=1$.

The differential equation of projectile flight (2.2.12) for the Mayevski function (2.5.1) and the Mayevski-Ingalls function (2.5.2) can be written respectively

$$\frac{d\vec{v}}{dt}=\vec{g}-c\cdot h(y)A_M v^n\frac{\vec{v}}{v},\quad \frac{d\vec{v}}{dt}=\vec{g}-c\cdot h(y)Av^n\frac{\vec{v}}{v},$$

(2.5.5)

where A_m and A are given in table 1.

Example 1. Find the values of Mayevski's function (2.5.1),

$$f_M(v)=A_m\cdot v^n,$$

for (a) $v=102m/s$, (b) $v=680m/s$, and (c) $v=850m/s$.

Solution

(a) $f(v)_{may} = A_m \cdot v^n = 1.210888 \cdot 10^{-4} \cdot (102)^2 = 1.258$

(b) $f(v)_{may} = A_m \cdot v^n = 2.09422 \cdot 10^{-3} \cdot (680)^{1.7} = 136.86$

(c) $f(v)_{may} = A_m \cdot v^n = 5.7080241 \cdot 10^{-3} \cdot (850)^{1.55} = 198.19$

Example 2. For the Mayevski-Zaboudski standard projectile, find the value of the function of resistance (2.2.6),

$$f(v) = 4.732 \cdot 10^{-4} \cdot v^2 C(v/a),$$

if the value of the drag coefficient $C(v/a)$ of the standard projectile are the following:[15]

(a) $C(v/a) = 0.228$, for $M = (v/a) = 0.3$, $(v=102m/s)$
(b) $C(v/a) = 0.603$, for $M = (v/a) = 2$, $(v=680m/s)$
(c) $C(v/a) = 0.559$, for $M = (v/a) = 2.5$, $(v=850m/s)$

Solution

Substituting in the above formula, we find the following values of the Mayevski-Zaboudski function of resistance:

(a) $f(102) = 4.732 \cdot 10^{-4} \cdot v^2 C(v/a) = 4.732 \cdot 10^{-4} (102)^2 \cdot (0.228) = 1.122$

(b) $f(680) = 4.732 \cdot 10^{-4} \cdot v^2 C(v/a) = 4.732 \cdot 10^{-4} (680)^2 \cdot (0.603) = 131.94$

(c) $f(850) = 4.732 \cdot 10^{-4} \cdot v^2 C(v/a) = 4.732 \cdot 10^{-4} (850)^2 \cdot (0.559) = 191.11$

Note: Comparing the above outcomes with the outcomes obtained in example 1, we see that they are approximately equal.

Example 3. Consider a projectile ballistics coefficient $C = m/d^2 = 0.225 \cdot lb/in^2$ that corresponds to $c = d^2 1000/m = 6.316 m^2 / kg$.

---

[15]   See table 1 section 2.1

Find the acceleration of the projectile using functions of resistance (2.5.1) and (2.5.2), for $v=100m/s=328\,ft/s$ and $v=550m/s=1804.50\,ft/s$.

Solution
(a) In SI units

For $v=100m/s$,

$$a_D=\frac{id^2}{m}1000A_m v^n=(6.3158)\cdot(1.12076\cdot10^{-4})(100)^2=7.0785m/s^2.$$

For $v=550m/s$,

$$a_D=\frac{id^2}{m}1000A_m v^n=(6.316)(3.154139\cdot10^{-4})\cdot(550)^2=602.63m/s^2\cdot$$

(b) In English units

For $v=328\,ft/s$,

$$a_{De}=\frac{\cdot1}{m/(i\cdot d^2)}Av^n=(0.225)^{-1}(4.6761669\cdot10^{-5})\cdot(328)^2=22.359\,ft/s=6.82m/s.$$

For $v=1804.50\,ft/s$,

$$a_{De}=\frac{\cdot1}{m/(i\cdot d^2)}Av^n=(0.225)^{-1}(1.3160125\cdot10^{-4})\cdot(1804.50)^2=1904.55\,ft/s^2=580.51m/s^2.$$

Example 4. The measured value of the Siacci function for a nonstandard projectile of caliber $d=76.2$ mm, mass $m=6.5$ kg, at speed $v=600$ m/s is $K(v)=65$. Find the form coefficient of the given projectile at the given speed.
Solution
Employing (2.2.18), $f(v)=i\cdot f_D(v)$, for the Siacci form coefficient we can find

$$i=\frac{K(v)}{K_D(v)}=\frac{K(600)}{(v-240)/3}=\frac{65}{(600-240)/3}=0.54167$$

2.6 Estimation of Siacci's Form Factor

The form factor of a projectile i depends on the projectile speed or, better to say, on the Mach number of the projectile. The form factor depends as well on the particular function of resistance in use. In other words, a projectile might have different form factors.

We will see some methods to determine experimentally the form factor i and the ballistics coefficient c when we are able to measure the projectile ballistics characteristics in two points on the trajectory of flight that are relatively close to each other, so close that the projectile trajectory can be assumed to be a straight line parallel to the x-axis.

(I) **Estimation of the Form Factor** when we measure the projectile speed in two relatively close points of the projectile trajectory

Consider the equation of projectile motion (2.2.13),

$$\frac{d\vec{v}}{dt}=\vec{g}-4.732 \cdot 10^{-4} c \cdot h(y)v^2 C_D(v/a)\frac{\vec{v}}{v}. \tag{2.6.1}$$

We assume that the projectile trajectory between two relatively close points with abscissa x_1 and x_2 is approximately a line parallel to the x-axis (for example, the projectile is launched with a relatively narrow launching angle). The speed of the projectile in those two points is respectively v_1 and v_2. Since the trajectory is almost a straight line parallel to the x-axis from (2.6.1), we obtain

$$\frac{dv}{dt}=-4.732 \cdot 10^{-4} c \cdot h(y)v^2 C_D(v/a). \tag{2.6.2}$$

Hence, we write

$$\frac{dv}{dx}=-4.732 \cdot 10^{-4} c \cdot h(y)v C_D(v/a). \tag{2.6.3}$$

Note as well that $h(y)$ is a constant since the projectile trajectory is approximately a horizontal line. Substituting in (2.6.3) the average Mach number ($\overline{M}=\overline{v}/a$) in the interval (x_1, x_2), we write

$$\frac{dv}{v}=-4.732 \cdot 10^{-4} c \cdot h(y) C_D (\overline{v}/a) \cdot dx \tag{2.6.4}$$

where $\overline{v}=(v_1+v_2)/2$. Integrating (2.6.4), we have

$$\ln(\frac{v_1}{v_2})=4.732 \cdot 10^{-4} c \cdot h(y) C_D (\overline{v}/a) \cdot (x_2-x_1) . \tag{2.6.6}$$

Hence, for the average ballistics coefficient c of the given projectile in the interval (x_1, x_2), we find

$$c=\frac{\ln(v_1/v_2)}{4.732 \cdot 10^{-4} h(y) C_D (\overline{v}/a) \cdot (x_2-x_1)} . \tag{2.6.7}$$

Since $c=1000 i d^2 /m$, from (2.6.7), for the average form factor i with respect to a particular drag coefficient $C_D (v/a)$, we obtain

$$i=\frac{m \cdot \ln(v_1/v_2)}{4.732 \cdot 10^{-1} d^2 h(y) C_D (\overline{v}/a) \cdot (x_2-x_1)} . \tag{2.6.8}$$

Actually, the form factor of a projectile i , as the formula (2.6.1) shows, depends on the projectile speed or, better to say, on the projectile Mach number. For a known drag coefficient, $C_D (v/a)$, employing (2.6.7) or (2.6.8), we can estimate an average value of the ballistics coefficient or the average form factor i in the given interval.

Solving (2.6.7) for the drag coefficient, we find a formula to estimate the average drag coefficient of the standard projectile

$$C_D (\overline{v}/a)=\frac{\ln(v_1/v_2)}{4.732 \cdot 10^{-4} c \cdot h(y)(x_2-x_1)} \tag{2.6.9}$$

Since the drag coefficient of a given projectile is $C(\bar{v}/a)=iC_D(\bar{v}/a)$ and its ballistics coefficient is $c = 1000 \cdot id^2 / m$, from (2.6.9), we find that the drag coefficient of the given projectile is

$$C(\bar{v}/a)=\frac{m \cdot \ln(v_1/v_2)}{4.732 \cdot 10^{-1} d^2 \cdot h(y)(x_2-x_1)}. \qquad (2.6.10)$$

Consider (2.2.6),

$$f_D(v)=4.732 \cdot 10^{-4} \cdot v^2 C_D(v/a), \text{ or } f(v)=4.732 \cdot 10^{-4} \cdot v^2 C(v/a),$$

Solving the obtained equation for $C_D(v/a)$, we find

$$C_D(\bar{v}/a) = f_D(\bar{v}) \div (4.732 \cdot 10^{-4} \bar{v}^2)$$

Then, substituting the above result in (2.6.7) and (2.6.8), we find respectively

$$c=\frac{\bar{v}^2 \ln(v_1/v_2)}{h(y)f_D(\bar{v}) \cdot (x_2-x_1)} \qquad (2.6.11)$$

and

$$i=\frac{m \cdot \bar{v}^2 \ln(v_1/v_2)}{d^2 h(y)f_D(\bar{v}) \cdot (x_2-x_1)}. \qquad (2.6.12)$$

Application for the Siacci Function

(a) For speeds of projectile greater than 256 m/s

For such speed of projectile, the Siacci function is $K_D(v)=(v-240)/3$. Substituting $K_D(v)$ for $f_D(v)$ in (2.6.11) and (2.6.12), we obtain

$$c=\frac{3\bar{v}^2 \ln(v_1/v_2)}{h(y)(\bar{v}-240) \cdot (x_2-x_1)} \qquad (2.6.13)$$

and

$$i=\frac{3m\cdot\bar{v}^2\ln(v_1/v_2)}{d^2h(y)(\bar{v}-240)\cdot(x_2-x_1)},$$
(2.6.14)

where

$$c=\frac{id^2}{m}\cdot1000 , \quad h(y)=(\frac{289.08-0.006328y}{289.08})^{4.4} .$$

(b) For speed of projectile less than 256 m/s

For such speed, the Siacci function is $K_D(v)=1.212\cdot10^{-4}v^2$. In a similar way, like for the projectile speed greater than 256 m/s, we obtain the following formulas:

$$c=\frac{1}{1.212\cdot10^{-4}h(y)(x_2-x_1)}\ln\frac{v_1}{v_2},$$
(2.6.15)

$$i=\frac{m}{1.212\cdot10^{-1}d^2h(y)(x_2-x_1)}\ln\frac{v_1}{v_2}.$$
(2.6.16)

In a similar way, we can find the ballistics coefficient and the form factor for Mayevski's function or other non-Siacci functions.

Second Approach
Consider (2.2.12),

$$\frac{d\vec{v}}{dt}=\vec{g}-c\cdot h(y)f_D(v)\frac{\vec{v}}{v}.$$
(2.6.17)

Since the trajectory is almost a horizontal line, we have

$$\frac{dv}{dt}=-c\cdot h(y)f_D(v).$$
(2.6.18)

We write (12.6.8) as

$$\frac{dv}{dx}=-c\cdot h(y)\frac{f_D(v)}{v}.$$
(2.6.19)

Substituting in (2.6.19) the Siacci function ,

$$K_D(v) = \begin{cases} 1.212 \cdot 10^{-4} v^2 & for \quad v \le 256 \\ (v-240)/3 & for \quad v > 256 \end{cases} \tag{2.6.20}$$

for $f_D(v)$, we write

$$\frac{dv}{dx} = -1.212 \cdot 10^{-4} c \cdot h(y) \cdot v \tag{2.6.21}$$

and

$$\frac{dv}{dx} = -c \cdot h(y) \frac{(v-240)}{v} \tag{2.6.22}$$

respectively, when the speed of projectile is less than 256 m/s or greater than 256 m/s.

Integrating (2.6.21) and (2.6.22), we obtain

$$\ln \frac{v_1}{v_2} = 1.212 \cdot 10^{-4} c \cdot h(y)(x_2 - x_1) \tag{2.6.23}$$

and

$$(v_1 - v_2) + 240 \cdot \ln \frac{v_1 - 240}{v_2 - 240} = \frac{c \cdot h(y)}{3}(x_2 - x_1) . \tag{2.6.24}$$

Solving for c, we find

$$c = \frac{1}{1.212 \cdot 10^{-4} h(y)(x_2 - x_1)} \ln \frac{v_1}{v_2} \tag{2.6.25}$$

and

$$c = 3 \cdot [(v_1 - v_2) + 240 \cdot \ln \frac{v_1 - 240}{v_2 - 240}] \div [h(y)(x_2 - x_1)], \tag{2.6.26}$$

respectively, when the speed of projectile is less than 256 m/s or greater than 256 m/s.

Substituting $c=1000 \cdot id^2/m$ in the above equations, we are able to find the form factor of the projectile i.

II. Estimation of the Form Factor when we measure the projectile speed as a function of time in two relatively close points on the trajectory.

Using (2.6.20) to integrate (2.6.18), we find an approximate value of the ballistics coefficient

$$c = \frac{1}{1.212 \cdot 10^{-4} h(y)t}(\frac{1}{v} - \frac{1}{v_0}) \qquad (2.6.27)$$

and

$$c = \frac{3}{h(y) \cdot t} \cdot \ln(\frac{v_0 - 240}{v - 240}), \qquad (2.6.28)$$

respectively, when the speed of projectile is less than 256 m/s or greater than 256 m/s.

Hence, for the form factor corresponding to Siacci law we have

$$i_1 = \frac{m}{d^2 h(y) 1.212 \cdot 10^{-1} t}(\frac{1}{v} - \frac{1}{v_0}), \qquad (2.6.29)$$

and

$$i_2 = \frac{3m}{1000 d^2 t \cdot h(y)} \cdot \ln\frac{v_0 - 240}{v - 240}, \qquad (2.6.30)$$

respectively when the speed of projectile is less than 256 m/s or greater than 256 m/s.

Comment: The above formulas show that the measured BC of a projectile depends by the measured velocity in two close points.

Thus, experimentally we are able to find BC as a function of velocity interval (or average velocity).

Thus, for the BC of Russian rifle of example 2 (below), we find the following BC function: 4.835 (735 - 640m/s); 4.164 (640 - 557m/s); 4.011 (557 - 485); 3.883 (485 - 424m/s); 3.852 (424 - 373 m/s); 3.858 (373 - 332 m/s); 3.995 (332 - 300); 4.323 (300 - 276 m/s); 5.342 (276 - 258).

III. Estimation of the Siacci Form Factor Using Non-Siacci Form Factors

In many problems of exterior ballistics, we need to use the drag coefficient of a projectile that is obtained experimentally to find the Siacci form factor of that projectile. In some other cases, we need to use as well some average form factors that are already determined experimentally for non-Siacci functions. For example, we need to use a non-Siacci form factor that is related to Mayevski's function, Ingall's functions, G-function, etc., to find the corresponding Siacci form factor. We have to modify a non-Siacci form factor, to transform it into a Siacci form factor, in order to use it with the Siacci function to solve exterior ballistics problems.

Consider the drag coefficient $C(v/a)$ of a particular projectile that is obtained experimentally—for example, a table of values as a function of Mach number. We denote i_n the form factor of the given projectile related to a non-Siacci drag coefficient $C_n(v/a)$. For example, $C_n(v/a)$ is the drag coefficient of the Mayevski standard projectile. Employing (2.1.5) for the drag coefficient of the given projectile, we write

$$C(\frac{v}{a})=i_n C_n(\frac{v}{a}).$$

(2.6.31)

For the acceleration of the same projectile, we can write (2.2.10),

$$a_D=(4.7302 \cdot 10^{-4}) \cdot \frac{d^2}{m} 1000 \frac{\rho}{\rho_{0N}} v^2 C(\frac{v}{a}).$$

(2.6.32)

On the other hand, employing (2.2.7), we can express the acceleration of the given projectile through the Siacci function

$$a_D = \frac{id^2}{m} 1000 \frac{\rho}{\rho_{0N}} K_D(v),$$
(2.6.33)

where i is the Siacci form coefficient of the projectile and

$$K_D(v) = \begin{cases} 1.212 \cdot 10^{-4} v^2 & for \quad v \leq 256 m/s \\ (v-240)/3 & for \quad v > 256 m/s \end{cases}$$
(2.6.34)

is the Siacci function. From (2.6.32) and (2.6.33), we write

$$\frac{id^2}{m} 1000 \frac{\rho}{\rho_{0N}} K_D(v) = (4.7302 \cdot 10^{-4}) \cdot \frac{d^2}{m} 1000 \frac{\rho}{\rho_{0N}} v^2 C(\frac{v}{a}).$$
(2.6.35)

From (2.6.35), we find the Siacci form coefficient expressed through a non-Siacci drag coefficient

$$i = (4.7302 \cdot 10^{-4}) \cdot \frac{v^2}{K_D(v)} C(\frac{v}{a}).$$
(2.6.36)

Substituting (2.6.31) in (2.6.36), we find the Siacci form coefficient expressed as a function of a non-Siacci form coefficient and drag coefficient of the non-Siacci standard projectile

$$i = (4.7302 \cdot 10^{-4}) \cdot \frac{v^2}{K_D(v)} i_n C_n(\frac{v}{a}).$$
(2.6.37)

The formulas (2.6.36) and (2.6.37) allow us to find the Siacci form factor when we know the drag coefficient of the given projectile or when we know the non-Siacci form coefficient of a given projectile and the drag coefficient of a non-Siacci standard projectile $C_n(v/a)$.

Considering (2.6.34), from (2.6.36), we find the Siacci form coefficient

$$i = 3.9043 \cdot C(\frac{v}{a})$$
(2.6.38)

for speed $v < 256 m/s$ and

$$i=0.0014196\frac{v^2}{v-240}\cdot C(\frac{v}{a})$$

(2.6.39)

for speed of projectile $v>256m/s$.

Substituting (2.6.34) into (2.6.37), we obtain two other formulas to find the Siacci form coefficient respectively

$$i=3.9043\cdot i_n C_n(\frac{v}{a})$$

(2.6.40)

for speed $v<256m/s$ and

$$i=0.0014196\frac{v^2}{v-240}i_n\cdot C_n(\frac{v}{a})$$

(2.6.41)

for projectile speed $v>256m/s$.

Example 1. Evaluate the average value of the drag coefficient $C(\bar{v}/a)$ for a $m=7.9gram$ bullet of a Russian semiautomatic rifle of caliber $d=7.62mm$ if the initial speed of the bullet is $v_0=735m/s$ and the speed of the projectile at $x=100m$ from the launching point is $v=640m/s$. The atmospheric conditions are standard. The speed of sound in standard conditions is $a_{0N}=340.83m/s$.

Solution
The average speed is

$$\bar{v}=(v_1+v_2)/2=(735+640)/2=687.5$$.

The average Mach number is

$$\overline{M}=\bar{v}/a=(687.5)/(340.83)=2.017135$$.

Substituting in (2.6.10), we find that the Siacci drag coefficient of the given bullet in the interval (0, 100 m) is

$$C(\bar{v}/a)=\frac{m{\cdot}\ln(v_1/v_2)}{4.732{\cdot}10^{-1}d^2{\cdot}h(y)(x_2-x_1)}=\frac{(0.0079)\ln(735/640)}{4.732{\cdot}10^{-1}(0.00762)^2(1)(100-0)}=0.3979$$

Example 2. Estimate the form factor of the bullet given in exercise 1.

Solution

Substituting the Siacci function of resistance $K_D(v)=(v-240)/3$ in (10) instead of $f_D(\bar{v})$, we find the coefficient of form

$$c=\frac{\bar{v}^2\ln(v_1/v_2)}{h(y)f_D(\bar{v}/a){\cdot}(x_2-x_1)}=\frac{3\bar{v}^2\ln(v_1/v_2)}{h(y)(\bar{v}-240)(x_2-x_1)}=$$

$$=\frac{3(0.0079)(687.5)^2\ln(735/640)}{(1)(687.5-240)(100-0)}=4.3855$$

Using (2.6.11), we find the coefficient of form

$$i=\frac{m{\cdot}c}{1000d^2}=\frac{(0.0079)(4.3855)}{1000(0.00762)^2}=0.59667\,.$$

Example 3. Find the form factor and the ballistics coefficient of a bullet of a Russian semiautomatic rifle of caliber $d=7.62mm$ using the following data obtained from the range table of the rifle: mass $d=7.9g$ and initial speed of bullet $v_0=735m/s$. At the range $x=100m$, the speed of projectile is $v=640m/s$.

Solution

Substituting the given numeric values in (2.6.26), we find that

$$c=3{\cdot}[(735-640)+240{\cdot}\ln\frac{735-240}{640-240}]\div[(1)(100-0)]=4.384271151.$$

Since $c=1000{\cdot}id^2/m$, the value of the form factor is

$$i=\frac{c{\cdot}m}{d^2h(y)1000}=\frac{(4.38427)(0.0079)}{(0.00762)^2(1000)}=0.5965055\,.$$

Note: The value obtained for the form factor $i=0.5965055$ in this example is quite equal to the value $i=0.59667$ obtained in example 2.

Example 4. Find the ballistics coefficient and the form factor of a bullet of a Russian semiautomatic rifle of caliber $d=7.62mm$ using the following data: mass $m=7.9g$ and initial speed of bullet $v_0=735m/s$. At a range of $x=100m$, the speed of impact and time of flight are respectively $v=640m/s$ and $t=0.14s$.

Solution

We can consider $h(y)=\rho/\rho_0=1$. Since the speed of the projectile during flight is greater than 256 m/s, to find an average value of the ballistics coefficient, we use (28). Substituting in (28), we get

$$c_2=\frac{3}{h(y)\cdot t}\ln(\frac{v_0-240}{v-240})=\frac{3}{(1)\cdot(0.14)}\ln\frac{735-240}{640-240}=4.5663.$$

The form factor is

$$i_2=10^{-3}c\frac{m}{d^2}=10^{-3}(4.5663)\frac{(7.9\cdot10^{-3})}{(7.62\cdot10^{-3})^2}=0.6213$$

It is different from the value $i=0.5965055$ obtained in example 3.

The different value of the form factor obtained in example 1 from the present value is related to the low accuracy of measurement of time of flight or with the rounding values of the time of flight.

Example 5. The value of the drag coefficient of a 0.30 M2 ball bullet caliber $d=0.308in=0.00782m$, for the Mach number $M=v/a=2.2$ is $C(v/a)=0.331$

Mass of the bullet is 0.00972 kg, and launching speed $v_0=853.43m/s$.[16] The density of air at sea level is $\rho_0=1.205kg/m^3$. The speed of sound is a = 341.46m/s,

(a) Estimate the Siacci form factor.
(b) Find an average form factor and an average ballistics coefficient.

---

[16] McCoy, Robert L. *Modern Exterior Ballistics*, p. 101. Schiffer Publishing, Ltd., 1999.

(c) Express the ballistics coefficient as a function of projectile speed,
$c=c(v)$.

Solution
(a) Form Factor

The speed of the 0.30 M2 ball bullet that corresponds to the Mach
number $M=v/a=2.2$ is

$$v=M{\cdot}a=(2.2)(341.46)=751.21m/s.$$

Employing (2.6.39), we find that the Siacci form factor that
corresponds to the speed $v=751.21m/s$ is

$$i=0.0014196\frac{v^2}{v-240}{\cdot}C(\frac{v}{a})=0.0014196\frac{(751.21)^2}{(751.21-240)}(0.331)=0.518702.$$

(b) Average Siacci Form Factor

In a similar way, we find the Siacci coefficients that correspond to
Mach numbers from 1.4 to 2.5. The estimated results are presented in
the following table.

Table 1. Siacci form factor for 0.30 Ball M2 bullet

Mach $M=v/a$	1.4	1.6	1.8	2	2.2	2.5
v	478.004	546.336	614.628	682.92	751.212	853.65
i	0.58193	0.5519	0.53394	0.52467	0.5187	0.51417

From the above table, we find that the **average value** of the Siacci
form coefficient is

$$\bar{i}=0.537552\approx0.538.$$

The average ballistics coefficient is

$$c=\frac{id^2}{m}1000=\frac{(0.538)(0.00782)^2}{0.00972}1000=3.384773m^2/kg\,.$$

(c) Ballistics Coefficient as a Function of the Projectile Speed, $c=c(v)$

Using the data of table 1 and a Minitab program, we find a second-degree regression equation that expresses the form coefficient as a function of the given projectile speed

$$i=0.913405-0.0009944v+0.00000062v^2\,. \tag{1}$$

Substituting the diameter of the projectile

$d=0.308in=(0.308)(0.0254)=0.0078232m$ and the mass $m=0.00972kg$ in

$$c=\frac{id^2}{m}1000\,,$$

we find the ballistic coefficient of 0.30 M2 ball bullet as a function of projectile speed

$$c(v)=(0.913405-0.0009944v+0.00000062v^2)\cdot(6.2913992)\,. \tag{2}$$

2.7 The Systems of Differential Equations of Projectile in Air

The differential equation of projectile motion (2.2.12),

$$\frac{d\vec{v}}{dt}=\vec{g}-c\cdot h(y)f_D(v)\frac{\vec{v}}{v}\,, \tag{2.7.1}$$

where

$$c=\frac{id^2}{m}1000 \tag{2.7.2}$$

and

$$h(y)=(\frac{289.08-0.006328y}{289.08})^{4.4}\,, \tag{2.7.3}$$

is a fundamental equation exterior ballistics uses to study the projectile flight in air and to construct the ballistics and range tables.

The function of resistance $f_D(v)$ we use in (2.7.1) is the approximate Siacci's function

$$K_D(v) = \begin{cases} 1.212 \cdot 10^{-4} v^2 & \text{for } v \leq 256 m/s \\ (v-240)/3 & \text{for } v > 256 m/s \end{cases} \qquad (2.7.4)$$

Using the Siacci function (2.7.4), the differential equation (2.7.1) can be written

$$\frac{d\vec{v}}{dt} = \vec{g} - c \cdot h(y) K_D(v) \frac{\vec{v}}{v} \qquad (2.7.5)$$

Initial conditions are

$$\vec{r} = 0\vec{i} + 0\vec{j} + 0\vec{k}, \quad \vec{v} = v_0 \cos\alpha_0 \vec{i} + v_0 \sin\alpha_0 \vec{j} + 0 \cdot \vec{k} \quad \text{when } t=0 \qquad (2.7.6)$$

Projecting the differential equation (2.7.5) on x-, y-, and z-axes, we obtain the following system of six differential equations:

$$\begin{cases} \dfrac{dv_x}{dt} = -ch(y) K_D(v) \dfrac{v_x}{(v_x^2 + v_y^2)^{1/2}} \\[2ex] \dfrac{dv_y}{dt} = -ch(y) K_D(v) \dfrac{v_y}{(v_x^2 + v_y^2)^{1/2}} - g, \\[2ex] \dfrac{dv_z}{dt} = 0 \end{cases} \qquad (2.7.7)$$

where

$$\left\{ v_x = \frac{dx}{dt}, \quad v_y = \frac{dy}{dt}, \quad v_z = \frac{dz}{dt}. \right. \qquad (2.7.8)$$

Integrating the last equation of (2.7.8), $dv_z/dt=0$, considering that at $t=0$ the component of velocity along z-axis is zero, we find that the component of the velocity of the projectile along z-axis at any moment t is zero,

$$v_z(t) = dz/dt = 0.$$

Integrating the above equation for the initial conditions $z=0$ when $t=0$, we find that at any time t, the z-coordinate of the projectile is zero, i.e., $z(t)=0$. That means that the trajectory of projectile is all the time in the xy-plane determined by the initial velocity vector $\vec{v}_0$ and the x-axis.

Thus, the system of the six differential equations (2.7.7) and (2.7.8) is reduced to a system with four differential equations

$$
\begin{cases}
\dfrac{dv_x}{dt} = -ch(y)K_D(v)\dfrac{v_x}{(v_x^2+v_Y^2)^{1/2}} \\
\dfrac{dv_y}{dt} = -ch(y)K_D(v)\dfrac{v_y}{(v_x^2+v_y^2)^{1/2}} - g \\
\dfrac{dx}{dt} = v_x \\
\dfrac{dy}{dt} = v_y
\end{cases}
\qquad (2.7.9)
$$

with initial conditions

$$x=0, \ y=0, \ v_x = v_0\cos\alpha_0, \ v_y = v_0\sin\alpha_0 \text{ when } t=0. \quad (2.7.10)$$

Solving the system of differential equations (2.7.9) (employing the methods of numerical integration) for a given projectile lunched in air with a given initial velocity, we are able to find the parametric equations and all the elements of the projectile trajectory, as for example the coordinates and the velocity at any moment, the time of flight to the target, the coordinates of the maximum altitude of flight, coordinates of impact on the ground, etc.

In some particular cases, which are shown in the following chapters, we are able to solve the system of differential equations of projectile motion analytically, i.e., to obtain formulae to calculate the projectile trajectory and its elements.

Introducing the Variable p

To simplify the solution of (2.7.9), we need to express the y-component of velocity v_y through the x-component of velocity v_x. For that reason, exterior ballistics introduces a new variable p that is a function of time t and is defined by the equation

$$v_y = pv_x \qquad (2.7.11)$$

where $p = \tan\alpha$.

The Differential Equations of Variable Time t

Differentiating both sides of the equation (2.7.11) with respect to time t, we write

$$\frac{dv_y}{dt} = v_x \frac{dp}{dt} + p \frac{dv_x}{dt}. \qquad (2.7.12)$$

Substituting the first two equations of system (2.7. 9) in (2.7.12), we can write (2.7.9) as follows:

$$\begin{cases} \dfrac{dv_x}{dt} = -ch(y)K_D(v)\dfrac{v_x}{(v_x^2+v_y^2)^{1/2}} \\[2mm] \dfrac{dp}{dt} = -\dfrac{g}{v_x} \\[2mm] \dfrac{dx}{dt} = v_x \\[2mm] \dfrac{dy}{dt} = v_y \end{cases}, \qquad (2.7.13)$$

where

$$p = \frac{v_y}{v_x} = \tan\alpha. \qquad (2.7.14)$$

The Differential Equations of Variable x

We can transform (13) into a system of differential equations of variable x. Considering the second equation of (13), we have

$$\frac{dv_X}{dt}=\frac{dv_X}{dx}\cdot\frac{dx}{dt}=v_X\frac{dv_X}{dx}\;,\;\frac{dp}{dt}=\frac{dp}{dx}\cdot\frac{dx}{dt}=v_X\frac{dp}{dx}\;,\;\frac{dy}{dt}=\frac{dy}{dx}\cdot\frac{dx}{dt}=v_X\frac{dx}{dt}\;.(2.7.15)$$

Employing (2.7.15) and (2.7.14), we write (2.7.9) as

$$\begin{cases} \dfrac{dv_x}{dx}=-ch(y)\cdot\dfrac{K_D(v)}{v} \\[2mm] \dfrac{dp}{dx}=-\dfrac{g}{v_x^2} \\[2mm] \dfrac{dt}{dx}=\dfrac{1}{v_x} \\[2mm] \dfrac{dy}{dx}=p \end{cases}\;, \qquad\qquad (2.7.16)$$

where

$$p=v_y/v_x=\tan\alpha$$

and

$$K_D(v)=\begin{cases}1.212\cdot10^{-4}v^2 & for \quad v\le256m/s \\ (v-240)/3 & for \quad v>256m/s\end{cases} \qquad (2.7.17)$$

The Differential Equations of Variable p

To simplify the solution of (2.7.9) we change the variable of integration from t to $p=\tan\alpha$. Using the chain rule, and considering as well the second equation of (2.7.13), we can write

$$\frac{dv_x}{dt}=\frac{dv_x}{dp}\cdot\frac{dp}{dt}=(-\frac{g}{v_x})\frac{dv_x}{dp}\;,\quad\frac{dx}{dt}=\frac{dx}{dp}\cdot\frac{dp}{dt}=-(\frac{g}{v_x})\frac{dx}{dp}\;,$$

$$\frac{dy}{dt}=\frac{dy}{dp}\cdot\frac{dp}{dt}=-(\frac{g}{v_x})\frac{dy}{dp}\;. \qquad\qquad (2.7.18)$$

Substituting (2.7.18) in (2.7.13) and taking into account that

$$v=(v_x^2+v_y^2)^{1/2}=v_x(1+v_y^2/v_x^2)^{1/2}=v_x(1+p^2)^{1/2}\;, \qquad (2.7.19)$$

we obtain the differential equations of projectile flight

$$\begin{cases} \dfrac{dv_x}{dp}=\dfrac{c \cdot h(y)}{g}K_D(v)\dfrac{v_x}{(1+p^2)^{1/2}} \\[2mm] \dfrac{dt}{dp}=-v_x/g \\[2mm] \dfrac{dx}{dp}=-v_x^2/g \\[2mm] \dfrac{dy}{dp}=-pv_x^2/g \end{cases} \qquad , \qquad (2.7.20)$$

where

$$p=v_y/v_x=\tan\alpha,$$

and

$$K_D(v)=\begin{cases} 1.212 \cdot 10^{-4}v^2 & for \quad v \le 256 m/s \\ (v-240)/3 & for \quad v>256m/s \end{cases}. \qquad (2.7.21)$$

In general, the solution of the differential equations of projectile motion (2.7.13), (2.7.16), and (2.7.20) for some given initial conditions can be solved numerically, for example, by using Runge-Kutta numerical methods of integration.

In some particular cases, we can get approximate analytical solutions that are close to the results obtained by numerical integration and coincide strongly with the practical data of shooting.

Example 1. Show that the horizontal component of the projectile velocity decreases along the trajectory of flight.

Solution
We use the differential equations (2.7.16) to prove the statement of exercise 1.

The first equation of (2.7.16)

$$\frac{dv_x}{dx}=-\frac{ch(y)}{g}\frac{K_D(v)}{v} \qquad (1)$$

shows that the derivative of the horizontal component of the velocity is negative.

It follows that the horizontal component of velocity decreases as x increases.

Example 2. Show that the projectile speed is asymmetric and that the terminal speed is lower than its initial speed on condition that gun and target are at the same level above sea or at sea level.

Solution

Consider two points on the trajectory $P_1(x_1,y)$ and $P_2(x_2,y)$ that have the same ordinates $y_1=y_2=y$. The potential energy of the projectile at the point P_1 is equal to the potential energy of the projectile at P_2. The change in kinetic energy due to the work done by drag force is

$$\frac{mv_2^2}{2} - \frac{mv_1^2}{2} = \int_{s_1}^{s_2} D\cos(\vec{D},\vec{v})ds$$

,

where $\vec{D}$ is the drag force. Since the angle between drag force and the velocity of projectile is a straight angle, we get

$$\frac{mv_2^2}{2} - \frac{mv_1^2}{2} = -\int_{s_1}^{s_2} Dds < 0$$

.

The last expression shows that the kinetic energy of the projectile at P_2 is less than its kinetic energy at P_1. Hence, it follows that $v_2 < v_1$.

The above inequality holds also for initial and terminal velocities $v_0 < v_{terminal}$ on condition that the gun and the target are at the same horizontal plane above or at sea level.

2.8 The Ingalls-Siacci Differential Equations of Projectile Flight

Using the approximate Ingalls-Siacci function of resistance (2.3.6), we write the system (2.7.16) in English units

$$\begin{cases} \dfrac{dv_x}{dx}=-\dfrac{1.42233731}{C}h(y)\cdot K_I(v) \\[2mm] \dfrac{dp}{dx}=-\dfrac{g}{v_x^2} \\[2mm] \dfrac{dt}{dx}=\dfrac{1}{v_x} \\[2mm] \dfrac{dy}{dx}=p \end{cases} \qquad , \qquad (2.8.1)$$

where

$$C=\frac{m}{id^2}, \qquad h(y)=e^{-0.00003159145y} \qquad\qquad (2.8.2)$$

and

$$K_I(v)=\begin{cases} 1.1260\cdot10^{-5}v^2 & for \quad v\leq840\,ft./s \\[2mm] (v-787.4016)/3 & for \quad v>840\,ft./s \end{cases}. \qquad (2.8.3)$$

We use the systems of equations (2.8.1) and those that can be derived from it (similar to the differential equations of section (2.7)) to solve the problems of exterior ballistics in English units.

Example 1. Find the ballistics coefficient C, the density function $h(y)$ at the altitude $y=1,200\,ft.$ for the projectile 7.62 mm M80. The mass, the caliber, and the Ingalls form factor of the given bullet are respectively $m=0.021lb.$, $d=0.30in.$, $i=0.57$. Find as well the density of air at the given altitude, considering that the density of air is $\rho_{0N}=0.0752257lb/in^3$ (Army Standard Metro).

Solution
Employing ((2.8.2), we find that the ballistics coefficient and the density function are respectively

$$C=\frac{m}{id^2}=\frac{0.021}{(0.57)(0.30)^2}=0.409357$$

and

$$h(y)=e^{-0.00003159145y}=e^{-0.000013159(1,200)}=0.9628.$$

The density of air at $y=1,200\,ft.$ is

$$\rho(y)=\rho_0 h(y)=(0.0752257)(0.9628)=0.07243lb./in.$$

Example 2. The average Ingalls' form factor for the projectile Caliber 0.30 Ball M2 is $i_{ing}=0.53755$ (See exercise 5, section 2.6). Find the Siacci Ballistics coefficient, if the projectile caliber is $d=0.308in.$ and the mass of projectile is $m=(150/7000)lb.$

We find that

$$C=\frac{m}{i_{ing}d^2}=\frac{(150/7000)}{(0.53755)(0.308)^2}=0.42022 \ \ lb/in^2$$

Substituting in (10) section 2.4, we find that the Ballistics coefficient is

$$c=\frac{1.422334331}{C}=\frac{1.42233731}{0.42022}=3.385m^2/kg.$$

2.9 Mayevski–Zaboudski's and Mayevski–Ingalls' Differential Equations of Projectile Flight

The vectorial differential equations (2.7.16) for Mayevski-Zaboudski's and Ingall's functions of resistance are respectively

$$\frac{d\vec{v}}{dt}=\vec{g}-c\cdot h(y)A_m v^n\frac{\vec{v}}{v}$$

$$(2.9.1)$$

and

$$\frac{d\vec{v}}{dt}=\vec{g}-\frac{1}{C}\cdot h(y)Av^{n}\frac{\vec{v}}{v}.$$

$$(2.9.2)$$

The Mayevski-Zaboudski differential equations and the Ingalls ones obtained from (2.9.1) and (2.9.2) are respectively

$$\begin{cases} \dfrac{dv_x}{dx}=-ch(y)\cdot A_m v^n \\ \dfrac{dp}{dx}=-\dfrac{g}{v_x^2} \\ \dfrac{dt}{dx}=\dfrac{1}{v_x} \\ \dfrac{dy}{dx}=p \end{cases}$$

$$(2.9.3)$$

and

$$\begin{cases} \dfrac{dv_x}{dx}=-\dfrac{0.422334331}{C}h(y)\cdot Av^n \\ \dfrac{dp}{dx}=-\dfrac{g}{v_x^2} \\ \dfrac{dt}{dx}=\dfrac{1}{v_x} \\ \dfrac{dy}{dx}=p \end{cases},$$

$$(2.9.4)$$

where the values of A_m, A, and n are displayed in table 1, section 2.7.

The integration of systems (2.9.3) and (2.9.4) is not so easy since the values of the parameters A_m, A, and n change continuously when the projectile passes from one interval of speeds to another (see table 1, section 2.3).

The Modified Mayevski-Ingalls Function

To simplify the solutions of (2.9.3) and (2.9.4) and to obtain analytical solutions of differential equations of projectile flight, we present hereafter the modified Mayevski-Zaboudski and Ingalls functions.

The average converting factor that relates the Mayevski-Zaboudski functions of resistance (2.9.1) and (2.9.2), respectively, with Siacci's functions of resistance (2.3.1) is "0.896".[17]

Using the approximate Siacci functions of resistance (2.3.2) instead of Siacci's functions (2.9.1), we can approximate the Mayevski-Zaboudski functions of resistance (2.9.1) with the approximate Siacci function (in SI units)

$$f(v)_{may} = (0.896) K_D(v), \tag{2.9.5}$$

where

$$K_D(v) = \begin{cases} 1.212 \cdot 10^{-4} v^2 & for \quad v \le 256 \\ (v - 240)/3 & for \quad v > 256 \end{cases}, \tag{2.9.6}$$

In English units, we can write

$$f(v)_{ing} = (0.896) K_{De}(v), \tag{2.9.7}$$

where

$$K_{De}(v) = \begin{cases} 1.1260 \cdot 10^{-5} v^2 & for \quad v \le 840\, ft./s \\ (v - 787.4016)/3 & for \quad v > 840\, ft./s \end{cases}. \tag{2.9.8}$$

We can write the differential equations of projectile flight using the approximate Mayevski-Zaboudski functions (2.9.5) and (2.9.7).

Thus, for example, we can write the differential equations of projectile flight analog to 2.7.16 respectively as

---

[17]   Cline, Donna. *Trajectories, Trajectories*. 1999–2004.
www.Angelfire.com/poetry/u31240468/Aeroballistivcs.pdf [Web site]

Note. The multiplication coefficient 0.896 that is an average factor that converts the Siacci function of resistance to Mayevski-Zaboudski's function of resistance.

Shapiro, J. M. *Vneshnaja Balistika*, p.48. Oborongiz 50'.

$$\begin{cases} \dfrac{dv_x}{dx}=-(0.896)ch(y)\cdot K_D(v) \\[2mm] \dfrac{dp}{dx}=-\dfrac{g}{v_x^2} \\[2mm] \dfrac{dt}{dx}=\dfrac{1}{v_x} \\[2mm] \dfrac{dy}{dx}=p \end{cases}$$

(2.9.9)

and

$$\begin{cases} \dfrac{dv_x}{dx}=-0.896\dfrac{1.42334331}{C}h(y)\cdot K_D(v) \\[2mm] \dfrac{dp}{dx}=-\dfrac{g}{v_x^2} \\[2mm] \dfrac{dt}{dx}=\dfrac{1}{v_x} \\[2mm] \dfrac{dy}{dx}=p \end{cases}$$

(2.9.10)

Differential equations of projectile flight (2.9.9) and (2.9.10) can be integrated relatively easy.

2.10 The Standard Functions of Resistance of Bullets of Small Arms

For many projectiles of small arms nowadays there are given the drag coefficients $C_D(v/a)$, in tabular forms, obtained experimentally using Doppler radars.

Since most of the ballisticians, professionals or amateurs, or some manufacturing companies, do not have the possibility to measure the drag coefficient $C_D(v/a)$ of their bullet using Doppler radar technology, it is possible to measure the drag coefficent (with a certain accuracy) using the chronograph measurements.

Employing the drag coefficient $C_D(v/a)$, measured experimentally, we can find the function of resistance $f_D(v)$ for a given bullet, and so we can avoid the measurement of the form coefficient that in this case remains equal to one ($i = 1$) all over the trajectory. In other words, we do not need to measure experimentally the BC, but we can calculate it using the caliber and the mass, i.e.:

$$c = 1000\frac{d^2}{m}.$$

Hereafter we will show the way to obtain the function of resistance $f_D(v)$ that has a form similar to (2.3.2), when the speed of the bullet is supersonic.

Function of Resistance of the 0.338 Lapua GB488 Scenar 16.2g (250 gr.) Bullet

The function that we will find can be used throughout the book to replace the Siacci's function of resistance (2.3.2) for the given Lapua bullet.

Consider the four first columns in the following table of drag coefficient $C_D(v/a)$ obtained using the data for **0.338 Lapua GB488 Scenar 16.2g**
(see: http://www.lapua.com/en/customer-center/lapua-ballistics/download-lapua-edition.html).
In the following graph it is drawn the drag coefficient as a function of Mach number.

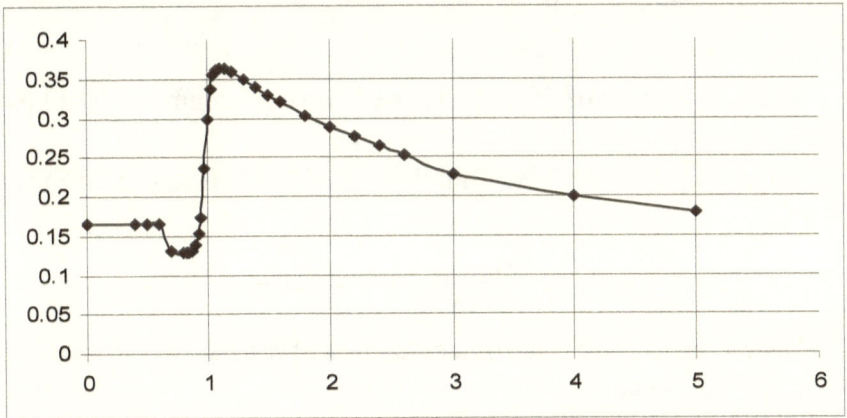

For the ICAO atmosphere (The International Civil Aviation Organization Atmosphere: pressure $p_{0N} = 760mm\ Hg$, temperature $t_{0N} = 15°\ C$, speed of sound $a_{0N} = 340.30m/s$, humidity 0%, density of air $\rho_{0N} = 1.2251kg/m^3$) we obtain the following relation:

$$f_D(v) = 4.811 \cdot 10^{-4} v^2 \cdot C_D(v/a) \qquad (2.10.1)$$

that is similar to the first equation of (2.2.6). The above formula can be easily obtained considering that $\rho_{0N} = 1.2251kg/m^3$.

The density function, for the ICAO atmosphere, is

$$h(y) = (\frac{288.15 - 0.006328y}{288.15})^{4.4}. \qquad (2.10.2)$$

(For more information see: G. Klimi, *Exterior ballistics: A New Approach*, section 1.3, Xlibris, 2010.)

Substituting in (2.3.6) the values of Mach number $M = (v/a)$, given in columns (1) and (3) of the following tableau and the corresponding projectile speed,

$$v = M \cdot a = M \cdot 340.30, \tag{2.10.3}$$

we obtain the values of $f_D(v)$ that are shown in columns (4) and (6) of the table.

For example, if the Mach number is $M = 1.5$, using (2.10.3) we find:

$$v = 340.30 \cdot M = 340.30 \cdot (1.5) = 510.45 m / s.$$

The corresponding value of the function of resistance (for $M = 1.5$, $v = 510.45 m / s$) is

$$f_D(v) = 4.811 \cdot 10^{-4} v^2 \cdot C_D(v/a) = 4.811 \cdot 10^{-4} (340.43)^2 C(1.5) = 41.36716,$$

since $C_D(1.5) = 0.33$.

The function of resistance (2.10.1) for the given Lapua bullet is presented in column (6) and (8) of the table.

Lapua GB488 Scenar 16.2g Bullet								
M = v/a	C_D(v/a)	M = v/a	C_D(v/a)		v	f_D(v)	v	f_D(v)
1.000	0.299	1.4	0.34		340.300	16.65829	476.420	37.12737
1.025	0.337	1.5	0.33		348.808	19.72590	510.450	41.36716
1.050	0.355	1.6	0.321		357.315	21.80551	544.480	45.78300
1.075	0.361	1.8	0.304		365.823	23.24253	612.540	54.87542
1.100	0.364	2.0	0.289		374.330	24.53839	680.600	64.40463
1.150	0.363	2.2	0.276		391.345	26.74617	748.660	74.42412
1.200	0.360	2.4	0.264		408.360	28.8818	816.720	84.71995
1.300	0.350	2.6	0.252		442.390	32.95445	884.780	94.90880
		3.0	0.228				1020.90	114.3240

Employing the data of the table (columns (5), (7) and the respective columns (6) and (8)) and the techniques of the linear regression we obtain the following function of resistance for the given Lapua GB488 Scenar 16.2g Bullet:

$$f_D(v) = 0.143706 v - 31.1521, \tag{2.10.4}$$

for the projectile speed in the interval: $[340m/s,\ 1021m/s]$.

The form coefficient of the **Lapua GB488 Scenar 16.2g** bullet (mass $m = 12.60g = 0.01260kg$, the diameter (caliber) $d = 0.388'' = 0.00859m$) is $i = 1$ while for the ballistics coefficient we find:

$$c = 1000\frac{d^2}{m} = 1000\frac{0.00859^2}{0.0162} = 4.5548 . \qquad (2.10.5)$$

Function of Resistance of Lapua GB528 Scenar 19.44 g, Caliber 8.59 mm bullet

The following tableau shows the values of the drag coefficient of the Lapua GB528 Scenar 19.44 g, Caliber 8.59 mm, obtained experimentally using the Doppler radar measurements (see http://en.wikipedia.org/wiki/External_ballistics).

Drag Coefficient of Lapua GB528 Scenar 19.44 g, 8.59 mm caliber						
Mach number	0.000	0.400	0.500	0.600	0.700	0.800	0.825
Drag coefficient	0.230	0.229	0.200	0.171	0.164	0.144	0.141
Mach number	0.85	0.875	0.9	0.925	0.95	0.975	1
Drag coefficient	0.137	0.137	0.142	0.154	0.177	0.236	0.306
Mach number	1.025	1.05	1.075	1.1	1.15	1.2	1.3
Drag coefficient	0.334	0.341	0.345	0.347	0.348	0.348	0.343
Mach number	1.4	1.5	1.6	1.8	2	2.2	2.4
Drag coefficient	0.336	0.328	0.321	0.304	0.292	0.282	0.27

Using the same method as for the Lapua GB 488 bullet and the formula (2.10.1), we obtain the function of resistance of the Lapua GB528 Scenar 19.44 g, 8.59 mm caliber bullet:

$$f_D(v) = \begin{cases} 0.15v - 35.6712 & 280 \leq v \leq 830 \\ 0.02364v - 1.27552 & 136 \leq v < 280 \\ 1.924 \times 10^{-4} v^2 & 0 \leq v < 136 \end{cases}. \qquad (2.10.7)$$

The following graph represents the drag coefficient of the Lapua GB488 bullet.

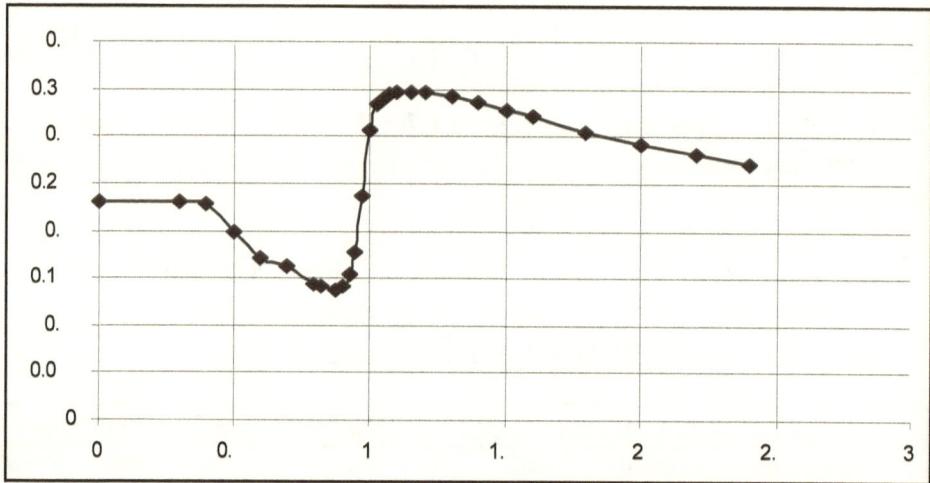

Another function of resistance for Lapua GB528 Scenar 19.44 g, 8.59 mm caliber when the speed of the bullet is in the interval [340 – 830] is

$$f_D(v) = 0.141v - 30.031, \qquad (2.10.8)$$

where $340 \leq v \leq 830$

The function of resistance (2.10.8) can be written:

$$f_D(v) = \frac{v - 213}{7.09} \qquad (2.10.9)$$

The function of resistance of Lapua bullet (2.10.8) or (2.10.9) can be used for the projectile trajectories that have the range till around 1500 meters.

The ballistics coefficient of the given Lapua bullet (since $i = 1$) is

$$c = 1000\frac{d^2}{m} = 1000\frac{(0.00859)^2}{0.01944} = 3.796.$$

NOTE

- To solve the ballistics problems related with the given Lapua bullet, we can use all the methods presented farther in this book, substituting (2.10.7) or (2.10.9) instead of Siacci's function (2.3.2), where the density function is given by (2.10.2).

- All the PC programs presented in this book and in the other two of my books: *Exterior Ballistics of Small Arms* (Xlibris, 2009) and *Exterior Ballistics: A New Approach* (Xlibris, 2010) can be modified to be used with the given Lapua function of resistance or with any other standard function of resistance.

- In Appendix F is given the PC program RPMELA.BAS that is obtained modifiyng RCPMET.BAS (see Exteriror Ballistics of Small Arms, p. 221- 234, XLIBRIS, 2009) by substituting in it the function of resistance (2.10.7). RPMELA.BAS can be used to solve ballistics problems related with the Lapua bullet presented above. The results obtained using the function of resistance (2.10.7) are correct for ranges till around 1500m – 2100m. For better results till 1500 meters we can use (2.10.8) instead of (2.10.7).

- In a similar way, modifying RPMELA.BAS, we can solve the ballistics problems related with any other bullet for which the ballistician (shooter) can find the standard function of resistance of the bullet in use.

2.11 A New Method to Derive the Standard Function of Resistance

Traditionally in exterior ballistics the solution of the exterior ballistics problems is based on the G-standard functions of resistance and in the form coefficient i that is introduced to adapt a known G-function (G_1, ...G_8, Siacci, etc.) to a particular projectile (see McCoy, R., *Modern Exterior Ballistics*, Schiffer Publishing Ltd., 1999).

The new method presented hereafter, eliminates the large errors exterior ballisticians (and shooters) tolerate predicting the ballistic trajectory for long range shooting using standard G-functions and an average ballistics coefficient.

The Doppler radars, or other modern equipments and methods, are used nowadays to obtain drag coefficients, for particular projectiles, in order to increase the accuracy of the prediction of the elements of the trajectory.

In section 2.3 we have already shown how to obtain a particular standard function of resistance $f_D(v)$ for a given bullet using the drag coefficient $C_D(v/a)$ that is measured by Doppler radars, chronographs, etc.

The simple piece-wise function of resistance or a linear function of resistance $f_D(v)$ that we have obtained, allow us to use the analytical methods, presented in this book (approximate, Siacci's, Euler's, etc.) to solve the differential equations of projectile flight.

Such analytical methods of solutions are not possible, or are too complex to be applied to solve the same differential equations of projectile flight using the drag coefficients $C_D(v/a)$ in tabular form.

Working on the idea to avoid the use of the known standard functions of resistance to solve exterior ballistics problems, and thus to eliminate the need to measure experimentally a fixed form coefficient i, I discovered a new and simple method to find theoretically the function of resistance $f_D(v)$ for any bullet.

The new method, introduced hereafter, is based on the determination of the standard function of resistance $f_D(v)$ for a given bullet (projectile) using the existing range tables of a given firearm (bullet) that are presented in the respective manuals.

Determining the standard function of resistance for any small arm (firearm), we reduce the large errors that ballisticians do to predict the ballistics elements of the trajectory, for long range and sniper shootings, by employing the already known standard G-functions of resistance, usually G_1 and G_7, and the BC given by the small arms manufacturers.

So, it will be convenient that the range tables of small arms, together with the form coefficient (BC) and other ballistics characteristics of small arms and respective bullets, to give as well the function of resistance $f_D(v)$ that is unique for the ballistics of the given bullet.

The theoretical determination of the functions of resistance, $f_D(v)$, for each bullet is the finest option to:

- Eliminate the need to employ (the expensive) Doppler radar measurements (or other experimental methods) to find the drag coefficient of each bullet.
- Increase the accuracy of the prediction of the ballistics elements of the projectile trajectories.
- Facilitate the analytical solutions (Siacci's, etc) of the differential equations of the projectile flight.

Derivation of the Standard Function of Resistance of a Bullet

We assume that the drag coefficient of a given bullet is unknown, but we know the velocity as a function of the range (for example using the range table of the given bullet or the data obtained by chronograph measurements.

We demonstrate the way we find the function of resistance of the **Lapua GB528 Scenar 19.44 g, Caliber 8.59 mm** bullet using the data in the "range" table 1, obtained employing the PC program RPMELA.BAS that uses the function of resistance (2.10.9).

In the third and the sixth row of the table (for comparison) are given the data obtained by radar measurements shown at http://en.wikipedia.org/wiki/External_ballistics).

Table 1. Velocity vs. Range: Lapua GB528 Scenar 19.44 g, Caliber 8.59 mm									
Range	0	100	200	300	400	500	600	700	800
Velocity	830	791	752	708	677	641	605	571	538
Radar velocity	830			711			604		
Range	900	1000	1100	1200	1300	1400	1500	1600	
Velocity	507	476	447	420	395	371	349	329	
Radar velocity	507			422			349		

In Table 2 are shown the values of the Drag Coefficient $C_D(\bar{v}/a)$ and the function of resistance $f_D(\bar{v})$ that are obtained using respectively the formulas:

$$C_D(\bar{v}/a) = \frac{\ln(v_1/v_2)}{4.811 \cdot 10^{-4} c \cdot h(y)(x_2 - x_1)} \qquad (2.11.1)$$

and

$$f_D(\bar{v}) = 4.811 \cdot 10^{-4} \bar{v}^2 \cdot C_D(\bar{v}/a), \qquad (2.11.2)$$

where:

- v_1 and v_2 are the speeds of the projectile in two consecutive ranges x_1 and x_2 respectively,
- the density function (at the sea level) is $h(y) = 1$,
- The average speed between two consecutive speeds is $\bar{v} = (v_1 + v_2)/2$,
- For each two consecutive ranges: $x_2 - x_1 = 100$,
- The speed of sound is $a = 340.30 m/s$,
- The BC of the given Lapua bullet is

$$c = 1000\frac{d^2}{m} = 1000\frac{(0.00859)^2}{0.01944} = 3.796. \qquad (2.11.3)$$

Note that formula (2.11.1) is similar (2.6.9) and is valid for ICAO atmosphere, while (2.11.2) is the formula (2.3.17), (see Example 1, section 2.6).

Substituting $c = 3.796$, $h(y) = 1$ and $x_2 - x_1 = 100$ in (2.12.1) we obtain the following simplified formula that can be used instead of (2.11.1) to estimate the drag coefficient of the given bullet:

$$C_D(\bar{v}/a) = 5.4757 \cdot \ln(v_1/v_2).$$

The values of the drag coefficient, estimated using the above formula, are presented in column 5 of table2 while the values of the function of resistance are calculated using (2.11.2) are shown in column 6.

Table 2. Calculated values of the Drag Coefficient and the Function of Resistance					
x	v	$\bar{v}$	$\bar{v}/a$	$C_D(\bar{v}/a)$	$f_D(\bar{v})$
0	830	810.5	2.3817	0.2635	83.2866
100	791	771.5	2.2671	0.2769	79.2806
200	752	730	2.1452	0.3301	84.6410
300	708	692.5	2.0350	0.2452	56.5624
400	677	659	1.9365	0.2992	62.5129
500	641	623	1.8307	0.3165	59.0997
600	605	588	1.7279	0.3167	52.6806
700	571	554.5	1.6294	0.3260	48.2189
800	538	522.5	1.5354	0.3250	42.6824
900	507	491.5	1.4443	0.3455	40.1516
1000	476	461.5	1.3562	0.3442	35.2685
1100	447	433.5	1.2739	0.3412	30.8437
1200	420	407.5	1.1975	0.3360	26.8459
1300	395	383	1.1255	0.3432	24.2229
1400	371	360	1.0579	0.3347	20.8706
1500	349	339	0.9962	0.3231	17.8661
1600	329				

Using column (3) and Column (6) of table 2 and the techniques of linear regression we obtain the following function of resistance for the given Lapua bullet:

$$f_D(v) = 0.1422v - 30.676.$$ (2.11.4)

A better formula is obtained if we eliminate from the data of table 2, the numbers in row (3) and (4), i.e. the points (692.5, 56.56) and (730, 84.6). In this case, using the linear regression we obtain the standard function of resistance for the given Lapua bullet:

$$f_D(v) = 0.1406v - 29.82,$$ (2.11.5)

where the projectile speed is $330m/s \le v \le 830m/s$.

The function of resistance (2.11.15) can be written:

$$f_D(v) = \frac{v - 212.09}{7.11}.$$ (2.11.6)

Note that the Lapua standard functions of resistance (2.11.4) or (2.11.5) are slightly different from the standard function of resistance (2.10.8) of the same Lapua bullet obtained in section (2.10) but, as we will see hereafter, the accuracy of the calculation of the ballistics characteristics of the trajectory is exceptional.

Accuracy of the Method

Using the PC program RPMELA.BAS that can be modified by substituting the function of resistance (2.10.7), i.e

$$f_D(v) = 0.15v - 35.6712,$$ (2.11.17)

with (2.11.6), we find the following data that are quite equal to the data given by "External Ballistics", http://en.wikipedia.org/wiki/External_ballistics".

Thus, the acuracy of the method described in the above example is remarcable.

Table 3. Lapua GB528 Scenar 19.44 g, Caliber 8.59 mm								
Range	0	100	200	300	400	500	600	700	800
Velocity	830	791	752	714	677	641	605	571	538
Drop	0	0.074	0.304	0.708	1.305	2.116	3.167	4.485	6.105
Time	0	0.12	0.25	0.39	0.53	0.69	0.85	1.02	1.20
Range	900	1000	1100	1200	1300	1400	1500	1600	
Velocity	507	476	447	420	395	371	349	329	
Drop	8.063	10.404	13.178	16.443	20.264	24.717	29.885	35.861	
Time	1.39	1.59	1.81	2.04	2.28	2.55	2.82	3.12	

As is shown in "External Ballistics", the data obtained using G_1 and G_7 functions of resistance, for long range shooting are very different from the data obtained by Doppler radar method as well as from the data we have obtained using the new and original approach (see http://en.wikipedia.org/wiki/External_ballistics").

The new methods to determine the function of resistance of a given bullet that are presented in sections 2.10 and 2.11, together with the method presented in section 3.6, introduce in exterior ballistics an inovative and acurate method to solve the exterior ballistics problems related with long range shooting.

Note: Substituting (2.11) in (2.12) we obtain the function of resistance in the form,

$$f_D(\bar{v}) = \frac{\bar{v}^2 \ln(v_1/v_2)}{c \cdot h(y)(x_2 - x_1)},$$
(2.11.18)

Considering that that $h(y) = 1$ the equation (2.1.18) can be written:

$$f_D(\bar{v}) = \frac{\bar{v}^2 \ln(v_1/v_2)}{c \cdot (x_2 - x_1)}.$$
(2.11.19)

Hence,

$$c \cdot f_D(\bar{v}) = \frac{\bar{v}^2 \ln(v_1 / v_2)}{(x_2 - x_1)}. \tag{2.11.20}$$

Considering (2.11.20), the differential equation (2.2.12) can be written:

$$\frac{d\vec{v}}{dt} = \vec{g} - h(y) f_{D1}(v) \frac{\vec{v}}{v}. \tag{2.11.21}$$

Where

$$f_{D1}(v) = \frac{\bar{v}^2 \ln(v_1 / v_2)}{(x_2 - x_1)} \tag{2.11.22}$$

For example, considering the function of resistance (2.11.15) of the Lapua GB528 Scenar 19.44 g, Caliber 8.59 mm bullet, and the corresponding ballistics coefficient $c = 3.796$ we have

$$f_{D1}(v) = 0.03846v - 7.855637. \tag{2.11.23}$$

Thus, we can write the differential equation (2.11.21) in the following form:

$$\frac{d\vec{v}}{dt} = \vec{g} - h(y) \cdot (0.03846v - 7.855637) \frac{\vec{v}}{v}. \tag{2.11.24}$$

Note that in the differential equation (2.11.24) it is not explicitly present the value of BC. The way we obtained (2.11.24), using (2.11.20), does not require even to know the mass and diameter of the bullet (i.e. the BC). We just need to know the velocity as a function of range, let's say every 100 meters or 50 meters.

The above differential equation that describes the flight of the bullet does not depend explicitly on the ballistics coefficient of the standard given bullet. The unique ballistics characteristics of a given bullet are represented by the corresponding unique function of resistance (2.11.23).

Differential equation (2.11.24) can be simplified further if the data are obtain at the sea level in ICAO atmosphere where the density function $h(y) = 1$.

In general, for any given bullet, the unique linear function of resistance, similar to (2.11.23) has the form:

$$f_{D1}(v) = E_1 \cdot v - F_1, \qquad\qquad (2.11.25)$$

where E_1 and F_1 are derived as in the above example.

Note: The topics 2.10 and 2.11 are written in New York on 07/04/2011.

3

Motion of High–Speed Projectiles

Introduction

In this chapter we introduce different approximate solution methods of differential equations that describe the projectile flight in air considering the projectile a point of mass.

The solutions of the systems of differential equations of the projectile flight, obtained using the model of a point-mass projectile, enable us to get approximate results for the elements of trajectories of projectiles flying in standard atmosphere for some particular cases, such as for the flight of projectiles when the time of flight is relatively short. The theoretical outcomes are close enough to practical results observed in infantry and artillery fire, such as in the flight of bullets, artillery shells, and fragments of antipersonnel ammunitions.

The approximate solution methods of ballistics problems developed here are simple mathematical models and fascinating applications of different techniques of calculus and differential equations.

To find the elements of trajectory of projectile flight, we use the systems of differential equations obtained in chapter 2 and the Siacci function of resistance given by equation (2.3.2).

The projectiles fired from artillery and infantry firearms in general have speeds much greater than 256 m/s. We study the motion of such projectiles, under the assumptions made in chapter 2 (section 2.2) for a point-of-mass projectile (figure 1, chapter 2):

- The projectile is considered a point of mass, without spinning motion, with axis along the tangent of the trajectory of flight.
- The projectile cross-sectional area is perpendicular to the projectile velocity $\vec{v}$.
- The earth does not rotate, and its surface is plane.
- The gravitational acceleration $\vec{g}$ is constant and has a magnitude of $g=9.80665m/s^2$.
- The flight is in standard atmosphere:
- At the ground level $y=0$: air density is $\rho_{0N}=1.205kg/m^3$, air temperature is $\tau_{0N}=289.08K$, atmospheric pressure is $p_{0N}=750$ mmHg, and speed of sound is $a_{0N}=340.83$.
- The temperature and the density change with height y above sea level.
- The projectile flight is in absence of wind.
- The projectile characteristics (mass, caliber, initial speed, etc.) are standard.

We now suppose that

- the projectile speed v at any time (from launching point till it hits the target) is greater than 256 m/s, i.e., $v>256m/s$.

For projectile speed $v \geq 256m/s$, the function of air resistance given in (2.3.2) has the form

$$K_D(v)=(v-240)/3.$$

We study the motion of projectiles launched in air with relatively high speeds (greater than 256 m/s).

3.1 The Differential Equations of Projectile Motion

In some special cases, the systems of differential equations of projectile flight presented in chapter 2 can be integrated analytically. First, we write the differential equations that describe the projectile flight.

(a) The independent variable is time t

Consider the system of differential equation (2.7.13),

$$
\begin{cases}
\dfrac{dv_x}{dt} = -ch(y)K_D(v)\dfrac{v_x}{(v_x^2 + v_y^2)^{1/2}} \\[2mm]
\dfrac{dp}{dt} = -\dfrac{g}{v_x} \\[2mm]
\dfrac{dx}{dt} = v_x \\[2mm]
\dfrac{dy}{dt} = v_y
\end{cases}
\qquad , \qquad (3.1.1)
$$

where

$$
K_D(v) = (v-240)/3 \qquad (3.1.2)
$$

Substituting (3.1.2) in the first differential equation of (3.1.1), we obtain the following system of differential equations that describes mathematically the projectile flight:

$$\begin{cases} \dfrac{dv_x}{dt} = -ch(y) \cdot \dfrac{(v-240)}{3} \dfrac{v_x}{v} \\[2mm] \dfrac{dp}{dt} = -\dfrac{g}{v_x} \\[2mm] \dfrac{dx}{dt} = v_x \\[2mm] \dfrac{dy}{dt} = v_y \end{cases} \qquad (3.1.3)$$

(b) The independent variable is the x coordinate of the projectile
The first differential equation of (3.1.3) can be written as

$$\frac{dv_x}{dx} \frac{dx}{dt} = -c \cdot h(y) \frac{(v-240)}{3} \cdot \frac{v_x}{v} . \qquad (3.1.4)$$

Considering the third equation of (3.1.3), we write (3.1.4) as

$$\frac{dv_x}{dx} = -c \cdot h(y) \frac{(v-240)}{3v} . \qquad (3.1.5)$$

Expressing in the same way the remaining three equations of (3.1.3) through x, we obtain an equivalent system of equations:

$$\begin{cases} \dfrac{dv_x}{dx} = -ch(y) \dfrac{v-240}{3v} \\[2mm] \dfrac{dp}{dx} = -\dfrac{g}{v_x^2} \\[2mm] \dfrac{dt}{dx} = \dfrac{1}{v_x} \\[2mm] \dfrac{dy}{dx} = p \end{cases} \qquad (3.1.6)$$

To obtain approximate solutions solving any of the above systems, we will consider $h(y)$ as a constant that is determined by

$$h(y)=h(\bar{y}) , \qquad (3.1.7)$$

where

- $\bar{y}=(2/3)y_m$, for field artillery[18]
- $\bar{y}=(1/2)y_m$, for antiaircraft artillery (uphill or downhill shooting)[19],

y_m is the maximum height of the trajectory,

and

$$h(\bar{y})=(\frac{289.08-0.006328\bar{y}}{289.08})^{4.4}$$

(3.1.8)

Considering (3.1.5), the equations of projectile motion (3.1.3) and (3.1.6) can be written, respectively, as

$$\begin{cases} \dfrac{dv_x}{dt} = -b\dfrac{v_x(v-240)}{v} \\[2mm] \dfrac{dp}{dt} = -\dfrac{g}{v_x} \\[2mm] \dfrac{dx}{dt} = v_x \\[2mm] \dfrac{dy}{dt} = v_y \end{cases}$$

(3.1.9)

and

---

[18]   Herrmann, Ernest E. *Exterior Ballistics*, p. 48. U.S. Naval Institute, The College Press, 1935.

[19]   Shapiro, J. M. *Vneshnaja Balistika*, p. 94. Oborongiz 50'.

$$\begin{cases} \dfrac{dv_x}{dx} = -b\dfrac{v-240}{v} \\[2mm] \dfrac{dp}{dx} = -\dfrac{g}{v_x^2} \\[2mm] \dfrac{dt}{dx} = \dfrac{1}{v_x} \\[2mm] \dfrac{dy}{dx} = p \end{cases} \qquad (3.1.10)$$

where

$$b = c \cdot h(\bar{y})/3 \,, \quad c = i\dfrac{d^2}{m} \cdot 1000 \,, \quad p = \tan\alpha \qquad (3.1.11)$$

while $h(\bar{y})$ is determined by (3.1.8).

For projectiles launched at relatively narrow angles at sea level, the density function $h(\bar{y})$ can be considered constant and equal to one, i.e., $h(\bar{y}) \approx 1$.

3.2 The Integration of Differential Equations of Projectile Motion When Launching Angle is Narrow (Method I)

The system of differential equations (3.1.3) and (3.1.6) can be integrated when the projectile trajectory is such that the y-component v_y of the projectile velocity at any moment during the flight is relatively small compared to x-component of projectile velocity v_x. In this case, we can consider that the projectile speed v is approximately equal to v_x,

$$v = (v_x^2 + v_y^2)^{1/2} = v_x(1 + v_y^2/v_x^2)^{1/2} \approx v_x. \qquad (3.2.1)$$

The above condition is satisfied if along the trajectory of flight,

$$v_y^2/v_x^2 = p^2 \ll 1, \qquad (3.2.2)$$

i.e., if

$$\tan^2\alpha \ll 1 . \tag{3.2.3}$$

It can be shown (see section 4.2) that the relative error we make substituting the projectile speed v with its horizontal component v_x is small for relatively narrow angles of impact. The projectile trajectory is relatively close to the ground or to the horizontal line that originates at the point where the projectile is launched.

As a result, for approximate solutions, we can consider the density function $h(y)$ to be constant and equal to the value it has at the launching point.

In standard conditions at sea level,

$$h(y)=h(\bar{y})=1$$

.

In practice, this scenario is observed during the fire from assaulting rifles, automatic guns, etc., or during the flight of fragments of antipersonnel ammunitions—for example, fragmentation grenades, antipersonnel mines, etc.

Summarizing, to find the solutions of (3.1.3) and (3.1.6), we assume the following:

- The projectile speed v at any time from launching point till it hits the target is greater than 256 m/s, $v > 256 m/s$.
- During the projectile flight $p^2=\tan^2\alpha \ll 1$ and $v \approx v_x$.
- The density function $h(y)$ is constant and equal to the value it has at the launching point.

Note: According to Alger, [Alger, R. Ph., *Exterior Ballistics,* p. 60, The Lord Baltimore Press, 1906] the density function does not change significantly if the time of flight is less than 12 seconds.

Consider the equations of projectile motion (3.1.3),

$$
\begin{cases}
\dfrac{dv_x}{dt}=-b\dfrac{v_x(v-240)}{v} \\[2mm]
\dfrac{dp}{dt}=-\dfrac{g}{v_X} \\[2mm]
\dfrac{dx}{dt}=v_X \\[2mm]
\dfrac{dy}{dt}=v_Y
\end{cases}
\qquad (3.2.4)
$$

Assuming that $v \approx v_x$, we can write the first differential equation of system (3.2.4) as

$$
\frac{dv_x}{dt}=-b\frac{v_x(v-240)}{v}=-b(v_x-240)
\qquad (3.2.5)
$$

Thus, the system of differential equations (3.2.4) can be written

$$
\begin{cases}
\dfrac{dv_x}{dt}=-b\cdot(v_x-240) \\[2mm]
\dfrac{dp}{dt}=-\dfrac{g}{v_x} \\[2mm]
\dfrac{dx}{dt}=v_x \\[2mm]
\dfrac{dy}{dt}=v_y
\end{cases}
\qquad (3.2.6)
$$

where

$$
b=c\cdot h(\bar{y})/3, \quad c = 1000\frac{d^2}{m}, \quad p=\tan\alpha .
$$

We integrate (3.2.6) for the following initial conditions:

$$x(0)=0, \ y(0)=0, \ v_x(0)=v_0\cos\alpha_0 \approx v_0, \ v_y(0)=v_0\sin\alpha_0 \approx 0,$$
$$p(0)=p_0=\tan\alpha_0 \ \text{at} \ t=0.$$

The Speed and the Angle of Flight
Integrating the first equation of (6), we obtain the x-component of velocity,

$$v_x = 240 + (v_{0x} - 240) \cdot e^{-b \cdot t} .$$
(3.2.7)

Integrating the second equation of (3.2.6),

$$\frac{dp}{dt} = -\frac{g}{v_x} ,$$
(3.2.8)

we can write

$$p = p_0 - g \int_0^t \frac{dt}{v_x} .$$
(3.2.9)

Substituting (3.2.7) in (3.2.9), we obtain the following formula to estimate the angle of projectile velocity at any time t,

$$p = p_0 - \frac{g}{240 \cdot b} \ln(\frac{240 \cdot e^{b \cdot t} + v_{0x} - 240}{v_{0x}}) ,$$
(3.2.10)

where $p = \tan \alpha$, $p_0 = \tan \alpha_0$.
For y-component of velocity, we have

$$v_y = p \cdot v_x$$

or

$$v_y = [(p_0 - \frac{g}{240 \cdot b} \ln(\frac{240 e^{b \cdot t} + v_{0x} - 240}{v_{0x}})] \cdot [240 + (v_{0x} - 240) \cdot e^{-b \cdot t}], \quad (3.2.11)$$

The Coordinates of Projectile in Flight. The Equation of Projectile Trajectory.
Substituting (3.2.7) in the third equation of (3.2.6) and integrating, we find the x-coordinate of the projectile

$$x = 240 t + \frac{v_{x0} - 240}{b} (1 - e^{-b \cdot t}) .$$
(3.2.12)

To find   the y-coordinate as function of the time of flight t, we consider the fourth equation of (3.2.6),

$$dy/dt = v_y$$

Substituting (3.2.11) on the right side of the above equation, we have

$$\frac{dy}{dt} = [(p_0 - \frac{g}{240 \cdot b} \ln(\frac{240e^{b \cdot t} + v_{0x} - 240}{v_{0x}})] \cdot [240 + (v_{0x} - 240) \cdot e^{-b \cdot t}].$$

Integrating, we find the y-coordinate of the projectile

$$y = p_0[240t + \frac{v_{x0} - 240}{b} \cdot (1 - e^{-b \cdot t})] -$$

$$- \frac{g}{240 \cdot b} \int_0^t (240 + (v_{x0} - 240) \cdot e^{-b \cdot t}) \cdot \ln(\frac{240e^{b \cdot t} + v_{x0} - 240}{v_{x0}}) \, dt \qquad \text{. (3.2.13)}$$

We can use (3.2.7), (3.2.10), (3.2.11), (3.2.12), and (3.2.13) to find the elements of the projectile trajectory at any time during the flight.

Note: In example 1 and example 2, we show that the results obtained in this section give satisfactory outcomes that are very close to the practice of shooting.

Example 1. A bullet of mass 7.9 g is fired from a 7.62 mm Russian rifle with initial speed 735 m/s. Find the launching angle α_0 if the range is 100 m. The form coefficient of the given bullet, estimated in example 2, section 2.6, is $i=0.59667$.

Solution

Time of Flight to the Target
    First, we will use equation (3.2.12) to find the time of flight to the target located 100 m from the rifle, and then we will use (3.2.13) to find the launching angle.

The ballistics coefficient is

$$c=i\frac{\cdot d^2}{m}\cdot 1000=(0.59667)\cdot\frac{0.00762^2}{0.0079}\cdot 1000=4.385478645$$

The value of parameter b is

$$b=ch(y)/3=(4.385478645)\cdot(1)/3=1.461826215$$

Substituting in (3.2.12),

$$x=240t+\frac{(v_{x0}-240)}{b}(1-e^{-b\cdot t})$$

and considering

$$v_{x0}=v_0\cos\alpha_0=v_0\approx 735 \text{ m/s,}$$

we can obtain

$$100=240t+\frac{(735-240)(1-e^{-1.461826215\cdot t})}{(1.461826215)}$$

Simplifying, we have

$$0.2953184273-0.7087642255\cdot t=(1-e^{-1.461826215t})$$

The last equation can be solved for t using a graphing calculator, for example, as an intersection of two curves

$$y_1=0.2953184273-0.7087642255\cdot t$$

and

$$y_2=(1-e^{-1.461826215t})$$

The solution of the above equation obtained using a TI-83 Plus is $t=0.1458$.

Launching Angle
Substituting $y=0$ in (3.2.13), we have

$$100.868987\,p_0 - 0.0279520515 \int_0^{0.1458} (240 + 495 \cdot e^{-1.46183 \cdot t}) \cdot \ln(0.3265306 \cdot e^{1.46138 \cdot t} +$$

$$+\, 0.67346993877)dt = 0$$

The value of the integral on the right side, calculated using a graphing calculator TI-83 Plus, is 3.5616253. Substituting in the above equation the obtained value of the integral, we have

$$100.868987\,p_0 - 0.0995563846 = 0$$

Hence we find

$$p_0 = \tan \alpha_0 = 0.000986987$$

and

$$\alpha_0 = \tan^{-1}(0.000986987) = 0.056550°\,.$$

Terminal Angle

Substituting in (3.2.10), we have

$$p_T = p_0 - \frac{g}{240 \cdot b}\ln(\frac{240 \cdot e^{b \cdot t} + v_{x0} - 240}{v_{x0}}) =$$

$$= 0.001030995 - 0.0279520515 \cdot \ln(0.326531 \cdot e^{1.4618262 \cdot (0.1458)} + 0.6734694) = -0.001057$$

Hence, for the terminal angle, we find

$$\alpha_T = \tan^{-1}(-0.001031) = -0.060574°\,.$$

Terminal Speed

$$v_{xT} = 240 + (v_{x0} - 240) \cdot e^{-b \cdot t} = 240 + (735 - 240)e^{-1.46183 \cdot (0.1458)} = 639.98$$

$$v_{yT} = p_T v_x = -(0.001031)(639.98) = 0.66 m/s$$

$$v_T = [(639.98)^2 + (0.66)^2]^{1/2} = 639.98 \; m/s.$$

Maximum Altitude. Coordinates of the Vertex

The maximum altitude of the projectile trajectory is at that point H where the projectile velocity is parallel to the x-axis,

$$p_H = \tan\alpha = \tan 0° = 0.$$

Substituting in equation (3.2.10), for the time when the projectile reaches the maximum height we can write

$$p_T = p_0 - \frac{g}{240 \cdot b} \ln(\frac{240 \cdot e^{b \cdot t} + v_{x0} - 240}{v_{x0}}) =$$

$$= 0.001031 - 0.02795205 \cdot \ln(0.326531 \cdot e^{1.46183 \cdot t} + 0.673469) = 0$$

Solving for t, we find that the projectile will be at the vertex of trajectory after $t = 0.0745$ sec.

To find the coordinates of the trajectory vertex, we employ equations (3.2.12) and (3.2.13).

- x-coordinate: Substituting in (3.2.12), we find

$$x = 240t + \frac{v_{x0} - 240}{b}(1 - e^{-b \cdot t})$$

$$= 240(0.0745) + \frac{(735 - 240)(1 - e^{-1.46183 \cdot (0.0745)})}{(1.46183)} = 52.82 \text{ m.}$$

- y-coordinate: Substituting in ((3.2.13), we find

$$y = (001031)[240(0.0745) + \frac{735 - 240}{(1.46183)}(1 - e^{-1.46183 \cdot (0.0745)})] -$$

$$- \frac{9.80665}{240 \cdot (1.46183)} \int_0^{0.0745} (240 + 495 \cdot e^{-(1.46183) \cdot t}) \cdot \ln(0.326531 \cdot e^{1.46183 \cdot t} + 0.67346939) dt =$$

$$= (0.001031) \cdot (52.82) - (0.027952) \cdot (0.95068753) = 0.027m$$

Example 2. (Known b, v_0, x) Find the launching angle α_0 needed to hit a tank that is in a horizontal distance of 500 m from antitank cannon 45 mm if a projectile of mass 0.98 kg is fired with the initial speed of 1,200 m/s. The coefficient of projectile form is 0.651. Meteorological conditions are normal.

Find as well the time of flight, the terminal angle, and the terminal speed.

Solution

First, we calculate b. We find that

$$c=\frac{id^2}{m}\cdot1000=\frac{(0.651)(0.045)^2}{0.98}\cdot1000=1.34518$$

and

$$b=c\cdot h(y)/3=(1.34518)(1)/3=0.4483934 \; .$$

Since the launching angle is relatively narrow, we consider $h(y)\approx h(\bar{y})=1$ and $v_{x0}=v_0\cos\alpha_0\approx v_0=1200$ m/s.

Time of Flight

Substituting in (3.2.12), we have

$$240t+\frac{(v_{x0}-240)}{b}(1-e^{-b\cdot t})=240t+\frac{(1200-240)}{0.0457234}\cdot(1-e^{-0.0457234\cdot t})=500 \; .$$

Solving the above equation (using a graphing calculator TI-83 Plus), we find

$$t=0.45077s \; .$$

Launching Angle

Substituting the above values in (3.2.13) and the value $y=0$ at the point of impact, we obtain

$$[499.994\,p_0-0.0911277\int_0^{0.45077}(240+960\cdot e^{-0.44839\cdot t})\cdot\ln(0.2\cdot e^{0.44839\cdot t}+0.8)dt=0 \; .$$

The value of the integral is 10.378458. Substituting and then solving the above equation for p_0, we find that

$$p_0=\tan\alpha_0=0.00189155$$

Hence,

$$\alpha_0=\tan^{-1}(0.00189155)=0.10837786°=6.50'$$

Terminal Angle

Substituting in equation (3.2.10), the value of the launching angle $\alpha_0=6.50°$ and the time $t=0.45077s$, we find that

$$\tan\alpha_T=p_0-\frac{g}{240\cdot b}\ln(\frac{240\cdot e^{b\cdot t}+v_{x0}-240}{v_{x0}})=$$

$$=0.00189155-\frac{g}{240\cdot(0.4483934)}\cdot\ln(\frac{240\cdot e^{(0.4483934\cdot(0.4577)}+(1200-240)}{1200})=-0.002102 \quad .$$

The terminal angle is

$$\alpha_T=\tan^{-1}(0.002102)=-0.1204°=-7.23' \quad .$$

Terminal Speed

The components of the terminal velocity are respectively

$$v_x=240+(v_{x0}-240)\cdot e^{-b\cdot t}=1024.32m/s$$

and

$$v_y=pv_y=(-0.002102)(1024.32)=-2.15m/s \quad .$$

The *y*-component of velocity is irrelevant. The projectile speed at the point of impact is

$$v=(v_x^2+v_y^2)^{1/2}=[1024.32^2+(-2.15)^2]^{1/2}=1024.32m/s \quad .$$

Using QBasic PC program ANGLEC.BAS (see appendix B).[20]

Solution Instructions

---

[20]   Request an electronic copy or a CD with all PC programs to the author at: *gklimi@pace.edu*, or iven24@aol.com.

Input:
Range = 500, Launching speed = 1,200, ballistics coefficient = 1.34518.

Results:
Launching angle = 0.102356 [Degree], Time of flight = 0.432, Terminal angle = -0.1078523, Terminal Speed = 1,114 m/s.

Comment: The results obtained using the above approximate methods are quite equal to the results obtained using the QBasic PC ANGLEC.BAS program where the differential equations of projectile flight are solved using Runge-Kutta methods (see chapter 8).

3.3 The Equation of Projectile Trajectory for Narrow Launching Angles (Method II)

In section 3.2, integrating the differential equations of projectile, we estimated with a satisfying accuracy the elements of projectile trajectory when the launching angle is narrow. For such launching angles, the trajectory is close to the ground, and changes in y-coordinate of the projectile are small for relatively great changes in horizontal distances x, and so we can approximate the values of y for horizontal distances using the Maclaurin series expansion.

Consider the system of differential equations (3.1.10),

$$
\begin{cases}
\dfrac{dv_x}{dx} = -b\,\dfrac{v-240}{v} \\[2ex]
\dfrac{dp}{dx} = -\dfrac{g}{v_x^2} \\[2ex]
\dfrac{dt}{dx} = \dfrac{1}{v_x} \\[2ex]
\dfrac{dy}{dx} = p
\end{cases}
$$

Assuming that $v \approx v_x$, we write the above system as

$$\begin{cases} \dfrac{dv_x}{dx}=-b\dfrac{v_x-240}{v_x} \\[2mm] \dfrac{dp}{dx}=-\dfrac{g}{v_x^2} \\[2mm] \dfrac{dt}{dx}=\dfrac{1}{v_x} \\[2mm] \dfrac{dy}{dx}=p \end{cases} \qquad , \qquad (3.3.1)$$

where

$$b=c\cdot h(\bar{y})/3, \quad c=i\dfrac{d^2}{m}\cdot 1000, \quad p=\tan\alpha. \qquad (3.3.2)$$

Equation of the Projectile Trajectory

The equation of projectile trajectory is a function of the horizontal distance x, i.e., $y=f(x)$. Expanding $f(x)$ in Maclaurin series we can write approximately

$$y=f(0)+f'(0)\cdot x+\dfrac{f''(0)}{2!}\cdot x^2+\dfrac{f'''(0)}{3!}x^3+\dfrac{f^{IV}(0)}{4!}x^4, \qquad (3.3.3)$$

where

$$f(0)=y_0=0, \quad f'(0)=[\dfrac{dy}{dx}]_{x_0}=p(0)=p_0,$$

$$f''(0)=[\dfrac{d^2y}{dx^2}]_{x=0}=[\dfrac{dp}{dx}]_{x=0}=[-\dfrac{g}{v_x^2}]_{x=0}=-\dfrac{g}{v_x^2(0)}=-\dfrac{g}{v_0^2}$$

$$f'''(0)=\dfrac{d(-g/v_x^2)}{dx}=\dfrac{2g\cdot(dv_x/dx)_{x=0}}{v_{0x}^3}=-2gb\dfrac{v_{0x}-240}{v_{0x}^4}=-2gb\cdot\dfrac{v_0-240}{v_0^4}$$

$$f^{IV}(0)=[d(-2gb\cdot\dfrac{v_x-240}{v_x^4})/dx]_{x=0}=-6gb^2\cdot(v_0-240)\dfrac{v_0-320}{v_0^6}.$$

For relatively narrow angles, we have considered $v \approx v_x$ at any point on the projectile trajectory. Substituting in (3.3.3) the above derivatives and considering the fourth equation of (3.3.1), we find the equation of projectile trajectory

$$y = p_0 \cdot x - \frac{g}{v_0^2} \cdot \frac{x^2}{2} - \frac{gb(v_0 - 240)}{v_0^4} \cdot \frac{x^3}{3} - \frac{g \cdot b^2 (v_0 - 240) \cdot (v_0 - 320)}{v_0^6} \cdot \frac{x^4}{4}. \quad (3.3.4)$$

Ignoring the last term as very small, we have a less accurate but still satisfying equation for the projectile trajectory

$$y = p_0 \cdot x - \frac{g}{2 \cdot v_0^2} \cdot x^2 - \frac{gb(v_0 - 240)}{3 v_0^4} x^3. \quad (3.3.5)$$

As we will illustrate later, equation (3.3.4) can be used to solve two main problems of exterior ballistics:

- Finding the range x when v_0, α_0, c are known
- Finding the launching angle α_0 when v_0, x, c are known

Differentiating y, given in (3.3.4), and employing the fourth equation of (3.3.1), we find p as a function of x

$$p = p_0 - \frac{g}{v_0^2} \cdot x - \frac{gb(v_0 - 240)}{v_0^4} x^2 - \frac{g \cdot b^2 (v_0 - 240) \cdot (v_0 - 320)}{v_0^6} x^3, \quad (3.3.6)$$

or using (3.3.5), we obtain a less accurate formula for p

$$p = p_0 - \frac{g}{v_0^2} \cdot x - \frac{gb(v_0 - 240)}{v_0^4} x^2. \quad (3.3.7)$$

Thus, to calculate the elements of a projectile trajectory, we use (3.3.5) or (3.3.6) and the equations obtained in section 3.2, i.e.,

$$x = 240t + \frac{(v_{x0} - 240)}{b}(1 - e^{-b \cdot t}), \quad (3.3.8)$$

$$v_x = 240 + (v_{x0} - 240) \cdot e^{-b \cdot t} ,$$ (3.3.9)

$$p = p_0 - \frac{g}{240 \cdot b} \ln(\frac{240 \cdot e^{b \cdot t} + v_{x0} - 240}{v_{x0}}) ,$$ (3.3.10)

where $p = \tan\alpha$, $p_0 = \tan\alpha_0$, and

$$v_y = [(p_0 - \frac{g}{240 \cdot b} \ln(\frac{240 e^{b \cdot t} + v_{x0} - 240}{v_{x0}})] \cdot [240 + (v_{x0} - 240) \cdot e^{-b \cdot t}]$$ (3.3.11)

or

$$v_y = p \cdot v_x .$$ (3.3.12)

The following illustration exercises confirm the satisfying accuracy of the method.

Note: The Projectile Drop

The drop $|y|$ of a projectile launched at $\alpha_0 = 0°$ can be found by substituting $p_0 = \tan\alpha_0 = 0$ in (3.3.5). The absolute value of the y coordinate is the drop of the projectile under the horizontal line (figure 11).

Example 1. (See example 2, section 3.2) Find the launching angle α_0 needed to hit a tank that is in a horizontal distance of 500 m from antitank cannon 45 mm if a projectile of mass 0.98 kg is fired with the initial speed of 1,200 m/s. The coefficient of projectile form is 0.651. The meteorological conditions are normal. Find as well the time of flight, the terminal angle, and the terminal speed.

Solution
For the given projectile, we have found in example 2 section 3.2 that $c = 1.34518$ and $b = 0.4483934$.
Since the launching angle is relatively narrow, we assume $h(y) \approx h(\bar{y}) = 1$ and $v_{x0} = v_0 \cos\alpha_0 \approx v_0 = 1200$ m/s.

Launching Angle

To find the launching angle, we substitute $y=0$ in the equation of projectile trajectory (3.3.5)

$$x \cdot (p_0 - \frac{g}{2 \cdot v_0^2} \cdot x - \frac{gb(v_0-240)}{3v_0^4} x^2)=0.$$

Substituting $x=500$ and all other given values, we find

$$p_0 = \frac{g}{2 \cdot v_0^2} \cdot x + \frac{gb(v_0-240)}{3v_0^4} x^2 =$$

$$= \frac{g}{2(1200)^2}(500) + \frac{g(0.4483934)(1200-240)}{3(1,200)^4}(500)^2 = 0.00187219$$

Hence, we write

$$p_0 = \tan\alpha_0 = 0.0018722 .$$

The launching angle is

$$\alpha_0 = \tan^{-1}(0.0018722) = 0.10727° = 6.44' .$$

For a better accuracy, substituting $y = 0$ in (3.3.4), we have

$$p_0 = \frac{g}{2 \cdot v_0^2} \cdot x + \frac{gb(v_0-240)}{3v_0^4} x^2 + gb^2(v_0-240)\frac{v_0-320}{4v_0^6 \cdot} x^3 .$$

Substituting all known values in the above equation and solving the obtained equation, we find a more accurate value for the launching angle

$$\alpha_0 = 6.496' ,$$

which is the same as the result of example 2, section 3.2.

All other elements can be found using the formulas obtained in section 3.2.

Example 2. (Known v_0, α_0, c) Calculate the range of fire of a 7.9 g bullet of a Russian 7.62 mm rifle launched at 735 m/s. Use the launching angle $\alpha_0 = 0.05728°$ found in example 1, section 3.2. Employ as an approximate value of the ballistics coefficient the value $c=4.5663$ found in example 4, section 2.6.

Solution

Estimate b.

$$b=c/(3)=(4.5663)/3=1.5221$$.

Range

To find the range, substitute $y=0$ in (3.3.5). We write

$$p_0 \cdot x - \frac{g}{2 \cdot v_0^2} \cdot x^2 - \frac{gb(v_0-240)}{3v_0^4} x^3$$

$$=x \cdot [\tan(0.05728) - \frac{g}{2(735)^2} \cdot x - \frac{g(1.5221)(735-240)}{3(735)^4} x^2] = 0$$.

Hence, we find $x=0$, the solution that represents the launching point, and the following second-degree equation with respect to x

$$9.997249 \cdot 10^{-4} - 9.07645 \cdot 10^{-6} x - 8.439145 \cdot 10^{-9} x^2 = 0 .$$

Solving the above equation, we find that the horizontal range is

$$x=100.67m .$$

Note: For the same shooting data, the range table of the given projectile gives a range of 100 m.

Example 3. Estimate an appropriate ballistics coefficient and the corresponding form coefficient for a 7.9 g bullet fired from a 7.62 mm Russian rifle considering the following experimental data: initial speed $v_0 = 735m/s$, terminal speed at a range of 300 m is $v_T = 485$ m/s, time of flight $t=0.50s$. Use the estimated ballistics coefficient to find the launching angle α_0.

Solution

First, we calculate an approximate ballistics coefficient using formula (2.6.28)

$$c = \frac{3}{h(y) \cdot t} \cdot \ln(\frac{v_0 - 240}{v - 240}).$$

Substituting $t = 0.50s$, $v_0 = 735 m/s$, and $v_T = 485$ m/s, we find:

$$c = \frac{3}{h(y) \cdot t} \cdot \ln(\frac{v_0 - 240}{v - 240}) = \frac{3}{0.50} \ln \frac{735 - 240}{485 - 240} = 4.219797.$$

We estimate

$$b = ch(y)/3 = (4.219797)/(3) = 4.219797$$

To find α_0, we substitute $y = 0$ in (3.3.4), and then factorizing x, we write

$$x[p_0 - \frac{g}{2 \cdot v_0^2} \cdot x - \frac{gb(v_0 - 240)}{3v_0^4} x^2 - gb^2(v_0 - 240)\frac{v_0 - 320}{4v_0^6} x^3] = 0.$$

Substituting the respective data, we have

$$p_0 = \frac{g}{2(735)^2}(300) + \frac{g(4.219797)(735 - 240)}{3(735)^4}(300)^2 +$$

$$+ \frac{g(4.219797)^2(735 - 240)(735 - 320)}{4(735)^6}(300)^3$$

We can write

$$p_0 = \tan\alpha_0 = 0.0035986449$$

The launching angle is

$$\alpha_0 = \tan^{-1}(0.0035986449) = 0.2062° = 12.37'$$

Note: The result is close to the value $\alpha_0 = 12.96'$ given in the range table of the 7.62 mm Russian rifle for the same range.

Example 4. Find an appropriate ballistics coefficient and the corresponding form coefficient for a 12.8 g bullet 8 x 57 S fired from a Mauser K98kconsidering the following experimental data: initial speed $v_0 = 755 m/s$, the speed at horizontal range 100 m is $v_T = 706$ m/s, and time of flight $t = 0.14s$. Use the obtained form coefficient to find the launching angle α_0. Air density is $\rho_0 = 1.225 kg/m^3$.[21]

Solution

First we calculate an approximate ballistics coefficient using formula (2.6.28)

$$c = \frac{3}{h(y) \cdot t} \cdot \ln(\frac{v_0 - 240}{v - 240}).$$

Since the air density is $\rho_0 = 1.225 kg/m^3 \neq \rho_{0N} = 1.205 kg/m^3$, we need to evaluate $h(y)$ for the given value of $\rho_0 = 1.225 kg/m^3$. We can write

$$h(y) = \rho / \rho_0 = \frac{\rho}{\rho_{0N}} \frac{\rho_{0N}}{\rho_0} = \frac{\rho}{\rho_{0N}} \frac{1.205}{1.225} = 0.9836735(1) = 0.9836735$$

since on the ground ($y = 0$), the density of air has not the standard value ($\rho_{0N} = 1.205 kg/m^3$), but it is $\rho = \rho_0 = 1.225 kg/m^3$.

Substituting $t = 0.14s$ and $v_T = 706$ m/s in the above formula, we find an approximate average ballistics coefficient

$$c = \frac{3}{h(y) \cdot t} \cdot \ln(\frac{v_0 - 240}{v - 240}) = \frac{3}{(0.9836735) \cdot (0.14)} \ln\frac{755 - 240}{706 - 240} = 2.1780.$$

The Siacci coefficient of form for the given projectile is

---

[21] Mori, Edoardo. *Balistics Teorica e Pratica.* http://www.earmi.it/balistica/coefball.htm [Web site]

$$i=10^{-3}c\frac{m}{d^2}=10^{-3}(2.1780)\frac{(0.0128)}{(0.008)^2}=0.4356.$$

For b, we have

$$b=ch(y)/3=(2.1780)(0.9836735)/3=0.714152.$$

To find the launching angle, we substitute $y=0$ in (3.3.4)

$$(p_0-\frac{g}{2\cdot v_0^2}\cdot x-\frac{gb(v_0-240)}{3v_0^4}x^2-gb^2(v_0-240)\frac{v_0-320}{4v_0^6}x^3)=0.$$

Substituting the given data and the value of $b=0.714152$, we have

$$p_0\frac{g}{2\cdot(755)^2}(100)-\frac{g(0.714152)(755-240)}{3(755)^4}(100)^2-g(0.714152)^2(755-240)\frac{755-320}{4(755)^6}(100)^3=0$$

Hence, we can obtain

$$p_0=\tan\alpha_0=0.000898714$$

Hence,

$$\alpha_0=\tan^{-1}(0.000898714)=0.05149215°=3.0895'=3'05".$$

Note 1: The range table given by Edoardo Mori gives the value $\alpha_0=3'\ 10"$. The estimated value is approximately the same as the value we have from the range table of the given rifle. The difference in value probably is the result of the time measurement, rounded to the nearest hundredth.

Note 2: Using QBasic PC program ANGLEC.BAS (see appendix B)

Input:
Range = 100, Launching speed = 755, Ballistics coefficient = 2.178

Outcomes:
Launching angle $\alpha_0=0.052°=3'\ 07"$,

Time of flight = 0.137, Terminal angle = -0.053, Terminal speed = 706.30 m/s

Comment: The results obtain by the approximate method of section 3.3 are quite equal to the results obtained by E. Mori or by using QBasic PC program ANGLEC.BAS.

Example 5. Consider a 0.30 M2 ball bullet caliber d=0.308"=0.00782 .
    The bullet is fired with a speed of 853.43 m/s at an angle α_0=16.8'=0.28° . Mass of the bullet is m=0.00972 kg.

(a) Find the elements of trajectory at the points on the trajectory with abscissa:

x_1=100m, x_2=200m, x_3=300m, x_4=400m , and x_5=500m .

(b) Find as well the coordinates of the vertex.

    Use the average form factor i=0.538 we found in example 5, section 2.6 for projectile speed in the interval (478–854 m/s). The density of air at sea level is ρ_0=1.205kg/m^3 .

Solution
    Since the launching angle α_0=16.8'=0.28° is relatively narrow, we use the equation of trajectory (3.3.4). The initial value of p_0 is

$$p_0=\tan(0.28°)=0.00488696$$

The ballistics coefficient is

$$c=\frac{id^2}{m}1000=\frac{(0.538)(0.00782)^2}{0.00972}1000=3.384773m^2/kg$$

We assume that $h(y)$=1.
We find as well that

$$b=ch(y)/3=(3.384773)(1)/3=1.1282576$$

We perform the calculations only for the point on the trajectory that has the abscissa $x_3=300m$. In a similar way, we can find the elements of trajectory for all other points.

(a) Elements of the Projectile Trajectory for $x_3=300m$

Substituting the above data and the value $x_3=300m$ in (3.3.4), we have

$$y=p_0 \cdot x - \frac{g}{v_0^2}\frac{x^2}{2} - \frac{gb(v_0-240)}{v_0^4}\frac{x^3}{3} - \frac{gb^2(v_0-240)\cdot(v_0-320)}{v_0^6 \cdot}\frac{x^4}{4}$$

$$=(0.004887)(300)-\frac{g}{2(853.44)^2}(300)^2-\frac{g(1.1282576)(853.44-240)}{3(853.44)^4}(300)^3 \; .$$

$$-g(1.1282576)^2(853.44-240)\frac{853.44-320}{4(853.44)^6}(300)^4=0.745m$$

Time of Flight (that corresponds to the point with abscissa $x_3=300m$)

$$v_{x0}=v_0 \cos\alpha_0=(853.44)\cos(0.28°)=853.423m/s$$

Employing (3.3.8), we can write

$$x=240t+\frac{v_{x0}-240}{b}(1-e^{-b \cdot t})$$

$$(300)=240 \cdot t+\frac{(853.423-240)}{(1.1282576)}(1-e^{-(1.1282576) \cdot t})^{\cdot}$$

The above equation can be written as

$$0.44142986+0.448212675t=e^{-1.12825786 \cdot t} \; .$$

Using a TI-83 Plus graphing calculator to solve the last equation, we find that the time of flight is $t=0.41s$.

Angle of flight for $x_3=300m$.
Employing (3.3.10), we find

$$p = p_0 - \frac{g}{240 \cdot b} \ln\left(\frac{240 \cdot e^{b \cdot t} + v_{x0} - 240}{v_{x0}}\right)$$

$$= (0.004887) - \frac{g}{240 \cdot (1.128257)} \ln(240 \cdot e^{(1.128257) \cdot (0.41)} + 853.43 - 240) = -0.00066284$$

Hence,

$$\alpha_T = \tan^{-1}(p) = \tan^{-1}(-0.00066284) = -0.0382° = -2.29'$$

Speed
   Using (3.3.9) and (3.3.12), we find that

$$v_x = 240 + (v_{x0} - 240) \cdot e^{-b \cdot t} =$$

$$= 240 + (853.43 - 240) \cdot e^{-(0.1128257) \cdot (0.41)} = 626.07 m/s$$

and

$$v_y = p \cdot v_x = (-0.0066284) \cdot (626.07) = -0.42 m/s.$$

The speed of projectile is

$$v = (v_x^2 + v_y^2)^{1/2} = (626.07^2 + (-0.42)^2)^{1/2} = 626.07 m/s.$$

The results obtained for other points of the trajectory are presented in the table 3.

Table 3

Coordinate x in meter	Coordinate y in meters	Time t in seconds	Speed v in meters per second	Angle α in minutes
100	0.42	0.12	774	11.69'
200	0.67	0.26	698	5.45'
300	0.75	0.41	626	- 2.29'
400	0.60	0.58	559	-11.93'
500	0.23	0.77	498	-24.07'

(b) The Coordinates of the Vertex

At the vertex of the trajectory, the projectile velocity is parallel to the x-axis. The angle of flight and its tangent are both zero: $\alpha_m=0$, $p_m=\tan\alpha_m=0$.

Substituting, $p=p_m=\tan\alpha_m=0$, we have

$$p_0-\frac{g}{v_0^2}\cdot x-\frac{gb(v_0-240)}{v_0^4}x^2-\frac{g\cdot b^2(v_0-240)\cdot(v_0-320)}{v_0^6}x^3=0.$$

Substituting in the above equation $p_0=\tan(0.28°)=0.00488696$ and all the other values and then solving for x, we find that the abscissa of the trajectory vertex is $x_m=274.86m$.

Substituting $x_m=274.86m$ in (3.3.5),

$$y=p_0\cdot x-\frac{g}{v_0^2}\cdot\frac{x^2}{2}-\frac{gb(v_0-240)}{v_0^4}\cdot\frac{x^3}{3}-\frac{g\cdot b^2(v_0-240)\cdot(v_0-320)}{v_0^6}\cdot\frac{x^4}{4},$$

we find the ordinate of the trajectory vertex, $y_m=0.73m$.

That means that the trajectory of the projectile will hit a standing person located inside the range of shooting at a point of his body that is not higher then $y_m=0.73m$ from the projection point (that is at $x_s=500m$ from the launching point).

3.4 Uphill Motion of a Projectile for Narrow Projection Angles. Curvilinear Coordinates

In situation of shooting against aircraft or parachutists, or firing uphill or downhill, in a mountain terrain, the projection angle, as well as the distances to the target, are relatively short. Such scenarios are also encountered in criminal justice cases of shootings from firearms, for example, in shooting from or at a high building, etc.

In such situations, it is necessary to find the projection angle, the time of flight to the target, and other elements of trajectory.

For example, during the fire of antiaircraft guns, where the target is on the increasing part of trajectory, it is necessary to know the time of flight of a fragmentation projectile to set the projectile off at the moment when the projectile is next to the aircraft, in such distances that the probability the fragments will hit and destroy the airplane is large.

For some firearms, such as infantry firearms, there are given the range tables in standard normal conditions when the rifle and the target are on the flat ground.

We will show approximate solutions of the equations of projectile flight (3.1.1) for downhill or uphill shootings when projection angle is narrow and the slant distance relatively short.

Curvilinear Coordinates

Consider the vector differential equation (2.3.5),

$$\frac{d\vec{v}}{dt}=\vec{g}-c{\cdot}h(y)K_D(v)\frac{\vec{v}}{v}.$$

$$(3.4.1)$$

For projectiles flying with speeds greater than 256 m/s,

$$K_D(v)=(v-240)/3 \quad \text{for} \quad v>256m/s.$$

$$(3.4.2)$$

We construct a new coordinative system with axes $o\bar{x}$ and $o\bar{y}$, where the first axis has the direction of the initial velocity $\vec{v}_0$ while the second axis is directed on the opposite direction of y-axis (figure 8).

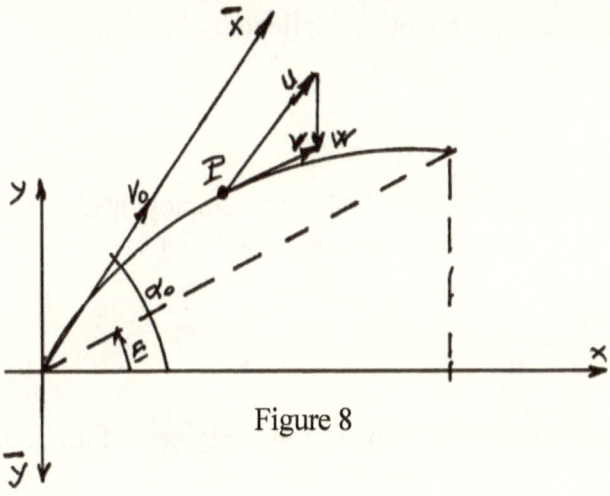

Figure 8

Let $\vec{u}$ and $\vec{w}$ be the components of velocity $\vec{v}$ in the new coordinate system. We can write

$$\vec{v}=\vec{u}+\vec{w}. \tag{3.4.3}$$

Substituting in (3.4.1), we can write

$$\frac{d\vec{u}}{dt}-\frac{dw}{dt}\vec{j}=-g\vec{j}-c{\cdot}h(y)K_D(v)\frac{\vec{u}}{v}+c{\cdot}h(y)K_D(v)\frac{w}{v}\vec{j}. \tag{3.4.4}$$

From (3.4.4), we obtain the following system of differential equations in curvilinear coordinates $\bar{x}$ and $\bar{y}$

$$\frac{du}{dt}=-c{\cdot}h(y)K_D(v)\frac{u}{v}, \tag{3.4.5}$$

$$\frac{dw}{dt}=g-c{\cdot}h(y)K_D(v)\frac{w}{v}, \tag{3.4.6}$$

$$\frac{d\bar{x}}{dt}=u, \quad \frac{d\bar{y}}{dt}=w. \tag{3.4.7}$$

We introduce a new variable $\bar{p}$ defined by the equation

$$\bar{p} = \frac{w}{u}.$$

(3.4.8)

By substituting (3.4.8) in (3.4.6) and employing (3.4.5), we obtain the following simple form of the equation (3.4.6)

$$\frac{d\bar{p}}{dt} = \frac{g}{u}$$

(3.4.9)

Thus we have obtained the following system of differential equations in curvilinear coordinates:

$$
\begin{cases}
\dfrac{du}{dt} = -c \cdot h(y) K_D(v) \dfrac{u}{v} \\[2mm]
\dfrac{d\bar{p}}{dt} = \dfrac{g}{u} \\[2mm]
\dfrac{d\bar{x}}{dt} = u \\[2mm]
\dfrac{d\bar{y}}{dt} = w
\end{cases}
$$

(3.4.10)

where

$$K_D(v) = (v-240)/3.$$

(3.4.11)

Substituting (3.4.11) into (3.4.10) we obtain the equations of projectile in the form:

$$\begin{cases} \dfrac{du}{dt} = -ch(y)\dfrac{(v-240)}{3}\dfrac{u}{v} \\[2mm] \dfrac{d\bar{p}}{dt} = \dfrac{g}{u} \\[2mm] \dfrac{d\bar{x}}{dt} = u \\[2mm] \dfrac{d\bar{y}}{dt} = w \end{cases} \qquad (3.4.12)$$

Initial conditions:

When $t=0$,

$\bar{x}(0)=0$, $\bar{y}(0)=0$, $u(0)=v(0)=v_0$, $w(0)=0$, $\bar{p}(0)=\bar{p}_0=0$, $\alpha(0)=\alpha_0$.(3.4.13)

Approximate Solutions

For a relatively short shooting distance to the target $\overline{D}=OT$, the projectile trajectory deviates slightly from the direction of initial velocity v_0 of the projectile. We assume that at any point $P(\bar{x},\bar{y})$ of the trajectory

$$\vec{v}\approx\vec{u} \text{ , and } v\approx u$$

and

$$K_D(v)=(v-240)/3\approx(u-240)/3 \qquad (3.4.14)$$

We can estimate somehow the difference between u and v or, better to say, the error we make assuming $K_D(v)\approx K_D(u)$.

We denote A the projection angle and E the angle of sight (elevation angle or depression angle).

The launching angle (figure 9) is

(Note: the angle in fig. 9 is (90 + alpha) and not (90 + alpha0)

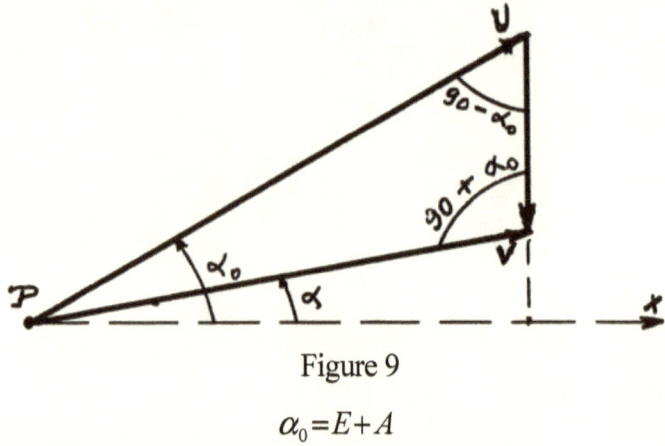

Figure 9

$$\alpha_0 = E + A$$

We construct the triangle of velocity vectors $\vec{u}$, $\vec{w}$, and $\vec{v}$ (figure 2). The angle between $\vec{v}$ and $\vec{w}$ is $(90° + \alpha)$. From the geometry of figure 9, we write[22]

$$u = v \frac{\cos\alpha}{\cos\alpha_0} = v \frac{\cos\alpha}{\cos(E+A)} . \qquad (3.4.15)$$

For short firing distances $\overline{D}$, when the shooting is uphill,

$$0 < \alpha < \alpha_0 = A + E \qquad (3.4.16)$$

and

$$|\alpha| > |A + E| \qquad (3.4.17)$$

when the shooting is downhill. Thus, for the uphill shooting,

$$\frac{\cos\alpha}{\cos(E+A)} > 1 , \qquad (3.4.18)$$

and as a result, from (3.4.15), we have $u > v$.

Considering (3.4.14) and (3.4.15), we can write the first equation (3.4.12) in the form

---

[22]   The speed u given in (3.4.15) is the Siacci pseudo-speed that is introduced in chapter 5.

$$\frac{du}{dt}=-ch(y)\frac{(u-240)}{3}\cdot\frac{\cos\alpha}{\cos\alpha_0}. \qquad (3.4.19)$$

To compensate somehow the error we make when we use (3.4.14), we multiply the right side of (3.4.19) by the quantity $(\cos\alpha_0/\cos\alpha)$ that is less than one, i.e., we write[23]

$$\frac{du}{dt}=-ch(y)\frac{u-240}{3}. \qquad (3.4.20)$$

Thus, the system of differential equations (3.4.12) can be written in the following form

$$\begin{cases} \dfrac{du}{dt}=-ch(y)\dfrac{u-240}{3} \\ \dfrac{d\overline{p}}{dt}=\dfrac{g}{u} \\ \dfrac{d\overline{x}}{dt}=u \\ \dfrac{d\overline{y}}{dt}=w \end{cases}. \qquad (3.4.21)$$

In the first differential equation of (3.4.21), the density function $h(y)$ is a function of y-coordinate of the projectile. The density function $h(y)$ does not allow us to obtain analytical solutions of (3.4.21). For that reason, we will assume the standard atmosphere and $h(y)$ to be a constant that is determined by

$$h(y)=h(\hat{y}),$$

where

$$h(\hat{y})=(\frac{289.08-0.006328\hat{y}}{289.08})^{4.4},$$

and

Siacci uses the same compensation variable factor.

$\hat{y}=(1/2)y_{max}$, for antiaircraft artillery; target is hit at the ascending points of the trajectory, while y_{max} is the maximum height of the projectile trajectory.

Denoting

$$B=\frac{c}{3}h(\hat{y})$$ (3.4.22)

we write

$$\begin{cases} \dfrac{du}{dt}=-B(u-240) \\ \dfrac{d\overline{p}}{dt}=-(-\dfrac{g}{u}) \\ \dfrac{d\overline{x}}{dt}=u \\ \dfrac{d\overline{y}}{dt}=w \end{cases}$$ (3.4.23)

Initial conditions (for $t=0$) are

$\overline{x}(0)=0$, $\overline{y}(0)=0$, $u(0)=v(0)=v_0$, $w(0)=0$, $\overline{p}(0)=\overline{p}_0=0$, $\alpha(0)=\alpha_0$.

The same system is also valid for downhill shooting.
System (3.4.23) has an identical form as system (3.2.6). Thus, the solution of (3.4.23) can be taken by formally substituting the curvilinear variables in the solutions of (3.2.6). Therefore, solving (3.4.23), we obtain

$$u=240+(v_0-240)\cdot e^{-B\cdot t}$$ (3.4.24)

$$\overline{p}=\frac{g}{240\cdot B}\ln(\frac{240\cdot e^{B\cdot t}+v_0-240}{v_0})\,,$$ (3.4.25)

where $\overline{p}=w/u\geq0$,

$$w=\frac{g}{240 \cdot B}\ln(\frac{240e^{B \cdot t}+v_0-240}{v_0}) \cdot (240+(v_0-240) \cdot e^{-B \cdot t}), \text{ or } w=p \cdot u, \quad (3.4.26)$$

$$\bar{x}=240t+\frac{(v_0-240)}{B}(1-e^{-B \cdot t}), \quad\quad (3.4.27)$$

$$\bar{y}=\frac{g}{240 \cdot B}\int_0^t (240+(v_0-240) \cdot e^{-B \cdot t}) \cdot \ln(\frac{240e^{B \cdot t}+v_0-240}{v_0})dt \quad\quad (3.4.28)$$

since $\bar{p}_0=w_0/u_0 \approx 0$.

Notes:

- The above formulas can be simplified substituting $\bar{p}_0=0$ since $\bar{p}_0=w_0/u_0=0$.
- The solution of the equations of projectile flight (3.2.4) can be obtained from the solution of equations (3.4.23) when the angle of projection is $A=0$.
- The value of $\bar{y}$, given in (3.4.28), can be interpreted as the "drop" of the projectile, flying in the direction of initial velocity $\bar{x}$ during the time of flight.

Elements of the Projectile Trajectory

Projectile Speed

Geometrically we can find some relationships for the elements of the trajectory.

From the triangle of velocities at any point on the trajectory or at the target T (figure 10), we obtain

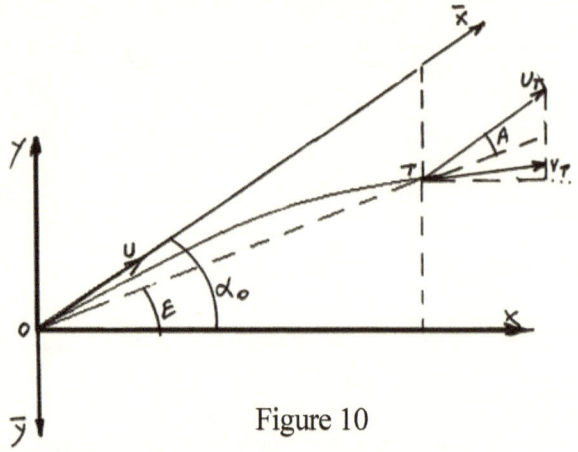

Figure 10

$$v^2 = u^2 + w^2 - 2uw \cdot \sin\alpha_0 \qquad (3.4.29)$$

Range

The distance to the target (figure 3) is

$$\overline{D} = \overline{x} \frac{\cos(A+E)}{\cos E} \qquad (3.4.30)$$

Projection Angle

Applying sine theorem (figure 3), we find that

$$\sin A = \frac{\overline{y}}{\overline{D}} \cos(A+E). \qquad (3.4.31)$$

Since the projection angle A in general is relatively narrow, we can consider

$$\sin A \approx A \ , \ \cos(E+A) \approx \cos E \ ,$$

and as a result (31) yields

$$A = \frac{\overline{y}}{\overline{D}} \cos(E). \qquad (3.4.32)$$

Terminal Angle

The angle α_T the projectile velocity forms with the horizontal plane at the point of impact to the target is

$$\alpha_T=\alpha_0-p_T\cos\alpha_0.\tag{3.4.33}$$

Terminal Speed
The speed of projectile at the target is

$$v_T=v_0\frac{\cos(A+E)}{\cos(A-E)}.\tag{3.4.34}$$

Cartesian Coordinates of Projectile
The Cartesian coordinates of projectile at any moment (see figure 3) are

$$x=\bar{x}\cos\alpha_0,\ y=\bar{x}\sin\alpha_0-\bar{y}.\tag{3.4.35}$$

Example 1. Find the projection angle A and time of flight t to the target of a bullet of an antiaircraft 14.5 mm when the firing distance and the angle of elevation of the target are respectively $D=1000m$ and $E=60^0$. The initial speed of the bullet is $v_0=945\,m/s$ while the ballistics coefficient is $c=1.64$.

Solution
Estimate B.

$$B=\frac{c\cdot h(\hat{y})}{3}$$

As a first approach we consider

$$\bar{x}\approx\bar{D}=1000\text{ m}$$

$$\bar{y}=y_m/2,\text{ where:}$$

$$y_m=D\sin A=(1000)\cdot\sin(60)=866.03m.$$

Thus

$$h(\hat{y}) = (\frac{289.08 - 0.006328 \cdot (433.01)}{289.08})^{4.4} = 0.96,$$

and

$$B = \frac{c \cdot h(\hat{y})}{3} = \frac{(1.62)(0.96)}{3} = 0.5178,$$

Substituting in (3.4.27),

$$\bar{x} = 240t + \frac{v_0 - 240}{B}(1 - e^{-B \cdot t}),$$

we have

$$1,000 = 240t + \frac{(945 - 240)}{0.5178}(1 - e^{-0.5178 \cdot t})$$

Simplifying we can write

$$0.73452 - 0.17629t = (1 - e^{-05178t})$$

Solving the above exponential equation for t (using a graphing calculator TI83+) we find the time of flight:

$$t = 1.335s$$

Employing equation (28) we find

$$\bar{y} = -\frac{g}{240 \cdot B} \int_0^t (240 + (v_0 - 240) \cdot e^{-B \cdot t}) \cdot \ln(\frac{240e^{B \cdot t} + v_0 - 240}{v_0})dt =$$

$$= -0.0789 \int_0^{1.335} (240 + 705e^{-0.5178 \cdot t}) \cdot \ln(0.253968e^{0.5178 \cdot t} + 0.746)dt$$

The integral on the right side of the above equation has the value 95.334. Thus the "drop" is

$$\bar{y} = -0.0789(95.334) = -7.52m.$$

Substituting in (32) we find that,

$$A = \frac{\overline{y}}{\overline{D}}\cos(A) = \frac{7.52}{1000}\cos(60°) = 0.003761 \text{ radian},$$

or

$$A = 0.003761(180/\pi) = 0.2155° = 12.93'.$$

Using (37) we find that the curvilinear coordinate $\overline{x}$ is

$$\overline{x} = \overline{D}\frac{\cos(E)}{\cos(E+A)} = 1,000\frac{\cos(60°)}{\cos(60.2155°)} = 1006.56 \text{ m}.$$

The results can be improved repeating the same operations as above considering x = 1006.56.

Note: The results obtained using the software QBasic PC program ANTAIRC.BAS (see appendix B) are the following:

Input Data:
Coordinates   of   the   Target: x_T=1000·cos(60)=500m, y_T=1000·sin(60)=866m, Projectile Speed v_0=945m/s, Ballistics Coefficient c=1.64.

Results:
Launching Angle is α_0=60.22168°, Time of Flight is t=1.355s, Impact Speed is v_T=579.06m/s, Impact Angle α_T=59.69°

Hence, we find that the projection angle is

$$A=\alpha_0-E=60.22168°-60°=0.22168°=13.3'.$$

Example 2. The projectile of an antiaircraft cannon explodes 6,000 m above the sea level. A projectile fragment with mass of 20 g is launched with an initial speed of 986 m/s and forms an angle of 15° with the x-axis.

Find the slant distance from the exploding point where the speed of the projectile fragment will be 316 m/s. The ballistics coefficient of the given projectile fragment is $c=90$.

Solution

We denote $\overline{D}$ the distance of the point where the projectile fragment speed will be 316 m/s. We assume that $\overline{x}=\overline{D}$, $u=v$.

We have

$$h(\hat{y})=(\frac{288.9-0.006328 \cdot (6000)}{288.9})^{4.4}=0.537969$$

and

$$B = \frac{c \cdot h(\hat{y})}{3} = \frac{(90)(0.537969)}{3} = 16.1388.$$

Substituting in

$$u=240+(v_0-240) \cdot e^{-B \cdot t},$$

we can write

$$(316)=240+(986-240) \cdot e^{-16.1388 \cdot t}.$$

Solving the above equation, we find the time of flight $t=0.142s$.

The $\overline{x}$ coordinate of projectile and at the same time an approximate estimation of slant distance is

$$\overline{x}=240t+\frac{(v_0-240)}{B}(1-e^{-B \cdot t})=$$

$$=240(0.142)+\frac{(986-240)}{(1.6457)(9.80665)}(1-e^{-(1.6457) \cdot (9.80665) \cdot (0.142)})=75.52m.$$

3.5 Approximate Trajectory Equation in Curvilinear Coordinates

The differential equations (3.3.23),

$$\begin{cases} \dfrac{du}{dt}=-B(u-240) \\[2mm] \dfrac{d\bar{p}}{dt}=-\left(-\dfrac{g}{u}\right) \\[2mm] \dfrac{d\bar{x}}{dt}=u \\[2mm] \dfrac{d\bar{y}}{dt}=w \end{cases},$$

can be expressed through the variable $\bar{x}$

$$\begin{cases} \dfrac{du}{d\bar{x}}=-B\dfrac{(u-240)}{u} \\[2mm] \dfrac{d\bar{p}}{d\bar{x}}=-\left(-\dfrac{g}{u^2}\right) \\[2mm] \dfrac{dt}{d\bar{x}}=\dfrac{1}{u} \\[2mm] \dfrac{d\bar{y}}{d\bar{x}}=p \end{cases}. \qquad (3.5.1)$$

Initial conditions for $\bar{x}(0)=0$ are the following:

$\bar{y}(0)=0$, $u(0)=v(0)=v_0$, $w(0)=0$, $\bar{p}_0=w_0/u_0\approx 0$, $\alpha(0)=\alpha_0$, $t=0$.

In a similar way as in section 3.3, we can obtain the following equations for the trajectory in curvilinear coordinates for projectiles flying with speeds greater than 256 m/s.

The equation of projectile trajectory in curvilinear coordinates is

$$\bar{y}=\dfrac{g}{2u_0^2}\bar{x}^2+\dfrac{gB(u_0-240)}{3u_0^4}\bar{x}^3+gB^2(u_0-240)\dfrac{u_0-320}{4u_0^6}. \qquad (3.5.2)$$

Neglecting the last term, we obtain another approximate formula, but less accurate than (3.5.2),

$$\bar{y}=\frac{g}{2\cdot u_0^2}\cdot\bar{x}^2+\frac{gB(u_0-240)}{3u_0^4}\bar{x}^3 .$$ (3.5.3)

Differentiating $\bar{y}$ with respect to $\bar{x}$ and employing the fourth equation of (3.5.1), we find $\bar{p}$ as a function of $\bar{x}$

$$\bar{p}=\frac{g}{u_0^2}\cdot\bar{x}+\frac{gB(u_0-240)}{u_0^4}\bar{x}^2+gB^2(u_0-240)\frac{u_0-320}{\cdot u_0^6}\bar{x}^3$$ (3.5.4)

or, less accurately,

$$\bar{p}=\frac{g}{u_0^2}\cdot\bar{x}+\frac{gB(u_0-240)}{u_0^4}\bar{x}^2 .$$ (3.5.5)

The formulas obtained in this chapter can be used to solve two main problems of exterior ballistics:

- Finding the slant range $\overline{D}$ when v_0,α_0,c are known.
- Finding the launching angle α_0 when $v_0,\overline{D}, c$ are known.

We can employ the results of this section together with those obtained in 3.4 to solve different problems of exterior ballistics.

In the following example 1, we solve the problem of exercise 1 in section 3.4 using the approximate approach presented in this section.

The Drop of a Projectile
A projectile fired horizontally drops continuously under the horizontal line that has the direction of the launching speed (figure 11).

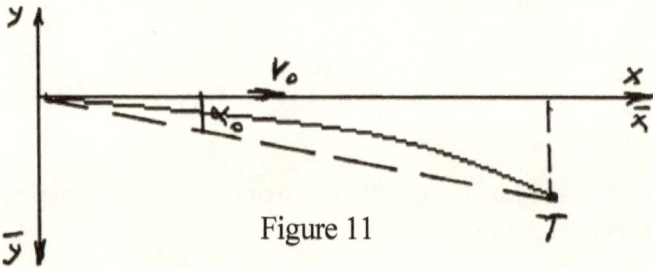

Figure 11

The $o\bar{x}$ coincides with ox. We find the drop $\bar{y}$ of the projectile substituting x for $\bar{x}$ in (3.5.2) or (3.5.3).

Example 1. Find the projection angle A and time of flight t to the target of a bullet of an antiaircraft 14.5 mm when the firing distance and the angle of elevation of the target are respectively $D=1000m$ and $E=60^0$. The initial speed of the bullet is $v_0=945\,m/s$ while the ballistics coefficient is $c=1.64$.

Solution

In example 1 in section 3.4, we found $B=0.496368$.
We consider

$$\bar{x}\approx\overline{D}=1000\,\text{m}.$$

Substituting in (3.5.3), we have

$$\bar{y}=\frac{g}{2\cdot u_0^2}\cdot\bar{x}^2+\frac{gB(u_0-240)}{3u_0^4}\bar{x}^3$$

$$=\frac{g}{2(945)^2}(1,000)^2+\frac{g(0.496368)(945-240)}{3(945)^4}(1,000)^3=6.925$$

The projection angle is

$$A=\frac{\bar{y}}{D}\cos(E)=\frac{6.93}{1000}\cos(60°)=0.003465=0.1985°=11.91'.$$

Improving Results

$$\bar{x}=\overline{D}\frac{\cos(E)}{\cos(A+E)}=1,000\frac{\cos(60°)}{\cos(60.199°)}=1006.6m$$

Substituting again in (3.5.5), we find

$$\bar{y}=\frac{g}{2\cdot u_0^2}\cdot\bar{x}^2+\frac{gB(u_0-240)}{3u_0^4}\bar{x}^3$$

$$=\frac{g}{2(945)^2}(1{,}006.6)^2+\frac{g(0.496368)(945-240)}{3(945)^4}(1{,}006.6)^3=7.02m$$

Projection angle is

$$A=\frac{\bar{y}}{D}\cos(E)=\frac{7.02}{1006.1}\cos(60°)=0.0034887=0.20°=12',$$

$$\bar{x}=D\frac{\cos(E)}{\cos(E+A)}=1{,}000\frac{\cos(60°)}{\cos(60.20°)}=1006.1m.$$

Note: There is no use to continue the "improving" procedure since the results we will obtain are quite equal to the result already obtained.

Example 2. A 7.9 g bullet is fired with initial speed 735 m/s. Determine the projection angle needed to hit a target located on a hill at a slant distance 300 meters from a 7.62 mm Russian rifle. The launching angle is 65°.

Use the data obtained in example 3 section 3.3: $c=4.219797$.

Solution

We assume that $\bar{x}\approx\overline{D}=300$. We find

$$B=\frac{c\cdot h(\bar{y})}{3}=\frac{(4.219797)(1)}{3}=1.406599$$

Substituting in (3.5.2), we can write

$$\bar{y}=\frac{g}{2\cdot u_0^2}\cdot\bar{x}^2+\frac{gB(u_0-240)}{3u_0^4}\bar{x}^3+gB^2(u_0-240)\frac{u_0-320}{4v_0^6}\cdot\bar{x}^4=\frac{g}{2(735)^2}(300)^2+$$

$$\frac{g(1.406599)(735-240)}{3(735)^4}(300)^3+g(1.406599)^2(735-240)\frac{735-320}{4(735)^6}(300)^4=1.0786m$$

The projection angle is

$$A=\frac{\bar{y}}{D}\cos(E)=\frac{1.078653}{300}\cos(65°)=0.00151953=0.08706°=5.224'$$

Improving Results

$$\bar{x}=\bar{D}\frac{\cos(E)}{\cos(E+A)}=300\frac{\cos(65°)}{\cos(65.08706°)}=300.98m$$

Substituting again in (3.5.2), we find the drop of the projectile from the $\bar{x}$ axis (direction of initial velocity)

$$\bar{y}=\frac{g}{2\cdot u_0^2}\bar{x}^2+\frac{gB(u_0-240)}{3u_0^4}\bar{x}^3+gB^2(u_0-240)\frac{u_0-320}{4v_0^6}\bar{x}^4=\frac{g}{2(735)^2}(300.98)^2+$$

$$\frac{g(1.406599)(735-240)}{3(735)^4}(300.98)^3+g(1.406599)^2(735-240)\frac{735-320}{4(735)^6}(300.98)^4=1.087m$$

The projection angle is approximately

$$A=\frac{\bar{y}}{D}\cos(E)=\frac{1.087}{300}\cos(65°)=0.001531=0.0877°=5.26'.$$

Note 1: Comparing the result $A=5.26'$ obtained in this example for the projection angle that corresponds to the slant distance 300 m with the value $\alpha_0=12.37'$ of the projection angle determined in example 3 in section 3.3 (when the rifle and the target are at sea level), we see a significant decrease in the value of the projection angle when the angle of sight increases from 0 to 65°.

Example 3. Consider exercise 2. Employ the approximate formula (1.9.6),

$$\sin(2A+E)=\sin2\alpha_0\cdot\cos^2 E+\sin E,$$

to estimate the projection angle A.

Solution
   Substituting, we have

$$\sin(2A+65°)=\sin[2\cdot(0.2062°)].\cos^2(65°)+\sin(65°)$$

and

$$\sin(2A+65°)=0.907593 \; .$$

Hence, we find

$$2A+65°=\sin^{-1}(0.907593)=65.17485874°$$

and

$$A=0.0874294°=5.25' \; .$$

Note: Using formula (3.5.6), we find the same result as the one obtained in exercise 2 using the approximate method II.

Exercise 4. Estimate the drop of the bullet fired horizontally by a Russian rifle 7.62 mm at a horizontal distance of 100 m. The mass of the bullet is 7.9 g. The ballistics coefficient is $c=4.5663$ (see exercise 4, section 2.7).

Solution

We assume that $\bar{x} \approx OT = \overline{D} = 100m$.

$$B=\frac{c\cdot h(\bar{y})}{3}=\frac{(4.5663)(1)}{3}=1.5221$$

As in exercise 2, for the problem posed above, we have (figure 4):

$$\bar{y}=\frac{g}{2\cdot u_0^2}\bar{x}^2+\frac{gB(u_0-240)}{3u_0^4}\bar{x}^3+gB^2(u_0-240)\frac{u_0-320}{4v_0^6}\bar{x}^4=\frac{g}{2(735)^2}(100)^2+$$

$$\frac{g(1.5221)(735-240)}{3(735)^4}(100)^3+g(1.5221)^2(735-240)\frac{735-320}{4(735)^6}(100)^4=0.0999m\approx0.10m$$

.

The y-coordinate of the impact point T at the horizontal distance 100 m is $y=-\bar{y}=-0.10m$. The bullet has dropped $0.10m$.

Exercise 5. A Sierra 0.257" caliber, 117 grain Spitzer Boat Tail bullet is fired horizontally with initial speed 2,900 ft/s. Estimate the drop of the bullet at 350 yd. The ballistics coefficient is $c=3.3593$.

Note 1: The ballistics coefficient $c=3.3593$ corresponds to the Siacci form coefficient $i=0.59768$ that is estimated using the data obtained from the ballistics tables of the given bullet. That form coefficient is valid for launching angles $0°$, or approximately $0°$).

Solution

We assume that $\bar{x} \approx OT = \overline{D} = 350 yard = 320m$ (figure 4).

$$B = \frac{c \cdot h(\bar{y})}{3} = \frac{(3.3593)(1)}{3} = 1.11977$$

The $\bar{y}$ coordinate of projectile at $\bar{x} \approx OT = \overline{D} = 320m$ is

$$\bar{y} = \frac{g}{2 \cdot u_0^2} \bar{x}^2 + \frac{gB(u_0 - 240)}{3u_0^4} \bar{x}^3 + gB^2(u_0 - 240)\frac{u_0 - 320}{4v_0^6} \bar{x}^4 = \frac{g}{2(884)^2}(320)^2 +$$

$$\frac{g(1.11977)(884 - 240)}{3(884)^4}(320)^3 + g(1.11977)^2(884 - 240)\frac{884 - 320}{4(884)^6}(320)^4 = 0.7935 \,.$$

The y-coordinate of the impact point T (figure 4) at the horizontal distance 320 m is $y = -\bar{y} = -0.7935m = -31.24inch$. The bullet has dropped 31.24 in. under the horizontal line.

Comment. The result obtained in this example is quite equal to the result 31.36 in. given in *Sierra's Exterior Ballistics*.[24]

Note 2: Using the PC program Sizero.Bas, we find that the drop of the bullet is

$$y = -0.7989m = -31.83inch \,.$$

---

[24]   R. Hayden, T. Almgren, K. Thomas, W. T. McDonald. Sierra's Exterior *Ballistics*, 5th ed., Table in figure 5.0-3. *http://www.exteriorballistics. com/ebexplained/5th/50.cfm*. {Web site]

Using the PC Program Sizero.Bas

The program "Sizero.Bas" (ref. Klimi, G. Exterior Ballistics of Small Arms, Xlibris, 2009) estimates the coordinates of the projectile trajectories of the Sierra 0.257" caliber, 117 grain Spitzer Boat Tail Bullet, is fired horizontally with Initial Speed 2900 ft./s. It can be modified and used for any other projectile on conditions that we know the ballistics coefficient as a function of projectile speed, $c=c(v)$. It also estimates the ballistics coefficient for any given launching angle.

The program can be used for the trajectory of the given bullet for any given launching angle.

We solve the problem for two values of the x-coordinate: x = 350 yard = 320m, and X = 656 yard = 600m.

Solution of the problem

Input Data
Input: Initial y-coordinate [m] = 0:
Input: Initial x-coordinate [m] = 0:
Input: Launching Speed [m/s] = 883.92,
Input: Launching Angle [Degree] = 0,
Input: Abscissa x of a point [m] =320,
Input: Abscissa X of another point = 600
Input: Integration Step ho = 0.01

Output
Abscissa [m]: x =320
Corresponding Ordinate: y= - 0.7989:
Corresponding Speed: v = 639.92;
Time of Flight [s]: t= 0.425
Corresponding Angle = - 0.3204948 degree

Abscissa X = 600m;
Corresponding Ordinate Y = -3.532
Corresponding Speed = 465;
Time = 0.939
Corresponding Angle = -.8560606

Ballistics Coefficient is $= 3.359304$ (corresponds to launching angle $0°$).

The Following Tables Illustrate the Techniques of Sections 3.4 and 3.3
and
Are Prepared by Prof. James Lewis (Marquette University)

Trajectory
270 Winchester, 130 grain projectile SuperX, Silvertip
Muzzle velocity: 3060 ft/s
Scope height: 3.556cm
$B=1.338029$ & $\alpha_0 = 30.07913$
Inclination Angle: $30°$
Method of Computation: Section 3.4
Author Prof. James Lewis

Horizontal Range (Yards)	Drop (cm)	Elevation, y' (Inches)
0	0	-1.4
10	0.0634	-0.761
20	0.2556	-0.174
30	0.5797	0.363
40	1.0387	0.845
50	1.6357	1.274
60	2.3741	1.646
70	3.2572	1.962
80	4.2886	2.220
90	5.4717	2.417
100	6.8101	2.554
110	8.3075	2.628
130	11.7949	2.582
140	13.7927	2.459
150	15.9652	2.267
160	18.3167	2.005
170	20.8514	1.671
180	23.5736	1.262
190	26.4878	0.779
200	29.5983	0.218

210	32.9909	-0.455
220	36.4272	-1.144
230	40.1550	-1.948
240	44.0981	-2.837
250	48.2614	-3.813
260	52.6501	-4.877
270	57.2693	-6.032
280	62.1240	-7.280
290	67.2197	-8.622
300	72.5618	-10.062

Trajectory
270 Winchester, 130 grain projectile SuperX, Silvertip
Muzzle velocity: 3060 ft/s
Scope height: 3.556cm
$b = 1.375684^* \ \& \ p_0 = 1.381218.10^{-3}$
Inclination Angle: 0°
Method of Computation: Section 3.3
Author Prof. James Lewis

Horizontal Range (yards)	Drop (cm)	Elevation (inches)	P (MOA)
0	0	-1.4	4.748
5	0.0118	-1.156	4.570
10	0.0474	-0.921	4.390
20	0.1191	-0.452	4.025
30	0.4328	-0.079	3.652
32.0809	0.4957	0	3.574- 1st crossing
40	0.7747	0.284	3.272
50	1.2189	0.606	2.883
60	1,7673	0.888	2.488
70	2.422	1.127	2.082
80	3.1860	1.324	1.670
90	4.0606	1.477	1.249
100	5.0486	1.585	0.818
118.4614	7.1781	1.664	0- vertex

130	8.7160	1.632	-0.527
150	11.7721	1.424	-1.472
160	13.4912	1.244	-1.959
170	15.3412	1.013	-2.456
180	17.3248	0.730	-2.964
190	19.4448	0.392	-3.482
200	21.7040	0	-4.010-2nd crossing
210	24.1053	-0.448	-4.550
220	26.6515	-0.953	-5.101
230	29.3457	-1.517	-5.662
240	32.1909	-2.140	-6.236
250	35.1901	-2.823	-6.821
260	38.3465	-3.569	-7.417
270	41.6631	-4.377	-8.026
280	45.1433	-5.250	-8.647
290	48.7904	-6.189	-9.281
300	52.6076	-7.194	-9.926

* The value of b was found from the muzzle velocity of 3060ft/s and a downrange value at 500 yards of 1708ft/s, as reported in the web site *Winchester.com*.

`The p_0 value was established by setting the second crossing at 200 yards.

The following examples (prepared by Prof. James Lewis) illustrate the way the data in the above tables are calculated.

Example 6. Narrow Angle Calculation of a 22 Caliber Projectile
(Method of Section 3.3 – See table 1, Example 3, section 5.6)

The following projectile calculation, for a 40 grain 22 caliber CCI cartridge, include a muzzle velocity of 1235ft/s (376.432m/s), and at 100 yards (91.441m) a velocity decrease to 1036ft/s (315.777m/s).

The Leupold 4-power scope on the rifle firing this 22 caliber projectile mounts h=1.2 inches (3.048cm) above the bore. Thus at zero range, the projectile strikes the target at $y_0 = -1.2 inches = -3.048 cm$.

Calculation of $b = ch(y)/3$

The integration of the first of equations (3.2.6) yields:

$$b = \{[(v_0 - v_x) - 240\ln[(v_0 - 240)/(v_x - 240)]\}/x \,.\, (1)$$

With $\quad v_0 = 376.432 m/s$, $\quad v_x = 315.777 m/s \quad$ and $\quad$ range $x_T = 91.441 m$, we have

$$b = \{[(376.432 - 315.777) - 240\ln[(376.42 - 240)/(315.777 - 240)]\}/(91.441)\,.$$

Hence,

$$b = 2.206727928 \,.$$

Note: The above value of "b" is used in the calculations of the data presented in the first table of Example 3, section 5.6.

Calculation of $p_0 = \tan\alpha_0$

Since all elevations will be referenced with respect to the *line of sight*, rather than to the *bore line* (x-axis), the Maclaurin expansion of y(x) includes an additional term in equation (3.3.4), namely, at x = 0, y is the negative of the scope height h above the bore, i.e. y = - h. Thus (3.3.4) becomes

$$y(x) = p_0 x - gx^2/2v_0^2 - gb(v_0 - 240)\cdot x^3/(3v_0^4) -$$

$$. \,(2)$$

$$- gb^2(v_0 - 240)\cdot(v_0 - 320)\cdot x^4/(4v_0^6) - h$$

With $v_0 = 376.432 m/s$, $g = 9.80665 m/s^2$, we find

$$g/(2v_0^2) = 3.460320116\cdot 10^{-5},$$

$$gb(v_0 - 240)/(3v_0^4) = 4.901370030\cdot 10^{-8}$$

and

$$gb^2(v_0 - 240) \cdot (v_0 - 320) / (4v_0^6) = 3.230598809 \cdot 10^{-11}.$$

With these values and the downrange line of sight, y(x) — the second crossing point — set equal to zero at 70 yards (64m), $p_0 = \tan\alpha_0$ calculates as

$$\tan\alpha_0 = 3.460320116 \cdot 10^{-5} \cdot (64) + 4.901370030 \cdot 10^{-8} \cdot (64)^2 +$$

$$+ 3.230598809 \cdot 10^{-11}(64^3) + 0.03048 / 64 = 2.900381 \cdot 10^{-3}$$

(Note: The above value of $p_0 = \tan\alpha_0$ is presented in the first table of example 3, section 5.6.)

Hence,

$$\alpha_0 = \tan^{-1}(2.900381 \cdot 10^{-3}) = 0.166179124° = 9.971\,MOA.$$

The first crossing point is located either by trial and error or via a graphing calculator as 12.28m or 13.43 yards. This is the distance used to sight-in the rifle.

The elevation at any other distance calculates from equation (2), the drop being

$$drop = [y(x) - p_0 x + h] = -gx^2 / 2v_0^2 - gb(v_0 - 240) \cdot x^3 / (3v_0^4) -$$

$$(3)$$

$$- gb^2(v_0 - 240) \cdot (v_0 - 320) \cdot x^4 / (4v_0^6)$$

Thus the drop (with respect to the line of sight) at 80 yards (73.153m) becomes 0.0204647cm,

The elevation,

$$y(x) = p_0 x + drop - h$$
,

calculates as -0.0229554m or -0.903inches.

Example 7. Narrow Angle Calculation (Method of Section 3.3)

The following example illustrates the way used to find the data for the trajectory range 500 yards as well as the data obtained for zeroing the Winchester projectile at 300 yards (see tables above).

This calculation, for a 130 grain 270 Winchester projectile (Silvertip) with a muzzle velocity of 3060ft/s (932.699m/s), includes at 500 yards (457.205m) a velocity of 1708ft/s (520.604m/s).

Calculation of $b = ch(y) / 3$

The integration of the first of equations (3.2.6) yields:

$$b = \{[(v_0 - v_x) - 240\ln[(v_0 - 240) / (v_x - 240)]\} / x. \qquad (1)$$

With $v_0 = 932.699m / s$, $v_x = 376.42m / s$ and range $x_T = 457.206m$, we have

$$b = \{[(932.699 - 457.206) - 240\ln[(932.699 - 240) / (520.604 - 240)]\} / (457.206)$$
.

Hence,

$$b = 1.375684.$$

The value of p_0 will be set by establishing *the second crossing point* at 200 yards (188.882m). For this rifle, the 2½ -10 power Leupold scope, located above the bore line, yields h of 1.4 inches (3.556cm). In equation (2) above we find

$$g / (2v_0^2) = 5.636471 \cdot 10^{-6},$$
$$gb(v_0 - 240) / (3v_0^4) = 4.116197 \cdot 10^{-9}$$

and

$$gb^2 (v_0 - 240) \cdot (v_0 - 320) / (4v_0^6) = 2.991163 \cdot 10^{-12}.$$

Thus $p_0 = \tan \alpha_0$ follows from equation (2) above

$$\tan \alpha_0 = 5.636471 \cdot 10^{-6} \cdot (457.206) + 4.116197 \cdot 10^{-9} \cdot (457.206)^2 +$$

$$+ 2.991163 \cdot 10^{-12} (457.206)^3 + 0.03556 / 457.206 = 3.80112 \cdot 10^{-3}$$

Hence,

$$\alpha_0 = \tan^{-1}(0.0380112 \cdot 10^{-3}) = 0.2177871° = 13.067 \, MOA.$$

The first crossing point is located either by trial and error or via a graphing calculator as 29.33513m or 32.0809 yards, as indicated in the above trajectory table for the 270 Winchester projectile.

The elevation at any other downrange distance, say for 300 yards (274.323m) follows from

$$y(x) = p_0 x - gx^2 / 2v_0^2 - gb(v_0 - 240) \cdot x^3 / (3v_0^4) -$$

$$- gb^2 (v_0 - 240) \cdot (v_0 - 320) \cdot x^4 / (4v_0^6) - h$$

with the scope height of 3.556cm, or 0.03556m

$$y(x) = p_0 x - drop - h.$$

The drop being, as above,

$$drop = [y(x) - p_0 x + h] = - gx^2 / 2v_0^2 - gb(v_0 - 240) \cdot x^3 / (3v_0^4) -$$

$$- gb^2 (v_0 - 240) \cdot (v_0 - 320) \cdot x^4 / (4v_0^6)$$

Thus, at 300 yards, we find a drop of 52.6076 cm, an elevation relative to the line of sight of -7.194 inches (-0.18272m), and the value of arctan(p) of -9.926MOA or -0.079137912°.

3.6 Approximate Polynomial Equations of the Projectile Trajectory

We have already shown the way we derive the standard function of resistance of a bullet $f_D(v)$ using the drag coefficient $C_D(v / a)$ that is available nowadays [section 2.10 and 2.11]. The drag coefficient is usually obtained using chronographs, Doppler radars or wind tunnels.

Since the drag coefficient is not available for any individual bullet or projectile, in exterior ballistics we use the G-functions (G_1, G_2 ... G_7, etc.), determined for certain standard projectiles, to solve the exterior ballistics problems related with a non-standard projectile employing the form coefficient (i).

In general, the ballistics elements of the projectiles obtained using the G-functions are acceptable approximations, but in some particular cases (for example for long range or sniper shootings with small arms) the accuracy is not satisfactory.

Working on the idea to "eliminate" the use of the Standard Functions of Resistance (G1,..., G7, Siacci's, Ingall's), and thus, to "eliminate" the need to measure a fixed-form coefficient (or a set of fixed-form coefficients), we have revived the old approximate method of expansion in series[25] to solve differential equations of exterior ballistics, at least for long range shooting with small arms.

In fact, the contemporary method we are introducing hereafter is an old method, but applied in an innovative way. The original application of the development in series we are using to find the elements of the trajectory for small arms, following Piton-Bressant (or Helie) and Duchene's approach[26], eliminates the large errors exterior ballisticians (and shooters) have tolerated with predictions of the ballistics trajectory for long range shooting using standard G-functions.

---

[25] Cranz, C., Becker, K, Handbook of Ballistics, p. 121 - 124, London, 1921.
[26] Cranz, C., Becker, K, Handbook of Ballistics, p. 124, p.126, London, 1921.

The series expansion method also avoids the use of Doppler radar measurements in ballistics of small arms, or other expensive experimental methods. It is based on the existing range tables of small arms, or on experimental data that every shooter can obtain with the exterior ballistics equipments that are in the market.

The original methods we are introducing add to the already existing exterior ballistics techniques a new technique to solve the problems of exterior ballistics without:

- Solving the differential equations that describe the projectile flight,
- Using the drag coefficient or the function of resistance,
- Measuring the form coefficient (i) and the ballistics coefficient, BC.

Practically, in the following two sections, we will show a method that can be used to find the standard function of resistance $f_D(v)$ that corresponds to a given bullet (form coefficient $i = 1$):

- Using the range tables that are already produced, for ranges till 1000 meters or over.
- Using the data from the firing tests (let's say the drop of the bullet) with small arms (ranges $0 - 1000$ or more), i.e. shooting on the targets located let say 50 or 100 meters from each others, without chronograph measurements of velocity and time; this is the most difficult case, but very interesting since every short/long range shooter gathers such data.

Thus, every shooter will have his respective bullet function of resistance, i.e. a bullet "fingerprint" that will increase the prediction accuracy of the respective small arm, especially for long ranges.

Series Solutions of Differential Equations of Projectile Trajectory

Consider the system of differential equations of the projectile flight [see equation (2.7.16)]

$$
\begin{cases}
\dfrac{dv_x}{dx} = -ch(y) \cdot \dfrac{f_D(v)}{v} \\[2ex]
\dfrac{dp}{dx} = -\dfrac{g}{v_x^2} \\[2ex]
\dfrac{dt}{dx} = \dfrac{1}{v_x} \\[2ex]
\dfrac{dy}{dx} = p
\end{cases}
\tag{3.6.1}
$$

with respect to a Cartesian coordinate system with the origin at the muzzle of the firearm, where:

- $f_D(v)$ is the standard function of resistance of the given bullet (projectile).

For projectiles the function of resistance has the form[27]

$$
f_D(v) = \begin{cases} A \cdot v^2 & for \quad v \le 256m/s \\ E \cdot v - F & for \quad v > 256m/s \end{cases}
\tag{3.6.2}
$$

- c is the ballistics coefficient $c = 1000 \cdot d^2 / m$, where i is the form coefficient of the projectile;
- $p = \tan\alpha$, α is the angle of flight, i.e. the angle that the projectile velocity forms with x-axis (horizon).
- $v_x = v\cos\alpha$ is the horizontal component of the projectile velocity $\vec{v}$.
- t is the time of flight.
- $g = 9.80665$.

---

[27]   Klimi, G, Exteriror Ballistics: A New Approach, chapter 1, Xlibris, 2011.

- $h(y)$ is the density function that will be considered constant.

We denote:

- α_0 the departure angle; x_T the range; α_T the terminal angle; t_T time of flight to the point of impact on the ground (x-axis), v_0 the departure speed and v_T the terminal speed.

The departure point of the bullet is at the origin of the coordinates. The projectile trajectory is in a standard atmosphere[28].

We will assume that the function of resistance $f_D(v)$ corresponds to the given trajectory, i.e. we consider that the form coefficient $i = 1$ and $c = 1000 \cdot d^2 / m$.

For low trajectories, like the trajectories of small arms, we can consider that the density function is constant, $h(y) = h$ where h is the value of the function at the firing location. The equation of projectile trajectory can be presented as a function of the horizontal distance x, i.e. $y = F(x)$, that is a solution of the system of differential equations (3.6.1). Assuming that the function $F(x)$ and its derivatives are finite and continuous, we can express $F(x)$ in Maclauren series:

$$y = F(0) + F^{(1)}(0) \cdot x + \frac{F^{(2)}(0)}{2!} \cdot x^2 + \frac{F^{(3)}(0)}{3!} x^3 + \frac{F^{(4)}(0)}{4!} x^4 + \cdots \tag{3.6.3}$$

Using the system of differential equations (3.6.1), the equation (3.6.3) can be written

$$y = p_0 \cdot x - \frac{g}{2v_0^2 \cos^2 \alpha_0} x^2 (1 + \frac{2}{3} \frac{hcf_D(v_0)/v_0}{v_0 \cos \alpha_0} \cdot x + \cdots) \tag{3.6.4}$$

---

[28] We are considering the flat fire trajectory in standard atmosphere only to develop and present the method, but the method can be applied for whatever atmosphere, inclined firing angle, etc.

Considering the time t, the slope of the trajectory tangent, $p = \tan\alpha$ and the horizontal component of the projectile velocity $(v_x = v \cdot \cos\alpha)$ as functions of the horizontal range, we obtain the following elements of the trajectory:

$$\tan\alpha = \tan\alpha_0 - \frac{g}{v_0^2 \cos^2\alpha_0} x \cdot (1 + \frac{chf_D(v_0)}{v_0^2 \cos\alpha_0} x + \cdots)$$

(3.6.5)

$$v = \frac{v_0 \cos\alpha_0}{\cos\alpha}(1 - \frac{chf_D(v_0)}{v_0^2 \cos\alpha_0} x + \cdots)_,$$

(3.6.6)

$$t = \frac{x}{v_0 \cos\alpha_0}(1 + \frac{chf_D(v_0)}{v_0^2 \cos\alpha_0} x + \cdots)_,$$

(3.6.7)

The Derivatives of the Polynomial Function $y = F(x)$

To find the derivatives of the polynomial function $y = F(x)$ we will consider only the flight of projectile for velocity greater than 256m/s, i.e. we assume that the function of resistance has the form:

$$f_D(v) = E \cdot v - F, \text{ where } v > 256m/s.$$

(3.6.8)

• We have:

$$F(0) = y_0, \quad F^{(1)}(0) = [\frac{dy}{dx}]_{x=0} = p_{x=0} = p(0) = \tan\alpha_0,$$

(3.6.9)

• Using the second equation of (3.6.1), for the second derivative of $F(x)$ at $x = 0$ we have:

$$F^{(2)}(0) = [\frac{d^2 y}{dx^2}]_{x=0} = [\frac{dp}{dx}]_{x=0} = [-\frac{g}{v_x^2}]_{x=0} = -\frac{g}{v_0^2 \cos^2\alpha_0},$$

(3.6.10)

- Considering (3.6.10) and using the first equation of (3.6.1) we find:

$$F^{(3)}(0) = [\frac{d(-g/v_x^2)}{dx}]_{x=0} = [\frac{2g \cdot (dv_x/dx)}{v_x^3}]_{x=0} = -2ghc \cdot [\frac{f_D(v)/v}{v^3 \cos^3 \alpha}]_{x=0},$$

(3.6.11)

since

$$dv_x / dx = -h \cdot c \cdot f_D(v) / v = -h \cdot c \cdot (Ev - F) / v$$

(3.6.12)

Thus,

$$F^{(3)}(0) = -2ghc \cdot [\frac{f_D(v)/v}{v^3 \cos^3 \alpha}]_{x=0} = -2ghc \cdot \frac{(E \cdot v_0 - F)/v_0}{v_0^3 \cdot \cos^3 \alpha_0}.$$

(3.6.13)

Continuing in the same way, we determine the following coefficients of the polynomial (3.6.3):

$$F^{(4)}(0) = 2gh^2c^2 \cdot [\frac{(dv_x dx) \cdot (3Ev - 4F)}{v_x^5}]_{x=0} = -2gh^2c^2 \cdot [\frac{(3Ev - 4f) \cdot (Ev - F)/v)}{v^4 \cos^4 \alpha}]_{x=0}$$

or

$$F^{(4)}(0) = -2gh^2c^2 \cdot \frac{(3Ev_0 - 4F) \cdot (Ev_0 - F)/v_0)}{v_0^5 \cos^5 \alpha}.$$

(3.6.14)

$$F^{(5)}(0) = -2gh^3c^3 \cdot \frac{(12E^2v_0^2 - 35EFv_0 + 24F^2) \cdot (Ev_0 - F)/v_0)}{v_0^7 \cos^7 \alpha}$$

(3.6.15)

$$F^{(6)}(0) = -2gh^4c^4 \cdot \frac{(60E^3v_0^3 - 282E^2Fv_0^2 + 413EF^2v_0 - 192F^3) \cdot (Ev_0 - F)/v_0)}{v_0^9 \cos^9 \alpha_0}$$

(3.6.16)

For approximate results, in the equation of the trajectory (3.6.3) we can consider some first terms of the expansion[29]. The error we make is estimated using the reminder of the series.

The above results can be applied for a known G-function of resistance or for any particular standard function of resistance of a bullet, $f_D(v)$ [30].

For relatively small angles of departure we can consider $v_x \approx v$, $\cos\alpha \approx 1$ at any point on the projectile trajectory, i.e. $v_{x=0} \approx v_0$, $\cos\alpha_0 \approx 1$, $\cos\alpha_T \approx 1$.

Note that $F(0)$, $F^{(1)}(0)$ and $F^{(2)}(0)$ do not depend by the function of resistance. They represent the coefficients of the second order parabola that describes the flight of projectile in absence of air resistance. The second order parabola,

$$y = p_0 \cdot x - \frac{g}{2v_0^2 \cos^2 \alpha_0} x^2 \quad ,$$

$$(3.6.17)$$

is obtained if we limit the Maclauren expansion to the third term.

Note. To illustrate the method we will use the data for the Lapua GB528 Scenar 19.4 g, Caliber 8.6 mm bullet given by http://en.wikipedia.org/wiki/External_ballistics.

Example 1. For the Lapua GB528 Scenar 19.4 g, Caliber 8.6 mm bullet we have the following function of resistance [see section 2.10]:

$$f_D(v) = 0.141v - 30.031 \text{ where } 340 \le v \le 830 . \quad (1)$$

---

[29]   It is interesting to note that, according to Cranz & Becker (page 123), the approximate equations of the projectile trajectories have been obtained using series expansion since the end of the 18th century.

[30]   Klimi, G., *Exterior Ballistics : A New Approach*, p. 43 - 52, Xlibris, 2010.

Find the coefficients of the series (3.6.3) and write the equation of the trajectory and the other ballistics characteristics.

Solution

From (1) we have $E = 0.141$ and $F = 30.031$ and $v_0 = 830 m/s$, $c = 3.796$.

$F(0)$, $F^{(1)}(0)$ and $F^{(2)}(0)$ are already given by (3.6.9) and (3.6.10).

Considering (3.6.13) we have:

$$\frac{F^{(3)}(0)}{3!} = -\frac{g}{2v_0^2 \cos^2 \alpha_0} \cdot \frac{2hc(Ev_0 - F)/v_0}{3v_0} \frac{1}{\cos \alpha_0} =$$

$$-\frac{g}{2v_0^2 \cos^2 \alpha_0} \cdot (3.195898815 \cdot 10^{-4}) \frac{1}{\cos \alpha_0} \qquad (2)$$

We denote:

$$A = 3.195898815 \cdot 10^{-4}, \qquad (3)$$

the coefficient in (2).

Using (3.6.14), (3.6.15) and (3.6.16) we obtain:

$$\frac{F^{(4)}(0)}{4!} = -\frac{g}{2v_0^2 \cos^2 \alpha_0} \cdot \frac{h^2 c^2 (3Ev_0 - 4F) \cdot (Ev_0 - F)/v_0}{6v_0^3} \frac{1}{\cos^3 \alpha_0} =$$

$$-\frac{g}{2v_0^2 \cos^2 \alpha_0} (1.01683644 \cdot 10^{-7}) \frac{1}{\cos^3 \alpha_0}$$

$$(4)$$

We denote:

$$B = 1.01683644 \cdot 10^{-7} \tag{5}$$

$$\frac{F^{(5)}(0)}{5!} = -\frac{g}{2v_0^2 \cos^2 \alpha_0} \cdot (3.056074173 \cdot 10^{-11}) \frac{1}{\cos^5 \alpha_0}.$$

We denote:

$$C = 3.056074173 \cdot 10^{-11}. \tag{6}$$

$$\frac{F^{(6)}(0)}{6!} = -\frac{g}{2v_0^2 \cos^2 \alpha_0} \cdot 3.056074173 \cdot 10^{-11} \frac{1}{\cos^5 \alpha_0}$$

We denote:

$$D = 8.27548897 \cdot 10^{-15} \tag{7}$$

Polynomial Equation of the Trajectory

In general, using the results of the above example we can write the algebraic equation of the trajectory of the projectile and the equations that describe the ballistics elements of the projectile, in the following forms:

Parabolic Trajectory

The equation of the trajectory (3.6.3) can be written:

$$y = \tan \alpha_0 \cdot x - \frac{gx^2}{2v_0^2 \cos^2 \alpha_0}(1 + \frac{A \cdot x}{\cos \alpha_0} + \frac{B \cdot x^2}{\cos^3 \alpha_0} + \frac{C \cdot x^3}{\cos^5 \alpha_0} + \frac{D \cdot x^4}{\cos^7 \alpha_0}) \tag{3.6.}$$

We say that the trajectory equation (3.6.17) is a parabola of the 6th degree order.

Angle of flight:

Differentiating (3.6.17) with respect to x, we find the angle of projectile flight at the point with abscissa x:

$$\tan \alpha = \tan \alpha_0 - \frac{gx}{v_0^2 \cos^2 \alpha_0} \cdot (1 + \frac{3}{2} \frac{A}{\cos \alpha_0} x +$$

$$+ 2 \frac{B}{\cos^3 \alpha_0} x^2 + \frac{5}{2} \frac{C}{\cos^5 \alpha_0} x^3 + 3 \frac{D}{\cos^7 \alpha_0} x^4)$$

(3.6.18)

Velocity

Differentiating the above equation with respect to x and then using the second equation of the system (3.6.1) we have:

$$v = \frac{v_0 \cos \alpha_0}{\cos \alpha} \cdot (1 + 3 \frac{A \cdot x}{\cos \alpha_0} + 6 \frac{B \cdot x^2}{\cos^3 \alpha_0} + 10 \frac{C \cdot x^3}{\cos^5 \alpha_0} + 15 \frac{D \cdot x^4}{\cos^7 \alpha_0})^{-1/2}$$

(3.6.19)

Time of flight

Using the third equation of (3.6.1), for the time of flight we obtain:

$$t = \frac{1}{v_0 \cos \alpha_0} \int_0^x \sqrt{1 + 3 \frac{Ax}{\cos \alpha_0} + 6 \frac{Bx^2}{\cos^3 \alpha_0} + 10 \frac{Cx^3}{\cos^5 \alpha_0} + 15 \frac{Dx^4}{\cos^7 \alpha_0}} \, dx$$

(3.6.20)

Departure Angle

Substituting $y = 0$ in equation (3.6.17) of example 2, for the departure angle we obtain:

$$\tan \alpha_0 = \frac{gx}{2v_0^2 \cos^2 \alpha_0} (1 + \frac{A \cdot x}{\cos \alpha_0} + \frac{B \cdot x^2}{\cos^3 \alpha_0} + \frac{C \cdot x^3}{\cos^5 \alpha_0} + \frac{D \cdot x^4}{\cos^7 \alpha_0})$$

(3.6.21)

The coefficients A, B, C, D that appear in the above equations for a any bullet can be obtain in the same way as the coefficients of the Lapua GB528 Scenar 19.4, Caliber 8.6mm bullet of found in example 1. For that we need to know the standard function of resistance $f_D(v)$ of the respective bullet.

For the Lapua GB528 Scenar 19.4 g, Caliber 8.6 mm, A, B, C and D are given respectively by (3), (5), (6) and (7)), i.e.

$$A = 3.195898815 \cdot 10^{-4}, \quad B = 1.01683644 \cdot 10^{-7},$$
$$C = 3.056074173 \cdot 10^{-11}, \quad D = 8.27548897 \cdot 10^{-15}$$

For less accuracy, or for smaller range of firing, we can consider 3rd, 4th or 5th degree order parabolas.

For example, for the given Lapua bullet the 4th degree parabola is

$$y = \tan \alpha_0 \cdot x - \frac{gx^2}{2v_0^2 \cos^2 \alpha_0}(1 + \frac{A \cdot x}{\cos \alpha_0} + \frac{B \cdot x^2}{\cos^3 \alpha_0}) \tag{3.6.21}$$

where

$$A = 3.195898815 \cdot 10^{-4}, \quad B = 1.01683644 \cdot 10^{-7}. \tag{3.6.22}$$

Example 2. For the Lapua bullet of Example 1 find the drop of the bullet and the respective velocity for the ranges 300, 600, 900, 1200 and 1500 meters, if the bullet is fired horizontally (departure angle $\alpha_0 = 0$). The departure speed of the Lapua bullet is $v_0 = 830m/s$.

Solution

Projectile Drop

Substituting in (3.6.17) the values of A, B, C, D for the Lapua bullet we obtain the trajectory equation of the given bullet:

$$y = \tan\alpha_0 \cdot x - \frac{gx^2}{2v_0^2 \cos^2\alpha_0}(1+ \frac{3.195898815 \cdot 10^{-4} x}{\cos\alpha_0} + \frac{1.01683644 \cdot 10^{-7} x^2}{\cos^3\alpha_0} +$$

$$+ \frac{3.056074173 \cdot 10^{-11} x^3}{\cos^5\alpha_0} + \frac{8.27548897 \cdot 10^{-15} x^4}{\cos^7\alpha_0})$$

(1)

Substituting in (1), $\alpha_0 = 0$, $v_0 = 830m/s$ and successively the values of x: 300, 600, 900, 1200 and 1500, we find the following corresponding values of the projectile drop:

$y_3 = -0.716m$, $y_6 = -3.205m$, $y_9 = -8.145m$, $y_{12} = -16.567m$,

$y_{15} = -30.0183m$.

(2)

Projectile velocity

For the Lapua bullet the equation (3.6.18) that can be used to find the projectile angle at a point that corresponds to the range x is:

$$\tan\alpha = \tan\alpha_0 - \frac{gx}{v_0^2 \cos^2\alpha_0} \cdot (1 + \frac{3}{2} \cdot \frac{3.195898815 \cdot 10^{-4}}{\cos\alpha_0} x +$$

(3)

$$+ 2\frac{1.01683644 \cdot 10^{-7}}{\cos^3\alpha_0} x^2 + \frac{5}{2} \cdot \frac{3.056074173 \cdot 10^{-11}}{\cos^5\alpha_0} x^3 + 3\frac{8.27548897 \cdot 10^{-15}}{\cos^7\alpha_0} x^4)$$

The velocity at a given range can be found by the equation:

$$v = \frac{v_0 \cos\alpha_0}{\cos\alpha} \cdot (1 + 3\frac{3.195898815 \cdot 10^{-4} x}{\cos\alpha_0} + 6\frac{1.01683644 \cdot 10^{-7} x^2}{\cos^3\alpha_0} +$$

(4)

$$+ 10\frac{3.056074173 \cdot 10^{-11} x^3}{\cos^5\alpha_0} + 15\frac{8.27548897 \cdot 10^{-15} x^4}{\cos^7\alpha_0})^{-1/2}$$

Substituting in (3) $\alpha_0 = 0$, $v_0 = 830m/s$ and successively the values of x: 300, 600, 900, 1200 and 1500, we find the following corresponding values of the angle of projectile (the angle that the velocity forms with the x-axis):

$\alpha_3 = -0.2849°$, $\alpha_6 = -0.6756°$, $\alpha_9 = -1.2243°$, $\alpha_{12} = -2.0072°$,

$\alpha_{15} = -3.1291°$.

$$(5)$$

Substituting in (4) $\alpha_0 = 0$, $v_0 = 830m/s$, and successively the values of x: 300, 600, 900, 1200 and 1500, and the corresponding values (5) of the projectile angle found above, we find the following corresponding values of the projectile velocity:

$v_3 = 713.88m/s$, $v_6 = 605.83m/s$, $v_9 = 508.79m/s$,

$v_{12} = 424.97m/s$, $v_{15} = 354.86m/s$.

$$(6)$$

Note: The above results are pretty accurate till around 1500meters. For comparison see Table 1 section (3.7).

Example 3. For the Lapua bullets of Example 1 find the angle of departure that is needed to hit a target located at the range 1,000 meters from the muzzle of the firearm. Shooting is in ICAO standard atmosphere. The initial velocity is $v_0 = 830m/s$. What is the departure angle if the range is 1200 meters?

Solution

Method I

Substituting $y = 0$ in equation (1) of example 2, for the departure angle we obtain:

$$\tan \alpha_0 = \frac{gx}{2v_0^2 \cos^2 \alpha_0}(1 + \frac{3.195898815 \cdot 10^{-4} x}{\cos \alpha_0} + \frac{1.01683644 \cdot 10^{-7} x^2}{\cos^3 \alpha_0} +$$
$$+ \frac{3.056074173 \cdot 10^{-11} x^3}{\cos^5 \alpha_0} + \frac{8.27548897 \cdot 10^{-15} x^4}{\cos^7 \alpha_0})$$

$$(1)$$

Range 1000 meters

As a first solution, we substitute on the right side of (1): $x = 1000$, $v_0 = 830$ and $\alpha_0 = 0$.

We find:

$$\tan \alpha_0 = 0.010501.$$

Hence:

$$\alpha_0 = \arctan(0.010501) = 0.60164°.$$

To improve the solution we substitute in (1) $x = 1000$, $v_0 = 830$ and $\alpha_0 = 0.60164°$.

We have:

$$\tan \alpha_0 = 0.010503$$

and

$$\alpha_0 = \arctan(0.010503) = 0.60176°$$

Range 1200 meters

In the same way as above we find that the launching angle is

$$\alpha_0 = \arctan(0.0138092) = 0.79116°.$$

The above value of the departure angle is equal to the value of the departure angle (for range 1200 meters) shown in table 2 of section 3.8.

Method II

We consider the trajectory equation for the Lapua bullet, i.e. the equation 1 of example 2:

$$y = \tan\alpha_0 \cdot x - \frac{gx^2}{2v_0^2 \cos^2\alpha_0}(1 + \frac{3.195898815 \cdot 10^{-4}x}{\cos\alpha_0} + \frac{1.01683644 \cdot 10^{-7}x^2}{\cos^3\alpha_0} +$$
$$+ \frac{3.056074173 \cdot 10^{-11}x^3}{\cos^5\alpha_0} + \frac{8.27548897 \cdot 10^{-15}x^4}{\cos^7\alpha_0})$$

(2)

and estimate the bullet drop at a certain distance, considering the departure angle $\alpha_0 = 0°$.

Bullet Drop at Range 1000 meters

Substituting in equation (2) $\alpha_0 = 0°$, $x = x_T = 1000m$, $v_0 = 830$ and $g = 9.80665$ we find the drop:

$$y_T = -10.501m.$$

Employing the principle of rigidity of the trajectory, i.e. rotating the trajectory counter clock wise till it is zeroed at $x_T = 1000m$, we find:

$$\tan\alpha_0 = \frac{|drop|}{x_T} = \frac{10.501}{1000} = 0.01051$$

Hence:

$$\alpha_0 = \arctan(0.010501) = 0.60164°.$$

Comment: Comparing the results obtained by method 1 and method 2 we see that we have an insignificant difference of:

$$\Delta\alpha_0 = 0.60164° - 0.60176° = -(1.1.2 \cdot 10^{-4})° = -0.0072'.$$

This approximation is another proof of the trajectory rigidity principle. We can say that the second method is much easier to use to find the departure angle.

Example 4. For the Lapua bullet of Example 1, use the second order parabola to find the drop of the bullet for the range 600 meters, if the bullet is fired horizontally (departure angle $\alpha_0 = 0$). The departure speed of the Lapua bullet is $v_0 = 830m/s$.

Solution

From equations (1) of example 2, we find the second order polynomial of the trajectory of flight

$$y = \tan\alpha_0 \cdot x - \frac{gx^2}{2v_0^2 \cos^2\alpha_0}(1 + \frac{3.195898815 \cdot 10^{-4}x}{\cos\alpha_0} + \frac{1.01683644 \cdot 10^{-7}x^2}{\cos^3\alpha_0}) \tag{1}$$

Substituting $v_0 = 830m/s$, $\alpha_0 = 0$ and $x = 600$ we find that the drop of projectile is

$$y = -3.161m$$

Note: The real drop ($y = -3.203m$, table 1 section 3.8) is slightly different from the drop obtained using the 4th order parabola.

The ballistics elements of the trajectory of a bullet calculated using a lower order of the parabola are less accurate then those obtained by the 6th order parabola, especially for relatively long ranges.

In all illustrations, the bullets are launched in ICAO atmosphere since the data we have used to find the parabolas are for the bullets that are thrown in ICAO atmosphere.

3.7 Rigidity Principle for Long Range Shooting

Consider the trajectory equation (3.6.17), i.e. the trajectory of a bullet launched with angle α_0 and departure speed v_0.

$$y = \tan\alpha_0 \cdot x - \frac{gx^2}{2v_0^2 \cos^2\alpha_0}(1 + \frac{A \cdot x}{\cos\alpha_0} + \frac{B \cdot x^2}{\cos^3\alpha_0} + \frac{C \cdot x^3}{\cos^5\alpha_0} + \frac{D \cdot x^4}{\cos^7\alpha_0}) \tag{3.7.1}$$

The drop of the bullet at x is

$$y - \tan\alpha_0 \cdot x = -\frac{gx^2}{2v_0^2 \cos^2\alpha_0}(1 + \frac{A\cdot x}{\cos\alpha_0} + \frac{B\cdot x^2}{\cos^3\alpha_0} + \frac{C\cdot x^3}{\cos^5\alpha_0} + \frac{D\cdot x^4}{\cos^7\alpha_0}). \quad (3.7.2)$$

Assuming that with the departure angle α_0 the trajectory (3.7.1) is zeroed at the point ($x = x_T$, $y = 0$). Substituting in (3.7.2), we find that the drop of the bullet at x_T is

$$\tan\alpha_0 \cdot x_T = \frac{gx_T^2}{2v_0^2 \cos^2\alpha_0}(1 + \frac{A\cdot x_T}{\cos\alpha_0} + \frac{B\cdot x_T^2}{\cos^3\alpha_0} + \frac{C\cdot x_T^3}{\cos^5\alpha_0} + \frac{D\cdot x_T^4}{\cos^7\alpha_0}). \quad (3.7.3)$$

If the bullet is launched with an angle $(E + \alpha_0)$, where E is the angle of sight, the drop of the bullet at the point (x_T, y_T) is

$$y_T - \tan(E + \alpha_0)\cdot x_T = \frac{gx_T^2}{2v_0^2 \cos^2(E + \alpha_0)}(1 + \frac{A\cdot x_T}{\cos(E + \alpha_0)} +$$

$$. \quad (3.7.4)$$

$$+ \frac{B\cdot x_T^2}{\cos^3(E + \alpha_0)} + \frac{C\cdot x_T^3}{\cos^5(E + \alpha_0)} + \frac{D\cdot x_T^4}{\cos^7(E + \alpha_0)})$$

For small projection angles α_0 and small angles of sight E we can consider $\cos\alpha_0 \approx 1$ and $\cos(E + \alpha_0) \approx 1$. For such angles the equation (3.7.3) and (3.7.4) can be written respectively:

$$\tan\alpha_0 \cdot x_T = \frac{gx_T^2}{2v_0^2}(1 + A\cdot x_T + B\cdot x_T^2 + C\cdot x_T^3 + D\cdot x_T^4)$$

$$(3.7.5)$$

and

$$y_T - \tan(E + \alpha_0)\cdot x_T = -\frac{gx_T^2}{2v_0^2}(1 + A\cdot x_T + B\cdot x_T^2 + C\cdot x_T^3 + D\cdot x_T^4)$$

$$(3.7.6)$$

From (3.75) and (3.7.6) we have:

$$y_T - \tan(E + \alpha_0) \cdot x_T = - \tan \alpha_0 \cdot x_T \qquad (3.7.7)$$

Hence,

$$\tan(E + \alpha_0) - \tan \alpha_0 = \frac{y_T}{x_T} . \qquad (3.7.8)$$

For small angles

$$\tan(E + \alpha_0) \approx E + \alpha_0, \quad \tan \alpha_0 \approx \alpha_0 \qquad (3.7.9)$$

Substituting (3.7.9) in (3.7.8) we can write

$$E \approx \frac{y_T}{x_T} \qquad (3.7.10)$$

Thus we can state, that for small sight angles that satisfy (3.7.10), we can consider that the drop of the projectile does not change if we rotate the trajectory that corresponds to the launching angle α_0 counter-clock or clock wise by a "small" angle of sight E.

This statement is the well known **Principle of Rigidity**.

How small is E in order that we can apply the principle of rigidity?

According to "Field Artillery, Vol. 6 – Ballistics and Ammunitions", chapter 6, section 1, 1992 DND Canada the rigidity of the trajectory is true if $|y_T| \leq 20 meters$, i.e. if

$$|E \cdot x_T| \approx |y_T| \leq 20m, \qquad (3.7.11)$$

or

$$|E| \leq 20 / x_T. \tag{3.7.12}$$

If the angle of sight is in degree than the rigidity criteria is.

$$|E°| \leq (\frac{20}{x_T}) \cdot \frac{180°}{\pi} \tag{3.7.13}$$

In the table below is given the range $x = x_T$ and the corresponding angle of sight E for which the principle of rigidity can be applied.

Table 1 – Rigidity Criteria (3.7.3)

Range (m)	200	400	600	800	1,000	1,200	1,400	1600
E (degree)	5.730	2.865	1.910	1.432	1.146	0.955	0.819	0.716

3.8 A Contemporary Method in Exterior Ballistics
(Analytical Estimation of the Ballistics Elements of the Trajectory of Bullet Flight)

George Klimi, James A. Boatright[31]

Introduction

The contemporary method we are introducing to solve the exterior ballistics problems is based on the expansion in series applied in an innovative way. The original application of the development in series we use to find the elements of the trajectory for small arms shooting, following Piton-Bressant and Duchene's approach, eliminates the

---

[31] James A. Boatright (BSc in Physics from Texas A&M University) is a retired aerospace engineer whose working career with Ford Aerospace Corporation and Lockheed EMSCO was mostly at NASA Johnson Space Center in Houston, Texas. Mr Boatright, amateur ballisticians, former benchrest rifle competitor and a former AIAA member, has been a gunsmith and a deputy sheriff, (E-mail: jim@theGuidedBullet.com)

relatively large errors exterior ballisticians tolerate with predictions of the ballistic trajectory for long range shootings using the known standard G- drag functions and a measured ballistics coefficient (or a set of ballistics coefficients).

The present method "eliminates" the need to measure the ballistics coefficient of a bullet and avoids the use of Doppler radar measurements, or other expensive experimental methods to find the trajectory of a bullet. The method is based on the existing range tables of small arms, or on experimental data that every shooter can obtain with the exterior ballistics equipments that are in the market.

The original method we present in the paper adds to the existing exterior ballistics methods a new technique to solve the problems of exterior ballistics of small arms without:

- Solving the differential equations that describe the projectile flight,
- Using a G-function of resistance,
- Measuring the ballistics coefficient.

The following approach is a work done together with James A. Boatright.

Duchene's Approximate Equation of the Projectile Trajectory[32]
4th order Parabola

According to Duchene, when v_0 and α_0 are known, using (1.6), we can consider that the equation of the projectile trajectory is:

$$y = \tan\alpha_0 \cdot x - \frac{gx^2}{2v_0^2 \cos^2\alpha_0}(1 + \frac{A \cdot x}{\cos\alpha_0} + \frac{B \cdot x^2}{\cos^2\alpha_0}) \qquad (3.8.1)$$

assuming that A and B are dependent on the projectile departure speed v_0, but are independent from the departure angle α_0.

---

[32] Cranz, C., Becker, K, *Handbook of Ballistics*, p. 126, London, 1921.

Hence, for the unknown constants, A and B, (when $y = 0$ and for the target distance x_T) we have:

$$\frac{v_0^2 \sin 2\alpha_0}{g \cdot x_T} = 1 + \frac{A}{\cos\alpha_0} x_T + \frac{B}{\cos^2\alpha_0} x_T^2 \tag{3.8.2}$$

where x_T is the range that corresponds to the departure angle α_0.

To determine A and B, using (3.8.2), we need to know two different horizontal ranges with their corresponding departure angles, or equivalently, we need to know the coordinates of two separated points of the trajectory well away from the origin of the coordinates.

The other elements of the trajectory of the same projectile can be estimated using the following approximate formulas:

Angle of Flight

$$\tan\alpha = \tan\alpha_0 - \frac{gx}{v_0^2 \cos^2\alpha_0} \cdot (1 + \frac{3}{2} \frac{Ax}{\cos\alpha_0} + 2\frac{Bx^2}{\cos^2\alpha_0}) \tag{3.8.3}$$

(The equation (3.8.3) is obtained differentiating (3.8.1) with respect to x)

Projectile Velocity

$$v = \frac{v_0 \cos\alpha_0}{\cos\alpha} \cdot \frac{1}{\sqrt{1 + 3Ax / \cos\alpha_0 + 6Bx^2 / \cos^2\alpha_0}} \tag{3.8.4}$$

(Equation 3.8.4 is obtained differentiating equation 3.8.3 with respect to x and using the second equation of (1.1)).

Time of Flight

$$t = \frac{1}{v_0 \cos\alpha_0} \int_0^x \sqrt{1 + 3Ax/\cos\alpha_0 + 6Bx^2/\cos^2\alpha_0}\ dx$$

(3.8.5)

Departure Angle
The departure angle can be determined using (3.8.1). When $y = 0$, for the departure angle we have:

$$\tan\alpha_0 = \frac{g\cdot x}{2v_0^2 \cos^2\alpha_0}(1 + \frac{A\cdot x}{\cos\alpha_0} + \frac{B\cdot x^2}{\cos^2\alpha_0})$$

(3.8.6)

Note: With regard to (3.8.5) we note that Duchene considers a different formula that takes into account an average value of time t.

Example 1. Approximate Equation of the Projectile Trajectory 4th order Parabola[33]

Determining the parameters A and B
Using Duchene's formula (3.8.2),

$$\frac{v_0^2 \sin 2\alpha_0}{g\cdot x_T} = 1 + \frac{A}{\cos\alpha_0}x_T + \frac{B}{\cos^2\alpha_0}x_T^2$$

(1)

we will show the way we determine the parameters A and B for the Lapua GB528 Scenar 19.4g, Caliber 8.6, considering the data in Table 1, below[34].

---

[33] We use the term nth order parabola, instead of nth order polynomial, to be consistent with Cranz's terminology.
[34] http://en.wikipedia.org/wiki/External_ballistics

We want to find the equation of Duchene's trajectory for the ranges till $x_T = 1500m$ ($y_T = 0$). We consider the following data from table 1, section 3 below:

Range: $x_T = 1500m$; drop: $\bar{y}_T = 30.035m$,

Range: $x_T = 1200m$; drop: $\bar{y}_T = 16.571m$ (2)

Based on the *principle of rigidity* of the projectile trajectory we find the respective departure angles:

$$\alpha_0 = \arctan(30.035/1500) = 1.1471°,$$

$$\alpha_0 = \arctan(16.571/1200) = 0.5791°. \tag{3}$$

Substituting in (1), $v_0 = 830m/s$, $g = 9.80665m/s^2$, $x_T = 1500m$, $\alpha_0 = 1.1471°$, we obtain a first linear equation:

$$1500.300670A + 2.250902101 \cdot 10^6 B = 0.874720526 \tag{4}$$

Substituting in (1), $v_0 = 830m/s$, $g = 9.80665m/s^2$, $x_T = 1200m$, $\alpha_0 = 0.5791°$, we obtain a second linear equation:

$$1200.114410A + 1.440274598 \cdot 10^6 B = 0.616474735 \tag{5}$$

Solving the system of linear equations (4) and (5), we find that the values of the parameters A and B, for the Lapua GB528 Scenar 19.4 g, Caliber 8.6 bullet, are:

$$A = 2.3644 \cdot 10^{-4}, \quad B = 2.3102 \cdot 10^{-7}.$$

Substituting the above values of A and B in (3.8.1) – (3.8.5), we find a set of 5 equations that allow us to find all the elements of the trajectory and solve exterior ballistics problems without employing the differential equations (1.1).

Thus the equation of the 4th order parabola is:

$$y = \tan\alpha_0 \cdot x - \frac{gx^2}{2v_0^2 \cos^2\alpha_0}(1 + \frac{2.3644\cdot 10^{-4}\cdot x}{\cos\alpha_0} + \frac{2.3102\cdot 10^{-7}\cdot x^2}{\cos^2\alpha_0}) \qquad (6)$$

For the projectile speed we have:

$$v = \frac{v_0 \cos\alpha_0}{\cos\alpha} \cdot \frac{1}{\sqrt{1 + 3\cdot 2.3644\cdot 10^{-4} x / \cos\alpha_0 + 6\cdot 2.3102\cdot 10^{-7} x^2 / \cos^2\alpha_0}} \qquad (7)$$

Note: To solve the linear equations (4) and (5) and to find all ballistics elements of the trajectory and solve a variety of exterior ballistics problems we can use a graphing calculator TI83+, HP, Casio, etc.

Approximate Equation of the Projectile Trajectory
5th order Parabola

Following Duchene's approach, we assume that the projectile flight is described by the following trajectory equation:

$$y = \tan\alpha_0 \cdot x - \frac{gx^2}{2v_0^2 \cos^2\alpha_0}(1 + \frac{A\cdot x}{\cos\alpha_0} + \frac{B\cdot x^2}{\cos^2\alpha_0} + \frac{C\cdot x^3}{\cos^3\alpha_0}) \qquad (3.8.7)$$

The constants A, B and C can be obtained using the equation:

$$\frac{v_0^2 \sin 2\alpha_0}{g\cdot x_T} = 1 + \frac{A}{\cos\alpha_0}x_T + \frac{B}{\cos^2\alpha_0}x_T^2 + \frac{C}{\cos^3\alpha_0}x_T^3 \qquad (3.8.8)$$

The other elements of the projectile trajectory are determined using the following formulas:

$$\tan\alpha = \tan\alpha_0 - \frac{gx}{v_0^2 \cos^2\alpha_0}\cdot(1 + \frac{3}{2}\frac{A}{\cos\alpha_0}x + 2\frac{B}{\cos^2\alpha_0}x^2 + \frac{5}{2}\frac{C}{\cos^3\alpha_0}x^3) \qquad (3.8.9)$$

$$v = \frac{v_0 \cos\alpha_0}{\cos\alpha} \cdot \frac{1}{\sqrt{1 + 3Ax/\cos\alpha_0 + 6Bx^2/\cos^2\alpha_0 + 10Cx^3/\cos^3\alpha_0}},$$

$$(3.8.10)$$

$$t = \frac{1}{v_0 \cos\alpha_0} \int_0^x \sqrt{1 + 3Ax/\cos\alpha_0 + 6Bx^2/\cos^2\alpha_0 + 10Cx^3/\cos^3\alpha_0}\ dx .$$

$$(3.8.11)$$

$$\tan\alpha_0 = \frac{gx^2}{2v_0^2\cos^2\alpha_0}(1 + \frac{A\cdot x}{\cos\alpha_0} + \frac{B\cdot x^2}{\cos^2\alpha_0} + \frac{C\cdot x^3}{\cos^3\alpha_0}).$$

$$(3.8.12)$$

Applications

Employing the approximate formulas obtained above and the data from the firing tests with small arms (or the existing range tables), i.e. for ranges 0 – 1000 or more, shooting on 3 targets located let say 200 meters or more from each other without employing ballistics chronograph or Doppler radar to measure the velocity and time, we are able to find the equation of the projectile trajectory with a satisfying accuracy.

The following example illustrates the use of the above formulae to predict the trajectory of the Lapua GB528 Scenar 19.4 g, Caliber 8.6 mm bullet with a great accuracy, which can be easily verified using the data of the 3rd table of "External Ballistics"[35] partially presented in following table1.

---

[35]   http://en.wikipedia.org/wiki/External_ballistics

Table 1 Lapua GB528 Scenar 19.4 g, Caliber 8.6 mm

Range (m)	0	300	600	900	1,200	1,500	1,800	2100
Velocity (m/s)	830	711	604	507	422	349	311	288
Time (s)	0.000	0.3918	0.8507	1.3937	2.0435	2.8276	3.7480	4.7552
Drop (m)	0.000	0.715	3.203	8.146	16.571	30.035	50.715	80.529

Example 2 **Equation of the Trajectory of the Lapua GB528 Scenar 19.4 g, Caliber 8.6 (ICAO Standard Atmosphere, 5th order Parabola)**

(a) Determining the unknown constants A, B, C.

From table 1, we select the following horizontal ranges (x_T) and the corresponding drops (y_T):

$(x_T = 900m , y_T = 8.146m),(x_T = 1200m , y_T = 30.035m),$
$(x_T = 1500m , y_T = 50.715m)$

Using the above data and the principle of the "rigidity of the trajectory" we find the respective departure angles:

$$\alpha_0 = \arctan(8.146/900) = 0.5186°,$$
$$\alpha_0 = \arctan(16.571/1200) = 0.7912°,$$
$$\alpha_0 = \arctan(30.035/1500) = 1.1471°.$$

Substituting in (3.8.8) the given ranges, the estimated corresponding departure angles and the departure velocity of the projectile, $v_0 = 830m/s$, we find three linear equations.

Solving the obtained system of linear equations we find:

$$A = 3.7961 \cdot 10^{-4}, \quad B = 1.6278 \cdot 10^{-8}, \quad C = 7.9523 \cdot 10^{-11}. \tag{1}$$

Substituting A, B and C in (3.8.7) – (3.8.11), we find a set of formulas that allow us to solve ballistics problem without:

- Solving the differential equations (1.1),
- Knowing the drag coefficient or the function of resistance,
- Measuring the form coefficient (i) and BC.

Thus, the equation of the trajectory of the GB528 Scenar 19.4 g, Caliber 8.6 mm bullet is:

$$y = \tan\alpha_0 \cdot x - \frac{gx^2}{2v_0^2 \cos^2\alpha_0}[1 + 3.7961 \cdot 10^{-4}(\frac{x}{\cos\alpha_0}) +$$

$$+ 1.6278 \cdot 10^{-8}(\frac{x}{\cos\alpha_0})^2 + 7.9523 \cdot 10^{-11}(\frac{x}{\cos\alpha_0})^3] \qquad (2)$$

while the projectile velocity along the trajectory is:

$$v = \frac{v_0 \cos\alpha_0}{\cos\alpha} \cdot (1 + 3\frac{3.7961 \cdot 10^{-4} \cdot x}{\cos\alpha_0} + 6\frac{1.6287 \cdot 10^{-8} \cdot x^2}{\cos^2\alpha_0} + 10\frac{7.4560 \cdot 10^{-11} \cdot x^3}{\cos^3\alpha_0})^{-1/2} \qquad (3)$$

where $v_0 = 830 m/s$.

Substituting the values of the constants A, B, C in (3.8.10) and (3.8.11) we are able to determine all the other ballistics elements of the trajectory at the impact point, or at any point on the trajectory as well as the maximum altitude of flight.

Using equation of trajectory (2) we can calculate the departure angle needed to hit a target located at a given range (from 0 – 1800m).

Verifying the Prediction Accuracy

Formula (2) of example 2 is pretty accurate till around 1800 meters (see table 2 below).

Graphing the above parabola, for example using a TI-84+, we can find the impact point for any given value of the range x_T.

Thus, using the table 1 for the distance $x_T = 1500m$, we find that the corresponding departure angle is

$$\alpha_0 = \arctan(drop/x) = \arctan(30.035/1500) = 1.1471°$$

Substituting the obtained value of the departure angle into the trajectory formula (2), we find that the y-coordinate of impact point at the range 1500 meters is

$$y_T = 8.0736 \times 10^{-5} m \approx 0.000,$$

i.e. the calculated trajectory intersects the center of the target.

The coordinates (range, y-coordinate) obtained using (2), for the LapuaGB528 Scenar 19.4 g, Caliber 8.6 mm are presented in the table below.

The second row of the table 2 corresponds to the 5th order parabola (formula (2), example 2), while the third row corresponds to the 4th order parabola (formula (1), example 1).

Table 2 Lapua GB528 Scenar 19.3 g, Caliber 8.6 mm

Departure Angle (degree)		0.1366	0.30586	0.51858	0.79116	1.1471	1.61388
Range (m)	0	300	600	900	1200	1500	1800
y-coordinate (m), 5th order P.	0	-0.001	- 0.002	0.000	0.000	0.000	-0.075
y-coordinate (m), 4th order P.	0	0.016	0.064	0.074	0.000	0.000	0.520

Note. Formula (2) of example 2 can be used for the inclined shooting as well. It can be modified to be used in non standard atmosphere, in high mountains, etc.

Comments

- As example 2 shows, there is no need for Doppler radars or ballistics chronographs measurements and no need to measure the form coefficient or BC of the bullet.

- What is needed to determine the equation of the trajectory of flight and the other equations, i.e. (3.8.7) – (3.8.11), are only the existing range tables of the small arms, or the experimental measurements of the drop in three different ranges, or the impact coordinates on three targets.

- For new bullets/firearms, or for experimenting with any individual small arm, we need to determine the coordinates of the impact of the projectile on the target in 3 different distances (or the drop of the bullet).

In the graph below are shown the 3rd, the 4th, the 5th degree parabolas and the ideal (vacuum) trajectory; the 5th (dash line) order parabola and the 4th degree parabola are two trajectories on top). The first form the bottom is the vacuum trajectory.

From the figure seems that the 4th and 5th degree parabolas are quite superimposed. The differences are visible around the vertex of the trajectories.

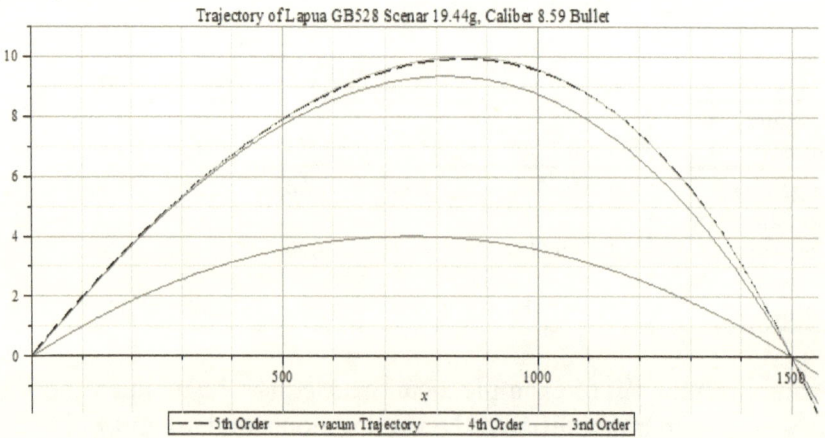

Trajectory of Lapua GB528 Scenar 19.44g, Caliber 8.59 Bullet

The question is:

Does the 5th (or 4th) degree parabola fit "perfectly enough" the real trajectory of the given Lapua bullet? In other words, does the geometric shape of the 5th (or 4th) order parabola and all the other elements of the trajectory (velocity, time of flight, angle of flight) present an acceptable good match for the real trajectory and the respective ballistics elements?

According to Cranz & Becker (page 122 of their Handbook on Ballistics), the function $y = F(x)$ **remains finite and continuous and its derivatives remains "so long as the projectile is over the muzzle horizon."**

That is, we should use the maximum range as one of our data points. Beyond the maximum range the 5th (4th) degree parabola is not a perfect fit.

- Thus, the 5th order trajectory of projectile flight does not fit the real trajectory below the horizon (x-axis). That is, the farthest data point used in the parabolic fit should be at very nearly the maximum range.

Indeed, we see that for distances greater than the range $x_T = 1500m$ the predicted drop is not that accurate.

Thus, in the last column of table 2 that corresponds to the range 1800m, are shown the relatively large errors obtained using 4th and 5th order parabolas.

Table 3 (below) shows as well the relatively large errors in predicted accuracy of the bullet velocity (row 3 and 4) at the range 1800m compared to the actual velocity (row 2).

- From table 3, obtained using formula (3) of example 2 (3rd row) and formula (2) of example 1 (2nd row), it can be seen that the 4th order parabola, with respect to the actual projectile speed (row 1), is not that accurate.

Table 3 Velocity Lapua GB528 Scenar 19.4 g, Caliber 8.6 mm

Range (m)	0	300	600	900	1,200	1,500	1,800
Lapua Velocity	830	711	604	507	422	349	311
Velocity (5th)	830	708.7	603.7	506.65	421.3	350.3	293
Velocity (4th)	830	717.6	598.3	499.5	423.2	364.7	319.2

Note. The predicted velocity obtained using 5th degree parabola is pretty accurate till around 1500m.

6th Order Parabola

Hereafter is shown the 6th order parabola that is obtained in a similar way as the 4th and 5th ones.

Equation of the Trajectory:

$$y = \tan\alpha_0 \cdot x - \frac{gx^2}{2v_0^2 \cos^2 \alpha_0}(1 + \frac{A \cdot x}{\cos\alpha_0} + \frac{B \cdot x^2}{\cos^2 \alpha_0} + \frac{C \cdot x^3}{\cos^3 \alpha_0} + \frac{D \cdot x^4}{\cos^4 \alpha_0})$$

$$(3.8.13)$$

The constants A, B, C and D can be obtained from the equation:

$$\frac{v_0^2 \sin 2\alpha_0}{g \cdot x_T} = 1 + \frac{A}{\cos\alpha_0}x_T + \frac{B}{\cos^2 \alpha_0}x_T^2 + \frac{C}{\cos^3 \alpha_0}x_T^3 + \frac{D}{\cos^4 \alpha_0}x_T^4$$

$$(3.8.14)$$

that is found from (3.8.13) substituting y = 0.

Substituting in (3.8.14) the departure angle and the corresponding horizontal ranges (4 points) starting from the largest range and then solving the obtained system of 4 linear equations we find the unknown values of A, B, C and D of a given small arm.

The other elements of the projectile trajectory are determined using the following formulas:

Angle of flight:

$$\tan\alpha = \tan\alpha_0 - \frac{gx}{v_0^2 \cos^2\alpha_0} \cdot (1 + \frac{3}{2}\frac{A}{\cos\alpha_0}x +$$

$$+ 2\frac{B}{\cos^2\alpha_0}x^2 + \frac{5}{2}\frac{C}{\cos^3\alpha_0}x^3 + 3\frac{D}{\cos^4\alpha_0}x^4)$$

$$(3.8.15)$$

(Hence we find the angle of flight α)

Velocity

$$v = \frac{v_0 \cos\alpha_0}{\cos\alpha} \cdot (1 + 3\frac{A \cdot x}{\cos\alpha_0} + 6\frac{B \cdot x^2}{\cos^2\alpha_0} + 10\frac{C \cdot x^3}{\cos^3\alpha_0} + 15\frac{D \cdot x^4}{\cos^4\alpha_0})^{-1/2}$$

$$(3.8.16)$$

Time of flight:

$$t = \frac{1}{v_0 \cos\alpha_0}\int_0^x \sqrt{1 + 3\frac{Ax}{\cos\alpha_0} + 6\frac{Bx^2}{\cos^2\alpha_0} + 10\frac{Cx^3}{\cos^3\alpha_0} + 15\frac{Dx^4}{\cos^4\alpha_0}}\ dx$$

$$(3.8.17)$$

Elements of the projectile trajectory at the impact point on the ground are obtained from equations (3.8.13), (3.8.15), (3.8.16) and (3.8.17) substituting the departure angle and the corresponding range.

The y-coordinate obtained from equation (3.8.13), substituting in it the value of the departure angle α_0 and the corresponding range x_T (from the range table/ experiments) will be used to verify the verity of the method.

Comment
It can be seen that for long range shooting the accuracy in predicted drop is very good for higher order parabolas (4th, 5th and 6th),

but the predicted accuracy of the calculation of the velocity (and time) is better the higher is the order of parabola.

So, for studies and accurate determination of all the elements of the trajectory we should use the 5th or 6th order parabolas at least for long range shootings.

Note: The coefficients of the 6th order parabola of the Lapua GB528 Scenar 19.4 g, Caliber 8.6 mm are:

$$A = 3.6605 \cdot 10^{-4}, \quad B = 5.167852 \cdot 10^{-8}, \quad C = 4.939870 \cdot 10^{-11},$$
$$D = 8.366851 \cdot 10^{-15}.$$

The above coefficients are obtained using the drop of the given bullet in four ranges, 1500m, 1200m, 900m and 600m (see table 1).

The Piton-Bressant Trajectory

The simplest parabola that describes the trajectory of projectile flight in presence of resistance of air, but the least accurate is the 3rd degree parabola that is known also as Piton Bressant formula.

It can be obtained from $(3.8.13) - (3.8.17)$ substituting B, C and D with zero.

$$y = \tan \alpha_0 \cdot x - \frac{gx^2}{2v_0^2 \cos^2 \alpha_0} \left(1 + \frac{A \cdot x}{\cos \alpha_0} \right). \tag{3.8.18}$$

The other elements of the projectile trajectory are determined using the following formulas:

Angle of flight:

$$\tan \alpha = \tan \alpha_0 - \frac{gx}{v_0^2 \cos^2 \alpha_0} \cdot \left(1 + \frac{3}{2} \frac{A}{\cos \alpha_0} x \right), \tag{3.8.19}$$

Using (3.8.19) we find the angle of flight α at any point on the trajectory and at the impact point.

Velocity

$$v = \frac{v_0 \cos\alpha_0}{\cos\alpha} \cdot (1 + 3\frac{A \cdot x}{\cos\alpha_0})^{-1/2}$$

(3.8.20)

Time of flight:

$$t = \frac{1}{v_0 \cos\alpha_0} \int_0^x (1 + 3Ax / \cos\alpha_0)^{1/2} dx$$

(3.8.21)

Firing at a certain range (for example at 600meters) we find the drop (or the y-coordinate of the impact on the target.

At the impact point, the horizontal range is $x = x_T$, and the y-coordinate is $y = y_T = 0$.

Substituting the above data in (3.8.18) we obtain the following equation:

$$\frac{v_0^2 \sin 2\alpha_0}{g \cdot x_T} = 1 + \frac{A}{\cos\alpha_0} x_T$$

(3.8.22)

that can be used to determine the parameter A.

The Piton-Bressant formula can be used by shooters (hunters) within the effective range of small arms.

For example, for the Lapua GB528 Scenar 19.4 g, Caliber 8.6 mm bullet and for ranges till around 600 meters, the value of A is:

$$A = 4.16649 \cdot 10^{-4}.$$

(3.8.23)

The above parameter A is obtained from the equation:

$$\frac{v_0^2 \sin 2\alpha_0}{g \cdot x_T} = 1 + \frac{A}{\cos \alpha_0} x_T$$

$$(3.8.24)$$

after substituting: range $x_T = 600m$; muzzle velocity, $v_0 = 830m/s$; departure angle, $\alpha_0 = \arctan(drop / x_T) = \arctan(3.203 / 600) = 0.305861°$, i.e. solving the equation:

$$600.008549 = 0.249993.$$

$$(3.8.25)$$

The formula of the trajectory of flight (3.8.18) gives good results for the effective distances of the small arms.

The velocity and time of flight, determined respectively by (3.8.20) and (3.8.21), represent approximate values.

For the Lapua GB528 Scenar 19.44 g, Caliber 8.59mm to predict the elements of the trajectory till 1500 meters we can substitute in (3.8.18)-(3.8.21) the following value of A:

$$A = 5.830 \cdot 10^{-4},$$

$$(3.8.26)$$

obtained in the same way as (3.8.23).

Zeroing the Rifle (Application of Piton-Bressant Formula)

Using formulas (3.8.18) and (3.8.24) we can set up the sight to zero the firearm within the effective range.

Indeed, since the departure angle is very small, in equation (3.8.24), we can substitute $\cos \alpha = 1$. Thus, for the departure angle we have:

$$\sin \alpha_0 = \frac{gx_T}{2v_0^2}(1 + Ax_T),$$

(3.8.27)

where A is given by (3.8.26).

Zeroing at 200 yards

For example, for the Lapua bullet GB528 Scenar 19.4 g, Caliber 8.6 mm, to zero the rifle at the range 200yards (182.88m) substituting in (3.8.24):

$$g = 9.80665m/s, \quad x_T = 182.88m, \quad v_0 = 830m/s,$$

$$A = 4.16649 \cdot 10^{-4}.$$

we can write:

$$\sin \alpha_0 = 0.0014008$$

Hence, for the angle of departure we find:

$$\alpha_0 = 0.0014009 radian = 0.080263°.$$

If the height of the sight is 1.5"=0.0381m, we find that to zero the rifle at 200yards, the rifleman should set up his sight considering the following angle[36]:

$$\alpha = \alpha_0 + \frac{h}{x_T} = 0.0014009 + \frac{0.0381}{182.88} = 0.001609 rad = 1.61mrad = 5.532 MOA.$$

Note. The rifle can be zeroed using the higher order parabolas, 4th or 5th. The approach is simple and similar to the one we just demonstrated. To improve the accuracy we use the iteration process as it is demonstrated in the following example.

---

[36] Klimi, G., Exterior Ballistics of Small Arms, p. 47, Xlibris, 2009

Departure Angle

One of the main tasks of exterior ballistics is the determination of the departure angle to hit a target located on the horizontal ground at a certain range y_T (when $y_T = 0$).

Once the departure angle is determined, we can find the projection angle, velocity and the time of flight.

For simplicity we will use the 4th degree parabola to show the way we determine the departure angle for small arms.

We use the equation of trajectory (1) of Example 1, for Lapua bullet GB528 Scenar 19.4g, Caliber 8.6 mm, i.e.

$$y = \tan\alpha_0 \cdot x - \frac{gx^2}{2v_0^2 \cos^2\alpha_0}(1 + \frac{2.3644 \cdot 10^{-4} \cdot x}{\cos\alpha_0} + \frac{2.3102 \cdot 10^{-7} \cdot x^2}{\cos^2\alpha_0}).$$

$$(3.8.28)$$

Substituting y = 0 we have:

$$\tan\alpha_0 \cdot x - \frac{gx^2}{2v_0^2 \cos^2\alpha_0}(1 + \frac{2.3644 \cdot 10^{-4} \cdot x}{\cos\alpha_0} + \frac{2.3102 \cdot 10^{-7} \cdot x^2}{\cos^2\alpha_0}) = 0.$$

$$(3.8.29)$$

Hence we find:

$$\tan\alpha_0 = \frac{gx}{2v_0^2 \cos^2\alpha_0}(1 + \frac{2.3644 \cdot 10^{-4} \cdot x}{\cos\alpha_0} + \frac{2.3102 \cdot 10^{-7} \cdot x^2}{\cos^2\alpha_0}).$$

$$(3.8.30)$$

For a given x (for example 900 meters) we solve the above equation with respect to α_0. One simple way to solve it is the following:

1. First Approximate Departure Angle

As a first approach, consider $\cos\alpha_0 \approx 1$, since the angle is small.

Substituting in (3.8.30): $x = 900$, $g = 9.80665$, $v_0 = 830$, $\cos\alpha_0 \approx 1$, we have:

$$\tan\alpha_0 = 0.008968$$

Hence for the departure we have first approximate value:

$$\alpha_0 = \arctan(0.008968) = 0.5138° \,. \tag{3.8.31}$$

2. Iteration process to improve the solution

Substituting in (3.8.30) all the above values and the angle given by (3.8.31) we have:

$$\tan\alpha_0 = 0.008969$$

and

$$\alpha_0 = \arctan(0.008969) = 0.5138° \,. \tag{3.8.32}$$

Repeating again the iteration process, we have:

$$\tan\alpha_0 = 0.0089686$$

and

$$\alpha_0 = \arctan(0.0089686) = 0.51385° \,. \tag{3.8.33}$$

The last result (3.8.33) is approximately the same as that given by (3.8.32). That means that (3.8.32) or (3.8.33) is the solution of (3.8.30).

Verification

Using the data from table 3 above, we find that the real departure angle is

$$\alpha_{0w} = \arctan(drop\,/\,x) = arac\tan(8.146\,/\,900) = 0.518576°\,. \quad (3.8.34)$$

The difference between (3.8.33) and (3.8.34) is

$$\Delta\alpha_0 = \alpha_0 - \alpha_{0w} = 0.51385° - 0.51858° = -0.00473° = -0.284'\,. \quad (3.8.35)$$

In the same way we find the departure angle for whatever parabola order.

Note. The equation (3.8.29) can be solved for the departure angle using any Math software or a graphing calculator.

3.9 Derivation of the Standard Function of Resistance Using 6th Order Parabola

The velocity of a bullet is found experimentally using Doppler radar measurements, or the ballistics chronographs (velocity meters). The PC programs that use the G-functions of resistance to determine the projectile velocity as a function of bullet range, or time of flight give approximate results, not accurate for relatively long ranges.

The contemporary method developed in section 3.9 allows us to find accurately the velocity of a bullet as a function of range using a fifth degree or a sixth degree parabola.

Once we find the velocity of the bullet as a function of range, let's say every 100 meters till 1500meters, we are able to find the standard function of resistance $f_D(v)$ of the respective bullet using the technique shown in section (2.11).

Method 1. We can measure the drop of the bullet at 4 ranges (4 targets)

We will illustrate the method using the 6th order parabola (formulas (3.8.13), (3.8.14), (3.8.15) and (3.8.16)), that is obtained using the drop of the Lapua GB528 Scenar 19.4 g, Caliber 8.6 mm bullet in 4 ranges (1500m, 1200m, 900m, 600m), i.e.

$$y = \tan \alpha_0 \cdot x - \frac{gx^2}{2v_0^2 \cos^2 \alpha_0}(1 + \frac{A \cdot x}{\cos \alpha_0} + \frac{B \cdot x^2}{\cos^2 \alpha_0} + \frac{C \cdot x^3}{\cos^3 \alpha_0} + \frac{D \cdot x^4}{\cos^4 \alpha_0})$$

(3.9.1)

Angle of flight:

$$\tan \alpha = \tan \alpha_0 - \frac{gx}{v_0^2 \cos^2 \alpha_0} \cdot (1 + \frac{3}{2} \frac{A}{\cos \alpha_0} x +$$

$$+ 2\frac{B}{\cos^2 \alpha_0} x^2 + \frac{5}{2} \frac{C}{\cos^3 \alpha_0} x^3 + 3\frac{D}{\cos^4 \alpha_0} x^4)$$

(3.9.2)

Velocity

$$v = \frac{v_0 \cos \alpha_0}{\cos \alpha} \cdot (1 + 3\frac{A \cdot x}{\cos \alpha_0} + 6\frac{B \cdot x^2}{\cos^2 \alpha_0} + 10\frac{C \cdot x^3}{\cos^3 \alpha_0} + 15\frac{D \cdot x^4}{\cos^4 \alpha_0})^{-1/2}$$

(3.9.3)

Time of flight:

$$t = \frac{1}{v_0 \cos \alpha_0} \int_0^x \sqrt{1 + 3\frac{Ax}{\cos \alpha_0} + 6\frac{Bx^2}{\cos^2 \alpha_0} + 10\frac{Cx^3}{\cos^3 \alpha_0} + 15\frac{Dx^4}{\cos^4 \alpha_0}} \, dx$$

(3.9.4)

where the coefficients for the Lapua GB528 Scenar 19.4 g, Caliber 8.6 mm bullet are:

$$A = 3.6605 \cdot 10^{-4}, \quad B = 5.167852 \cdot 10^{-8}, \quad C = 4.939870 \cdot 10^{-11},$$
$$D = 8.366851 \cdot 10^{-15}. \tag{3.9.5}$$

Considering that the departure angle is zero degree, $\alpha_0 = 0°$, using (3.9.2), (3.9.3) and the corresponding coefficients we find the velocity of the Lapua GB528 Scenar 19.4 g, Caliber 8.6 mm bullet as a function of range, in this example for ranges till 1500 meters.

In table 1 below is shown the velocity of bullet as well as the drop, time of flight and the terminal angle obtained using the formulas (3.9.1) till (3.9.4) when $\alpha_0 = 0°$, $v_0 = 830$, and the values of coefficients (3.9.5).

Table 1. Lapua GB528 Scenar 19.44 g, Caliber 8.59 mm								
Range	0	100	200	300	400	500	600	700	800
Velocity		830	786.59	746.51	708.68	672.42	637.32	603.22	570.08	537.95
Drop	0	-0.074	-0.306	-0.708	-1.319	-2.140	-3.203	-4.536	-6.171
Time	0	0.124	0.254	0.392	0.537	0.689	0.851	1.021	1.202
Term. Angle	0	-0.086	-0.182	-0.288	-0.406	-0.537	-0.683	-0.847	-1.030
Range	900	1000	1100	1200	1300	1400	1500	1600	
Velocity	506.94	477.16	448.7	421.7	396.2	372.3	349.8	328.8	
Drop	-8.145	-10.50	-13.29	-16.57	-20.40	-24.85	-30.02	35.99	
Time	1.393	1.597	1.813	2.043	2.288	2.545	2.826	3.122	
Term. Angle	-1.236	-1.469	-1.731	-2.028	-2.365	-2.746	-3.178	-3.667	

Comparing the results of the Table 1 above with the data given in Table 1 section 3.8 and with the data given in table 3 section 2.11, we see that they are approximately the same. That demonstrates the accuracy of the new methods we have introduced in sections (2.11), (3.8) and (3.9).

Method 2 The velocity of the bullet is measured at 4 ranges (4 targets)

We will show the method using the wikipedia data given in table 1 section 3.8, pretending that "we have measured" with great accuracy the velocity of the Lapua GB528 Scenar 19.44 g, Caliber 8.59 mm bullet at four ranges, i.e. using the following pairs (range, velocity) of data:

(600m, 604m/s), (900m, 507m/s), (1200, 422m/s) and (1500m, 349m/s).(3.9.6)

1. Deriving the set of ballistics equations related to the 6th order parabola

We need to determine the coefficients A, B, C, D employing the data given in (3.9.6) and using the equation (3.9.3), i.e.

$$v = \frac{v_0 \cos\alpha_0}{\cos\alpha} \cdot (1 + 3\frac{A \cdot x}{\cos\alpha_0} + 6\frac{B \cdot x^2}{\cos^2\alpha_0} + 10\frac{C \cdot x^3}{\cos^3\alpha_0} + 15\frac{D \cdot x^4}{\cos^4\alpha_0})^{-1/2}$$

$$(3.9.7)$$

where $\alpha_0 = 0°$, $v_0 = 830$.

The equation (3.9.7) can be written:

$$3A \cdot x + 6B \cdot x^2 + 10C \cdot x^3 + 15D \cdot x^4 = (\frac{830}{v_T \cdot \cos(a_T)})^2 - 1 \qquad (3.9.8)$$

As a first approach we consider the terminal angle $\alpha_T = 0$.

Substituting in (3.9.8) the sequence of pairs (3.9.6) we obtain the following system of equations:

$$1800A + 2.16 \cdot 10^6 B + 2.16 \cdot 10^9 \cdot C + 1.944 \cdot 10^{12} \cdot D = 81021 / 91204$$

$$(3.9.9)$$

$$2700A + 4.86 \cdot 10^6 B + 7.29 \cdot 10^9 C + 9.8415 \cdot 10^{12} D = 431851 / 257049$$
$$(3.9.10)$$

$$3600A + 8.64 \cdot 10^6 B + 1.728 \cdot 10^{10} C + 3.1104 \cdot 10^{13} D = 127704 / 44521$$
$$(3.9.11)$$

$$4500A + 1.35 \cdot 10^7 B + 3.375 \cdot 10^{10} C + 7.59375 \cdot 10^{13} = 567099 / 121801$$
$$(3.9.12)$$

Solving the above system of linear equations, for example using a TI-83Plus, we obtain the following values:

$$A = 3.884665046 \cdot 10^{-4}, \quad B = 3.793391251 \cdot 10^{-9},$$

$$(3.9.13)$$

$$C = 8.316809882 \cdot 10^{-11}, \quad D = 6.546647941 \cdot 10^{-16}$$

Substituting in (3.9.2), i.e. in

$$\tan \alpha = \tan \alpha_0 - \frac{gx}{v_0^2 \cos^2 \alpha_0} \cdot (1 + \frac{3}{2} \frac{A}{\cos \alpha_0} x +$$

$$+ 2\frac{B}{\cos^2 \alpha_0} x^2 + \frac{5}{2} \frac{C}{\cos^3 \alpha_0} x^3 + 3\frac{D}{\cos^4 \alpha_0} x^4),$$

$$(3.9.14)$$

the values of estimated coefficients (3.9.13), $\alpha_0 = 0$, $v_0 = 830$ and the sequence of pairs of data (3.9.6), we find respectively the following terminal angles:

$$\alpha_T = 0.68390°, \quad \alpha_T = 0.1.2357°, \quad \alpha_T = -2.02860°, \quad \alpha_T = -3.18106°.$$

$$(3.9.15)$$

Repeating again the substitutions in the equation (3.9.7), considering the terminal angles (3.9.5), we obtain a new system of equations similar to (3.9.9) – (3.9.12). The solution of the new system of the linear equations is:

$$A = 3.900159712 \cdot 10^{-4}, \quad B = 1.26370845 \cdot 10^{-9}, \qquad (3.9.16)$$

$$C = 8.439123385 \cdot 10^{-11}, \quad D = 6.994843301 \cdot 10^{-16}.$$

Thus, we found the coefficients, (3.9.16), that to be used in the formulas of the 6th order parabola, (3.9.1) – (3.9.4), to find all the elements of the Lapua GB528 Scenar 19.44 g, Caliber 8.59 mm bullet trajectory

In table 2 there are given the ballistics elements of the trajectory of the given bullet computed using the 6th degree parabola and the coefficients (3.9.6).

Table 2. Lapua GB528 Scenar 19.44 g, Caliber 8.59 mm								
Range	0	100	200	300	400	500	600	700	800
Velocity	830	785.00	745.04	707.94	672.46	637.91	604.03	570.82	538.4
Drop	0	-0.074	-0.307	-0.717	-1.323	-2.146	-3.210	-4.543	-6.178
Time	0	0.124	0.254	0.392	0.537	0.689	0.851	1.021	1.202
Term. Angle	0	-0.086	-0.183	-0.288	-0.407	-0.538	-0.684	-0.847	-1.030
Range	900	1000	1100	1200	1300	1400	1500	1600	
Velocity	507	476.82	448.05	420.8	395.2	371.3	349	328.3	
Drop	-8.152	-10.51	-13.30	-16.57	-20.41	-24.87	-30.02	36.03	
Time	1.393	1.597	1.813	2.044	2.288	2.551	2.829	3.125	
Term. Angle	-1.236	-1.468	-1.731	-2.030	-2.368	-2.751	-3.184	-3.676	

Comparing table 2 with wikipedia table 1 section 3.8 and with the data given in table 3 section 2.11, we see that the method 2 give excellent prediction results of the trajectory elements of the Lapua bullet.
The first method in practice is less expensive and can be easily used in practice.

The Standard Functions of Resistance

Using the velocities calculated in table 1 and table 2 above, and the method developed in section 2.11 we find the standard function of resistance for each of the methods described above:

Method 1

$$f_D(v) = 0.1491v - 34.4680 \, ,$$

(3.9.17)

Method 2

$$f_D(v) = 0.1494v - 34.562 \, ,$$

(3.9.18)

where the ballistics coefficient is

$$c = 1000\frac{d^2}{m} = 1000\frac{(0.00859)^2}{0.01944} = 3.796 \, ,$$

(3.9.20)

since the form coefficient is $i = 1$.

NOTE. Bullet Flight in Non-Standard Atmosphere and Inclined Firing

The parabolic theory in presence of air resistance, developed in sections 3.6 – 3.10, is appropriate for the flight of the bullets in standard atmosphere (in our examples we use the ICAO atmosphere).

In general, long range shooting with small arms is usually performed in non-standard atmosphere and over the sea level. To find the elements of the trajectories in non-standard atmosphere, in mountain or winter/summer shootings as well as in the inclined shooting among other methods we can use:

- Snell's law correction method employing the range tables obtained using the techniques of sections (3.6) and (3.7) [Klimi, G. Exterior Ballistics: A New Approach, section 4.1, Xlibris 2010];

- Different techniques of the Modern Theory of Corrections [Klimi, G. Exterior Ballistics: A New Approach, chapter 4, Xlibris 2010];
- Siacci's method for high-speed projectiles in non standard atmosphere employing instead of a G-function the standard functions of resistance (form coefficient $i = 1$) that is obtained by methods used in section (2.10), (2.11) or (3.9) [Klimi, G. Exterior Ballistics: A New Approach, section 3.2, 3.3, 3.4, Xlibris, 2010];
- Numerical integration of the system of differential equations (3.6.1) using the particular standard function of resistance of a given small arm (form coefficient $i = 1$) as in the PC program RPMELA.BAS. [See the PC programs included in Exterior Ballistics: A New Approach, Xlibris, 2010]

4

The Trajectory of Low-Speed Projectiles

Introduction

In this chapter, we study the projectile flight when the projectile speed all over the trajectory is less than 256 m/s. The drag function of a projectile that flies with speed less than 256 m/s is a function of the square of the projectile speed. Such projectiles are the ammunitions of some trench mortars, field cannons, hand grenades, game balls, etc.

The flight of low-speed projectiles is modeled by a system of differential equations that is solved relatively easy. We get approximate analytical solutions of differential equations of flight with a great practical accuracy.

The methods used in this chapter are basics for applications in military, sports, and other areas related with the motion of relatively slow projectiles, like in skydiving and parachute falling and in the motion of balls used in football, soccer, etc.

We assume that the projectile flight is in the standard atmosphere (see introduction of chapter 3) and the projectile characteristics are standard.

4.1 The Equations of Projectile Motion

We assume that

•

• the projectile speed v at any time from launching point till it hits the target is less than 256 m/s, $v<256m/s$.

For such projectile speeds, the Siacci function of resistance (2.3.2) is

$$K_D(v)=1.212\cdot10^{-4}v^2.$$

(4.1.1)

Considering (4.1.1), the system of differential equations of projectile flight (2.6.13) and (2.6.16) can be written, respectively, as

$$\begin{cases} \dfrac{dv_x}{dt}=-bh(y)v_x v \\[2mm] \dfrac{dp}{dt}=-\dfrac{g}{v_x} \\[2mm] \dfrac{dx}{dt}=v_x \\[2mm] \dfrac{dy}{dt}=v_y \end{cases},$$

(4.1.2)

and

$$\begin{cases} \dfrac{dv_x}{dx}=-bh(y)\cdot v \\[2mm] \dfrac{dp}{dx}=-\dfrac{g}{v_x^2} \\[2mm] \dfrac{dt}{dx}=\dfrac{1}{v_x} \\[2mm] \dfrac{dy}{dx}=p \end{cases},$$

(4.1.3)

where

$$b=1.212{\cdot}10^{-4}c \;,\quad c=\frac{id^2}{m}{\cdot}1000 \;,$$

$$h(y)=(\frac{289.08-0.006328y}{289.08})^{4.4} \;,\quad p=v_y/v_x=\tan\alpha \;. \qquad (4.1.4)$$

4.2 Approximate Solutions When Launching Angle is Narrow

In some particular cases, shown hereafter, we are able to obtain approximate analytic solutions of the system of differential equations of projectile motion, i.e. to obtain approximate formulae to calculate the projectile trajectory and its elements.

The systems of differential equations (4.1.2) and (4.1.3) can be integrated analytically when the projectile trajectory is such that the y-component, v_y of the projectile velocity at any moment during the flight is relatively low compared to the x-component v_x.

We can consider that the projectile speed v is approximately equal to its x-component v_x,

$$v=(v_x^2+v_y^2)^{1/2}=v_x(1+v_y^2/v_x^2)^{1/2}=v_x(1+p^2)^{1/2}\approx v_x \;. \qquad (4.2.1)$$

The above condition is satisfied all over the trajectory if

$$p^2=v_y^2/v_x^2<<1 \;. \qquad (4.2.2)$$

The approximation (4.2.1) is valid for the projectile trajectory relatively close to the ground or close to the horizontal line that originates at the point where the projectile is launched.

In practice, such scenario is observed during the fire from assault riffles, automatic guns, etc., or during the flight of fragments of antipersonnel projectiles—for example, fragmentation grenades, antipersonnel mines, etc.

For approximate solutions, we can consider $h(y)$ to be constant and equal to the value it has at the launching point. If the projectile is fired in standard conditions at sea level, then the value of the density function is $h(y) \approx h(0) = 1$.

To obtain approximate solutions for system (4.1.2), or system (4.1.3), we assume the following:

- The projectile speed v at any time from launching point till it hits the target is less than 256 m/s, $v < 256 m/s$. This condition is satisfied for projectiles fired with initial speeds less than 256 m/s.
- Condition (4.2.2) is true, i.e., $\tan^2 \alpha = p^2 \ll 1$.
- The density function $h(y)$ is constant and equal to the value it has at the launching point.

Substituting v_x for v in the first equation of (4.1.3), we obtain

$$\frac{dv_x}{dx} = -bv_x.$$
(4.2.3)

Thus, the system of the differential equations of projectile motion (4.1.3) can be written in the form

$$\begin{cases} \dfrac{dv_x}{dx} = -b \cdot v_x \\ \dfrac{dp}{dx} = -\dfrac{g}{v_x^2} \\ \dfrac{dt}{dx} = \dfrac{1}{v_x} \\ \dfrac{dy}{dx} = p \end{cases},$$
(4.2.4)

where $b = 1.212 \cdot 10^{-4} c$, $h(y) \approx h(y_0)$.

We solve (4.2.4) for the initial conditions:

$v_x = v_0 \cos\alpha_0$, $v_y = v_0 \sin\alpha_0$, $p = p_0 = \tan\alpha_0$, $y = 0$, $t = 0$ when $x = 0$. (4.2.5)

Integrating the first differential equation of (4.2.4), we obtain the x-component of the velocity

$$v_x = v_0 \cos\alpha_0 \cdot e^{-bx} . \tag{4.2.6}$$

Substituting (4.2.12) into the second equation of (4.2.4), we have

$$\frac{dp}{dx} = -\frac{1}{v_0^2 \cos^2\alpha_0 \cdot e^{-2bx}} . \tag{4.2.7}$$

Integrating, we find

$$p = \tan\alpha_0 + \frac{g(1-e^{2bx})}{2bv_0^2 \cos^2\alpha_0} . \tag{4.2.8}$$

By substituting $p = \tan\alpha$ in (4.2.8), we obtain the following equation that gives the angle of flight

$$\tan\alpha = \tan\alpha_0 + \frac{g(1-e^{2bx})}{2bv_0^2 \cos^2\alpha_0} . \tag{4.2.9}$$

Multiplying (4.2.6) and (4.2.8) and considering that $v_y = pv_x$, we find that the y-component of the velocity is

$$v_y = (\tan\alpha_0 + \frac{g(1-e^{2bx})}{2bv_0^2 \cos^2\alpha_0}) \cdot (v_0 \cos\alpha_0 \cdot e^{-bx}) . \tag{4.2.10}$$

Substituting (4.2.8) into the third equation of (4.2.4), we have

$$\frac{dy}{dx} = \tan\alpha_0 + \frac{1-e^{2bx}}{2bv_0^2 \cos^2\alpha_0} . \tag{4.2.11}$$

Integrating (4.2.11), we find the equation of the projectile trajectory

$$y=(\tan\alpha_0+\frac{g}{2bv_0^2\cos^2\alpha_0})x+\frac{g(1-e^{2bx})}{4b^2v_0^2\cos^2\alpha_0}. \qquad (4.2.12)$$

Integrating the last equation of system (4.2.4), we find the time as a function of the x-coordinate

$$t=\frac{e^{bx}-1}{bv_0\cos\alpha_0}. \qquad (4.2.13)$$

Summary. The solutions of the differential equations of the projectile flight (system [4.2.4]) are as follows:

$$\begin{cases} v_x=v_0\cos\alpha_0\cdot e^{-b\cdot x} \\ \\ \tan\alpha=\tan\alpha_0+\dfrac{g(1-e^{2bx})}{2bv_0^2\cos^2\alpha_0} \\ \\ v_y=(\tan\alpha_0+\dfrac{g(1-e^{2b\cdot x})}{2bv_0^2\cos^2\alpha_0})\cdot(v_0\cos\alpha_0 e^{-b\cdot x}), \\ \\ y=(\tan\alpha_0+\dfrac{g}{2bv_0^2\cos^2\alpha_0})x+\dfrac{g(1-e^{2b\cdot x})}{4b^2v_0^2\cos^2\alpha_0} \\ \\ t=\dfrac{e^{b\cdot x}-1}{bv_0\cos\alpha_0} \end{cases} \qquad (4.2.14)$$

where

$$b=1.212\cdot10^{-4}c, \quad c=1000\frac{id^2}{m}, \quad h(y)\approx h(0)=1. \qquad (4.2.15)$$

Solving the last equation of (4.2.20) for x, we obtain

$$x=\frac{\ln(1+bv_0\cos\alpha_0\cdot t)}{b}. \qquad (4.2.16)$$

Substituting (4.2.16) in all equations of (4.2.14), we can express the elements of the trajectory of flight through time t

$$
\begin{cases}
v_x = \dfrac{v_0 \cos\alpha_0}{1+bv_0\cos\alpha_0 \cdot t} \\[3mm]
\tan\alpha = \tan\alpha_0 - \dfrac{g}{v_0\cos\alpha_0}t - \dfrac{gb}{2}t^2 \\[3mm]
v_y = (\tan\alpha_0 - \dfrac{g}{v_0\cos\alpha_0}t - \dfrac{gb}{2}t^2)\cdot(\dfrac{v_0\cos\alpha_0}{1+bv_0\cos\alpha_0 \cdot t}) \\[3mm]
x = \dfrac{\ln(1+bv_0\cos\alpha_0 \cdot t)}{b} \\[3mm]
y = (\tan\alpha_0 + \dfrac{g}{2bv_0^2\cos^2\alpha_0})\cdot\dfrac{\ln(1+bv_0\cos\alpha_0 t)}{b} - \dfrac{g}{2bv_0\cos\alpha_0}t - \dfrac{g}{4}t^2
\end{cases}
\qquad .(4.2.17)
$$

The following examples illustrate two types of problems exterior ballistics solves when there are known: (I) b, α_0, v_0; (II) b, v_0, x.

Example 1. Find the launching angle α_0 needed to hit a tank that is in a horizontal distance of 300 m from a 82 mm antitank cannon if a projectile of mass 2 kg is fired with the initial speed 160 m/s. The coefficient of the projectile form is 0.56. Meteorological conditions are normal. Find also the time of flight and the terminal angle.

Solution

There are given b, v_0, x.

First let's estimate b. We find

$$
c = \frac{id^2}{m}1000 = \frac{(0.56)(0.082)^2}{2}1000 = 1.88272
$$

and

$$b=c \cdot 1.212 \cdot 10^{-4}=(1.883)(1) \cdot 1.212 \cdot 10^{-4}=2.2818566 \cdot 10^{-4}.$$

(a) Launching Angle

Since the initial speed is less than 256 m/s and the angle of fire is relatively narrow (target and gun are in the horizontal plane), we can use the fourth equation of (4.2.1) to find the launching angle α_0.

The y-coordinate of the point of impact is zero, i.e., $y=0$.

Thus, to find the launching angle α_0 we have to solve for α_0 the fourth equation of (4.2.14),

$$(\tan\alpha_0 + \frac{g}{2bv_0^2 \cos^2\alpha_0})x + \frac{g(1-e^{2bx})}{4b^2v_0^2 \cos^2\alpha_0} = 0. \tag{1}$$

Substituting the above data in (1), we have

$$(\tan\alpha_0 + \frac{g}{2(0.0002282)(160)^2 \cos^2\alpha_0}) \cdot (300) + \frac{g(1-e^{2(0.0002282)(300)})}{4(0.0002282)^2(160)^2 \cos^2\alpha_0} = 0 \tag{2}$$

Hence, we have

$$\tan\alpha_0 - \frac{0.06017564}{\cos^2\alpha_0} = 0.$$

We can write

$$\frac{\sin(2\alpha_0)-0.1203513}{2\cos^2\alpha_0} = 0$$

or

$$\sin(2\alpha_0)=0.1203513.$$

Solving the above equation, we find that

$$\alpha_0 = \frac{1}{2}\sin^{-1}(0.1203513) = 3.4562°$$

.

(b) Time of Flight

Substituting in the fifth equation of (4.2.20), we get

$$t = \frac{e^{bx} - 1}{bv_0\cos\alpha_0} = \frac{e^{(0002282)\cdot(300)} - 1}{(0.0002282)(160)\cos(3.4562)} = 1.94s .$$

(c) Terminal Angle

We denote α_T the terminal angle. Employing the third equation of (4.2.15), we find that

$$p = \tan\alpha_0 + \frac{g(1-e^{2bx})}{2bv_0^2\cos^2\alpha_0} = \tan 3.4562° + \frac{g(1-e^{2(0.0002282)(300)})}{2(0.0002282)(160)^2\cos^2(3.4562°)} = -0.063192 .$$

Since $p = \tan\alpha$, then

$$\tan\alpha_T = -0.06318$$

.

Hence, for the terminal angle, we get

$$\alpha_T = -3.615°$$

.

We note that the terminal angle in absolute value is slightly greater than the launching angle.

NOTE: In the following table, there are shown for comparison the results obtained using the PC program ANGLEC.BAS (see Appendix B).

Table 1

	Range	Initial angle	Time	Terminal Angle
Calculation	300	$\alpha_0 = 3.4562$	$t = 1.94 s$	$\alpha_T = -3.615$
Anglec.Bas	300	$\alpha_0 = 3.44922$	$t = 1.94 s$	$\alpha_T = -3.624$

Using the QBasic PC program ANGLEC.BAS

INPUT:
Range = 300, Launching Speed = 160, Ballistics Coefficient =1.88272, x-coordinate of a point on Trajectory = 150

RESULTS:
Launching Angle = 3.449219[Degree], Time of Flight = 1.944 [Sec.]
Terminal Angle =-3.6243 [Degree], Vertex (153.15, 154.22)

For $x = 150$ m, $y = 4.615$ m, Time = 0.962 s, Angle = 0.009685 degrees, Speed = 154.300 m/s

Example 2. (Known: b, v_0, α_0) Find the elements of the trajectory of a projectile of mass 22 kg launched with speed 183 m/s from a 122 mm cannon if the launching angle is 6°. The coefficient of the projectile form is $i = 0.645$. The cannon and the target are at the sea level.

Solution
    We first calculate the ballistics coefficient c and the parameter b, employing (21):

$$c = \frac{id^2}{m} \cdot 1000 = \frac{(0.645) \cdot (0.122)^2}{22} \cdot 1000 = 0.4363718182$$

and

$$b = c \cdot 1.212 \cdot 10^{-4} / g = (0.4363718182) \cdot (1.212 \cdot 10^{-4}) / 9.80665 = 5.39126 \cdot 10^{-6}.$$

(a) **Range** x_T

To find the horizontal distance to the target (the range) located on the x-axis, we substitute $y=0$ in the equation of the projectile trajectory (fourth equation of [4.2.20]). We have

$$(\tan\alpha_0 + \frac{g}{2bv_0^2\cos^2\alpha_0})x + \frac{g(1-e^{2bx})}{4b^2v_0^2\cos^2\alpha_0} = 0.$$

Substituting $c=0.4363718182$, $b=0.1805479190$, and $v_0=183m/s$, we can write

$$(\tan(6°) + \frac{g}{2\cdot(5.39\cdot10^{-6})(183)^2\cos^2(6°)})x + \frac{g(1-e^{2(5.39\cdot10^{-6})\cdot x})}{4(5.39\cdot10^{-6})^2(183°)^2\cos^2(6)} = 0.$$

Solving the obtained equation for x, using numerical methods, a graphing calculator TI-83 Plus, or any other math software, we find that the projectile will hit the ground at a horizontal distance of

$$x_T = 692.54m$$

from the launching point.

(b) The Terminal Angle α_T

Substituting in the third equation of (4.2.14), we have

$$\tan\alpha_T = \tan\alpha_0 + \frac{g(1-e^{2bx})}{2bv_0^2\cos^2\alpha_0} = \tan6° + \frac{g(1-e^{2(5.39\cdot10^{-6})\cdot(692.54)})}{2(5.39\cdot10^{-6})(183)^2\cos^2(6°)} = -0.10770.$$

Hence, solving for α_t, we find that the terminal angle is

$$\alpha_T = \tan^{-1}(-0.10770) = -6.147°.$$

(c) The Terminal Speed

Substituting $x_T = 692.54m$, $b=0.1805479190$, $v_0=183m/s$ in the first equation of (4.2.14), we get the x-component of the projectile velocity at the impact point,

$$v_x = v_0 \cos\alpha_0 \cdot e^{-bx} = (183)\cos(6°)\cdot e^{-(5.39\cdot10^{-3})\cdot(692.54)} = 175.45 m/s \ .$$

Using the second equation of (4.2.14) for the y-component of the velocity, we find

$$v_y = (\tan\alpha_0 + \frac{g(1-e^{2bx})}{2bv_0^2\cos^2\alpha_0})\cdot(v_0\cos\alpha_0\cdot e^{-bx}) = (\tan\alpha_t)\cdot v_x = (175.45)\cdot\tan(-6.147°) = -18.896 m/s$$

For the terminal velocity, we get

$$v_T = (v_x^2 + v_y^2)^{1/2} = [(175.45)^2 + (-18.896)^2]^{1/2} = 176.465 m/s \ .$$

(d) The Time of Flight to the Point of Impact

Substituting in the last equation of (4.2.14), we find that the time of flight to the impact point is

$$t = \frac{e^{bx}-1}{bv_0\cos\alpha_0} = \frac{e^{(5.39\cdot10^{-6})(692.54)}-1}{(5.39\cdot10^{-6})(183)\cos(6°)} = 3.88s \ .$$

(e) The Coordinates of the Trajectory Vertex

The vertex of the trajectory is located at the point where the velocity is tangent to the trajectory, i.e., when the velocity angle is zero. Substituting $\alpha=0$ in the third equation of (4.2.20), we have

$$\tan\alpha_0 + \frac{g(1-e^{2bx})}{2bv_0^2\cos^2\alpha_0} = \tan(6°) + \frac{1-e^{2(5.39\cdot10^{-6})x}}{2(5.39\cdot10^{-6})(183)^2\cos^2(6°)} = 0 \ .$$

Solving for x in the last equation, we find that the abscissa of trajectory vertex is

$$x_m = 348.38m \ .$$

Employing the equation of the trajectory of flight, i.e., the fourth equation of (4.2.14), we find that the y-coordinate of the vertex is

$$y_m = (\tan\alpha_0 + \frac{g}{2bv_0\cos^2\alpha_0})x + \frac{g(1-e^{2bx})}{4b^2v_0^2\cos^2\alpha_0} =$$

$$= (\tan 6° + \frac{g}{2\cdot(5.39\cdot10^{-6})(183)^2\cos^2(6°)})(348.38) - \frac{g(1-e^{2(5.39\cdot10^{-6})(348.38)})}{4(5.39\cdot10^{-6})^2(183)^2\cos^2(6°)} = 18.420m$$

(f) Equation of the Trajectory

Substituting in the fourth equation of (4.2.14), we can write the equation of trajectory in the form

$$y = 2.905044843 \cdot x - 26479.382 \cdot (e^{0.0001057404 \cdot x} - 1).$$

Note: The abscissa of the trajectory vertex ($x_m = 348.38$ meters) is greater than half of the horizontal distance ($x = 692.54$ m).

The obtained results are presented in the following table together with the results obtained employing the QBasic PC program RANGEC.BAS (see appendix C.)

Table 2

	Initial angle	Range	Time	Terminal angle	Terminal speed	x vertex	y vertex
Approximate method	$\alpha_0 = 6°$	692.54	3.88	$\alpha_T = -6.147°$	175.45	348.38	18.42
Rangec.Bas	$\alpha_0 = 6°$	693	3.88	$\alpha_T = -6.152°$	176.46	351	18.42

Note: Employing the RANGEC.BAS PC program

Solution Instruction

Input:
Initial x-coordinate $= 0$, Initial y-coordinate $= 0$, Launching angle $= 6$, Speed $= 183$, Ballistics Coefficient $= 0.43637$
Input: x-coordinate of a point on the trajectory, $x = 500$

Output:
Range = 693 m, Terminal Speed = 176.46, Terminal Angle = - 6.152°,
Time of Flight = 3.88 sec.

Trajectory Vertex (351, 18.42)
For x =500, y =14.88 m, Speed = 177.44 m/s, time = 2.78 sec.,
Angle = - 2.6864°

4.3 Projectile Motion When The Time of Flight Is Short

We will study the flight of a projectile when time of flight of the
projectile from the launching point to a given target is relatively short,
considering that the velocity of the projectile in flight at any moment is
less than 256 m/s.

The flight of the projectile is modeled by the system of equations
(4.1.2),

$$\begin{cases} \dfrac{dv_x}{dt}=-bh(y) \cdot v_x \cdot v \\[2mm] \dfrac{dp}{dt}=-\dfrac{g}{v_x} \\[2mm] \dfrac{dx}{dt}=v_x \\[2mm] \dfrac{dy}{dt}=v_y \end{cases} \qquad (4.3.1)$$

where

$$b=1.212 \cdot 10^{-4} c \quad c=\dfrac{id^2}{m} 1000 ,$$

$$h(y)=(\dfrac{289.08-0.006328y}{289.08})^{4.4} \quad p=v_y/v_x=\tan\alpha . \qquad (4.3.2)$$

The density function $h(y)$ is a function of projectile altitude "y".
To obtain approximate solutions, we assume the following:

- The density of atmospheric air is uniformly distributed with
  altitude.

- The density function $h(y)$ is considered to be a constant that is determined by

$$h(y)=h(\bar{y}),$$

where
$\bar{y}=(2/3)y_{max}$, for the field artillery; $\bar{y}=(1/2)y_{max}$, for the antiaircraft artillery; y_{max} is the maximum height of the trajectory.

- For projectiles launched at relatively narrow launching angles, the density function $h(y)$ can be considered constant and equal to one, i.e., $h(y)\approx1$.

We denote

$$B=b\cdot h(\bar{y}).$$

(4.3.3)

We solve (4.3.1) for the initial conditions

$$x=0,\ y=0,\ v_x=v_0\cos\alpha_0,\ v_y=v_0\sin\alpha_0 \text{ when } t=0.$$

The speed of projectile can be expressed as

$$v=(v_x^2+v_y^2)^{1/2} \qquad \text{or} \qquad v=v_x(1+p^2)^{1/2}$$

(4.3.4)

We integrate (4.3.1) using Taylor series (polynomials). Substituting (4.3.4) in the first equation of (4.3.1), we have

$$\frac{dv_x}{dt}=-Bv_x^2\cdot(1+p^2)^{1/2}.$$

(4.3.5)

Dividing both sides of (4.3.5) by v_x^2, we can write

$$\frac{d(1/v_x)}{dt}=B(1+p^2)^{1/2}.$$

(4.3.6)

If we denote

$$u=1/v_x ,$$
(4.3.7)

the differential equation (4.3.6) can be written as

$$\frac{du}{dt}=B(1+p^2)^{1/2} .$$
(4.3.8)

Substituting (4.3.7) in the second equation of (4.3.`1), we can write

$$\frac{dp}{dt}=-gu .$$
(4.3.9)

Substituting u given by equation (4.3.9) in (4.3.8), we find the second derivative of p

$$\frac{d^2p}{dt^2}=-gB(1+p^2)^{1/2} .$$
(4.3.10)

Differentiating (4.3.10), we obtain the third derivative of p

$$\frac{d^3p}{dt^3}=g^2B\frac{p\cdot u}{(1+p^2)^{1/2}} .$$
(4.3.11)

The function $p=p(t)$ has derivatives of any order in an interval containing $t=0$. So we can approximate p using Maclaurin series (polynomials)

$$p=p(0)+\frac{p'(0)}{1!}t+\frac{p''(0)}{2!}t^2+\frac{p'''(0)}{3!}t^3+... .$$
(4.3.12)

Differentiating (4.3.12) with respect to t, we can write

$$\frac{dp}{dt}=p'(0)+p''(0)t+\frac{p'''(0)}{2}t^2+...) .$$
(4.3.13)

Estimating the first, the second, and the third derivatives of p at $t=0$ by employing (4.3.9), (4.3.10), and (4.3.11), respectively; and considering (4.3.7), we can write (4.3.13) in the form

$$\frac{1}{v_x} = \frac{1}{v_x(0)} + B\frac{v(0)}{v_x(0)}t - gB\frac{v_y(0)}{2 \cdot v_x(0) \cdot v(0)}t^2 + \dots. \qquad (4.3.14)$$

For approximate solution, we can consider that the third term (in absolute value) is much lower than the second one, i.e.

$$B\frac{v(0)}{v_x(0)}t >> B\frac{v_y(0)}{2 \cdot v_x(0) \cdot v(0)}t^2.$$

The above condition is true when the time of flight of projectile to the target satisfies the condition

$$t << \frac{v^2(0)}{2 \cdot v_y(0)} = \frac{v_0}{2 \cdot \sin\alpha_0}. \qquad (4.3.15)$$

Assuming that restriction (4.3.15) is satisfied, we can neglect the third term in (4.3.14) and write approximately that

$$\frac{1}{v_x} = \frac{1}{v_x(0)} + B\frac{v(0)}{v_x(0)}t. \qquad (4.3.16)$$

Hence, we obtain the component of the velocity along the x-axis

$$v_x = \frac{v_x(0)}{1 + b \cdot v(0) \cdot t}.$$

Considering the initial conditions of the projectile flight, the last expression can be written in the form

$$v_x = \frac{v_0 \cdot \cos\alpha_0}{1 + B \cdot v_0 \cdot t}. \qquad (4.3.17)$$

Substituting (4.3.17) in the second equation of system (4.3.1), we have

$$\frac{dp}{dt} = -g\frac{1 + Bv_0 \cdot t}{v_0 \cdot \cos\alpha_0}.$$

Integrating, we get

$$p=\tan\alpha_0-\frac{g}{v_0\cos\alpha_0}\cdot(t+\frac{Bv_0t^2}{2}),\qquad(4.3.18)$$

where $p=\tan\alpha$.

Substituting (4.3.18) in the last equation of (4.3.2), we find that the y-component of the velocity is

$$v_y=[\tan\alpha_0-\frac{g}{v_0\cos\alpha_0}\cdot(t+\frac{Bv_0t^2}{2})]\cdot(\frac{v_0\cos\alpha_0}{1+Bv_0t}).\qquad(4.3.19)$$

Substituting v_X given by (4.3.17) in the third equation of (4.3.1) and integrating, we find

$$x=\frac{\cos\alpha_0}{B}\cdot\ln(1+Bv_0t)\qquad(4.3.20)$$

Substituting v_y given by (4.3.19) in the fourth equation of (4.3.1) and integrating, we have

$$y=(\frac{\sin\alpha_0}{B}+\frac{g}{2B^2v_0^2})\cdot\ln(1+Bv_0t)-\frac{g}{2Bv_0}t-\frac{g}{4}t^2.\qquad(4.3.21)$$

Solving (4.3.20) for t, we find the time of flight as a function of the x-coordinate of the projectile

$$t=\frac{e^{Bx/\cos\alpha_0}-1}{Bv_0}.\qquad(4.3.22)$$

Substituting (4.3.22) into (4.3.21), we obtain the equation of the trajectory of flight

$$y=(\tan\alpha_0+\frac{g}{2Bv_0^2\cos\alpha_0})\cdot x+\frac{g(1-e^{2Bx/\cos\alpha_0})}{4B^2v_0^2}.\qquad(4.3.23)$$

The solution of the system of differential equations (4.3.1) is shown in the following chart.

$$
\begin{bmatrix}
v_x = \dfrac{v_0 \cos\alpha_0}{1 + Bv_0 t} \\[3mm]
v_y = (\tan\alpha_0 - \dfrac{g}{v_0 \cos\alpha_0}(t + \dfrac{Bv_0 t^2}{2})) \cdot (\dfrac{v_0 \cos\alpha_0}{1 + Bv_0 t}) \\[3mm]
\tan\alpha = \tan\alpha_0 - \dfrac{g}{v_0 \cos\alpha_0}(t + \dfrac{Bv_0 t^2}{2}) \\[3mm]
x = \dfrac{\cos\alpha_0 \cdot \ln(1 + Bv_0 \cdot t|}{B} \\[3mm]
y = (\dfrac{\sin\alpha_0}{B} + \dfrac{g}{2B^2 v_0^2}) \cdot \ln(1 + Bv_0 t) - \dfrac{g}{2Bv_0} t - \dfrac{g}{4} t^2
\end{bmatrix}
\qquad (4.3.24)
$$

Note: For narrow launching angles α_0, the trajectory equation (4.3.23) has approximately the same form as the fourth equation of (4.2.18).

The following example demonstrates that the results obtained using (4.3.23) are quite the same as results obtained using trajectory equation of (4.2.18) for relatively narrow angles.

Example 1. (See example 1, section 4.1) Find the launching angle α_0 needed to hit a tank that is in a horizontal distance of 300 m from an 82 mm antitank cannon if a projectile of mass 2 kg is fired with the initial speed 160 m/s. The coefficient of the projectile form is $i=0.56$. Meteorological conditions are normal.

Determine as well the time of flight.

Solution

In example 1, section 4.1, we found

$$
c = \frac{id^2}{m}1000 = \frac{(0.56)(0.082)^2}{2} \cdot 1000 = 1.88272
$$

and

$$b=c\cdot1.212\cdot10^{-4}=(1.883)(1)\cdot1.212\cdot10^{-4}=2.2818566\cdot10^{-4}.$$

Since $h(\bar{y})\approx h(0)=1$, we have

$$B=bh(\bar{y})=b=0.0002282.$$

(a) Launching Angle

Since the initial speed is less than 256 m/s, we can use the trajectory equation of (4.3.23) to find the launching angle α_0. The y-coordinate of the target is zero, $y=0$. Thus, to find the launching angle α_0, we solve the trajectory equation for α_0, when $y=0$,

$$(\tan\alpha_0+\frac{g}{2Bv_0^2\cos\alpha_0})\cdot x+\frac{g(1-e^{2Bx/\cos\alpha_0})}{4B^2v_0^2}=0. \tag{1}$$

Substituting in (1), we have

$$(\tan\alpha_0+\frac{g}{2(0.0002282)(160)^2\cos\alpha_0})\cdot(300)+\frac{g(1-e^{2(0.0002282)(300)/\cos\alpha_0}}{4(0.0002282)^2(160)^2}=0 \tag{2}$$

Substituting

$$\cos\alpha_0=(1+\tan^2\alpha_0)^{-1/2},$$

in (2), and solving the obtained equation for $\tan\alpha_0$ (using a graphing calculator TI-83 Plus), we find that

$$\tan\alpha_0=0.06042. \tag{3}$$

Hence,

$$\alpha_0=3.47°$$

Note: This result is practically equal to the value $\alpha_0=3.46°$ we have found in example 1, section 4.2.

(b) Time of Flight

Substituting $\alpha_0 = 3.47°$ in equation (4.3.22), we get exactly the same value for the time of flight to the target as in example 1, section 4.2,

$$t = \frac{e^{Bx/\cos\alpha_0} - 1}{Bv_0} = \frac{e^{(0.0002282)\cdot(300)/\cos(3.47)} - 1}{(0.0002282)(160)} = 1.944s \,.$$

The following evaluation shows that restriction (4.315) on the time of flight is satisfied,

$$t = 1.944 <<= \frac{160}{2\cdot\sin(3.47)} = 1321.75 \,.$$

Example 2. The antitank cannon of example 1 (antitank cannon 82 mm) fires on a helicopter located at a distance $D=80\,m$ from the cannon. The angle of sight (angle of elevation) is $E=60°$.

Find the projection angle A and the coordinates of aiming point if the direction of helicopter flight is perpendicular to the firing plane (plane xy) in the direction of z-axis when the helicopter speed is 180 km/h. Launching speed of the projectile is 160 m/s. The form coefficient is $i=0.56$. Will the projectile hit the target?

Correct the initial data of shooting in order to hit the moving helicopter.

Metrological conditions are standard. The cannon is at sea level.

Solution

As a first approach, we consider the helicopter stationary (hovering) at the point with coordinates

$$x = D\cos A = (80)\cdot\cos(60) = 40$$

and

$$y = D\cos A = (80)\cdot\sin(60) = 69.282$$

The ballistics soefficient is

$$c=\frac{id^2}{m}\cdot1000=\frac{(0.56)(0.082)^2}{2}\cdot1000=1.88272$$

We have

$$\bar{y}=y(1/2)=69.3(1/2)=34.64m$$

We can write

$$h(\bar{y})=(\frac{289.08-0.006328y}{289.08})^{4.4}=(\frac{289.08-0.006328\cdot(34.64)}{289.08})^{4.4}=0.999667$$

For B, we find

$$B=bh(\bar{y})=1.212\cdot10^{-4}\,c\cdot h(\bar{y})=1.212\cdot10^{-4}(1.883\cdot)(0.999667)=0.00022811$$

Launching angle is

$$\alpha_0=60°+A \tag{1}$$

(a) Launching Angle α_0

To find the launching angle, we solve the equation of projectile trajectory (4.3.23),

$$y=(\tan\alpha_0+\frac{g}{2\,Bv_0^2\cos\alpha_0})\cdot x+\frac{g(1-e^{2B\cdot x/\cos\alpha_0})}{4B^2v_0^2}$$

in the same way as in example (1). Substituting, we can write

$$(\tan\alpha_0+\frac{g}{2(0.00022811)(160)^2\cos\alpha_0})\cdot(40)+\frac{1-e^{2(0.00022811)\cdot(40)/\cos\alpha_0}}{4(0.00022811)^2(160)^2}=69.282$$

If we substitute in the above equation

$$\cos\alpha_0=(1+\tan^2\alpha_0)^{-1/2}$$

and then solve the obtained equation for $\tan\alpha_0$ (using a graphing calculator TI-83 Plus), we find that the launching angle is

$$\alpha_0 = 60.45^0.$$

Substituting in (1) the last value, we find that the projection angle is

$$A = 60.45^\circ - 60^\circ = 0.45^\circ.$$

(b) Time of Flight

Substituting $\alpha_0 = 60.45^\circ$ in equation (4.3.22), we get the time of flight

$$t = \frac{e^{B \cdot x/\cos\alpha_0} - 1}{Bv_0} = \frac{e^{(0.00022811)\cdot(40)/\cos(60.45)} - 1}{(0.00022811)(160)} = 0.512 s.$$

The following evaluation shows that restriction (4.3.15) on the time of flight is satisfied,

$$t = 0.512 << \frac{v_0}{2 \cdot \sin\alpha_0} = \frac{160}{2\sin(60.45)} = 91.96.$$

During the time that the projectile covers the distance to the helicopter, the helicopter has deviated to the right from the plane of flight of the projectile at a distance,

$$d = v(helicopter) \cdot t = (180 \cdot \frac{1000}{3600}) \cdot 0.512 = 25.60 m.$$

The projection direction must be deviated to the right of firing plane xy, with an angle

$$\theta = \tan^{-1}(25.65/80) = 17.78^\circ.$$

The distance of flight of the projectile to reach the new position (where the projectile will hit the helicopter) is

$$D_1=(80^2+25.65^2)^{-1/2}=84m.$$

To hit the helicopter located at 84 m from the cannon, the projection angle should be changed. Repeating again the same calculations for the new distance 84 m, we find

$$x=D_1\cos A=(84)\cdot\cos(60)=42$$

and

$$y=D_1\cos A=(80)\cdot\sin(60)=72.75.$$

Substituting in the trajectory equation, we have

$$(\tan\alpha_0+\frac{g}{2(0.0002813)(160)^2\cos\alpha_0})\cdot(42)+\frac{1-e^{2(0.00022813)\cdot(42)/\cos\alpha_0}}{4(0.00022813)^2(160)^2}=72.75.$$

Solving the above equation, we find

$$\alpha_0=60.475^0.$$

Time of flight is

$$t_2=\frac{e^{B\cdot x/\cos\alpha_0}-1}{Bv_0}=\frac{e^{(0.00022811)\cdot(42)/\cos(60.475)}-1}{(0..00022811)(160)}=0.538s.$$

During $t=0.538s$, the helicopter has deviated,

$$d_2=v(helicopter)\cdot t=(180\frac{1000}{3600})\cdot 0.5328=26.64m.$$

The aiming direction will be deviated from the xy-plane with an angle of

$$\theta_2=\tan^{-1}(26.64/80)=18.42^\circ.$$

The distance of helicopter from the cannon at the time the projectile reaches the distance of 84 m is

$$D_2 = (84^2 + 26.64^2)^{-1/2} = 84.40m$$

It is obvious that the projectile will hit the target, considering the relatively large dimensions of the helicopter. For better accuracy, the calculations should be repeated one more time.

Summary: To hit the helicopter the projection angle must be $A_2 = 60.475° - 60° = 0.475°$ while the direction of fire should be rotated $\theta_2 = 18.42°$ to the right of the firing plane.

4.4 Projectile Trajectory for Low Speeds in Curvilinear Coordinates

For projectiles flying with speeds less than 256 m/s, when shooting is uphill or downhill with relatively narrow projection angles, we can obtain analytical solutions of ballistics problems using the curvilinear coordinates.

Curvilinear Coordinates
Consider the vector differential equation (2.6.5),

$$\frac{d\vec{v}}{dt} = \vec{g} - c \cdot h(y) K_D(v) \frac{\vec{v}}{v}.$$

(4.4.1)

For projectiles flying with speeds less than 256 m/s:

$$K_D(v) = 1.212 \cdot 10^{-4} v^2 \qquad \text{for} \qquad v < 256 m/s.$$

(4.4.2)

We construct a new coordinative system with axes $o\bar{x}$ and $0\bar{y}$, where the first axis has the direction of the initial velocity $\vec{v}_0$ while the second axis is directed on the opposite direction of y-axis (see figure 1 in section 3.4).

Let $\vec{u}$ and $\vec{w}$ be the components of velocity $\vec{v}$ in the new coordinate system. We can write

$$\vec{v} = \vec{u} + \vec{w}.$$

(4.4.3)

Substituting in (4.4.1), we write

$$\frac{d\vec{u}}{dt}-\frac{dw}{dt}\vec{j}=-g\vec{j}-c\cdot h(y)K_D(v)\frac{\vec{u}}{v}+c\cdot h(y)K_D(v)\frac{w}{v}\vec{j} \qquad (4.4.4)$$

From (4.4.4) we obtain the following system of differential equations in curvilinear coordinates $\bar{x}$, $\bar{y}$

$$\frac{du}{dt}=-c\cdot h(y)K_D(v)\frac{u}{v}, \qquad (4.4.5)$$

$$\frac{dw}{dt}=g-c\cdot h(y)K_D(v)\frac{w}{v}, \qquad (4.4.6)$$

$$\frac{d\bar{x}}{dt}=u, \quad \frac{dy}{dt}=w. \qquad (4.4.7)$$

Introducing a new variable $\bar{p}$ defined by

$$\bar{p}=\frac{w}{u}. \qquad (4.4.8)$$

Substituting (4.4.8) in (4.4.6) and employing (4.4.5), we obtain the following simple form of the equation (4.4.6):

$$\frac{dp}{dt}=\frac{g}{u}. \qquad (4.4.9)$$

Thus, we have the following system of differential equations in curvilinear coordinates

$$\left\{ \begin{array}{l} \dfrac{du}{dt}=-c\cdot h(y)K_D(v)\dfrac{u}{v} \\[2mm] \dfrac{dp}{dt}=\dfrac{g}{u} \\[2mm] \dfrac{dx}{dt}=u \\[2mm] \dfrac{d\bar{y}}{dt}=w \end{array} \right. \qquad (4.4.10)$$

Substituting (4.4.2) in (4.4.10), we obtain the differential equations of projectile motion in curvilinear coordinates

$$
\begin{cases}
\dfrac{du}{dt}=-c{\cdot}h(y)(1.212{\cdot}10^{-4}){\cdot}uv \\[2mm]
\dfrac{d\bar{p}}{dt}=\dfrac{g}{u} \\[2mm]
\dfrac{d\bar{x}}{dt}=u \\[2mm]
\dfrac{dy}{dt}=w
\end{cases}
\qquad (4.4.11)
$$

Initial conditions: when $t=0$,

$$\bar{x}(0)=0, \ \bar{y}(0)=0, \ u(0)=v(0)=v_0, \ w(0)=0,$$
$$\bar{p}(0)=\bar{p}_0=0, \ \alpha(0)=\alpha_0. \qquad (4.4.12)$$

The density function $h(y)$ is considered to be a constant and is determined by

$$h(y)=h(\hat{y}), \qquad (4.4.13)$$

where

$\hat{y}=(1/2)y_{max}$, for antiaircraft artillery (downhill or uphill shooting), and

y_{max} is the maximum altitude of the trajectory.

We denote

$$B=1.212{\cdot}10^{-4}c{\cdot}h(\hat{y}). \qquad (4.4.14)$$

Substituting (4.4.14) and

$$v=u\frac{\cos\alpha_0}{\cos\alpha} \qquad (4.4.15)$$

in the first equation of (4.4.11), we have[37]

$$\frac{du}{dt} = -Bu^2 \frac{\cos\alpha_0}{\cos\alpha}. \qquad (4.4.16)$$

For short uphill or downhill distances to the target, we can consider $\vec{v} \approx \vec{u}$, and $v \approx u$, the aiming angle very narrow, i.e., $\alpha \approx \alpha_0$. Equation (4.4.16) can be written in the form

$$\frac{du}{dt} = -Bu^2.$$

Thus, the system (4.4.11) can be written as follows:

$$\begin{cases} \dfrac{du}{dt} = -Bu^2 \\ \dfrac{d\overline{p}}{dt} = -\left(-\dfrac{g}{u}\right) \\ \dfrac{d\overline{x}}{dt} = u \\ \dfrac{d\overline{y}}{dt} = w \end{cases}. \qquad (4.4.17)$$

Expressing the above differential equations through the curvilinear coordinate $\overline{x}$, we have

$$\begin{cases} \dfrac{du}{d\overline{x}} = -Bu \\ \dfrac{d\overline{p}}{dx} = -\left(-\dfrac{g}{u^2}\right) \\ \dfrac{dt}{d\overline{x}} = \dfrac{1}{u} \\ \dfrac{d\overline{y}}{d\overline{x}} = \overline{p} \end{cases}. \qquad (4.4.18)$$

Initial conditions: when $\overline{x} = 0$,

---

[37]   See more details in section 3.3.

$$\bar{y}(0)=0, \ u(0)=v(0)=v_0, \ w(0)=0,$$
$$t(0)=0, \ \alpha(0)=\alpha_0, \ \bar{p}(0)=\bar{p}_0=0. \tag{4.4.19}$$

The system of equations (4.4.17) is similar to the system of equations (4.2.10). Adapting the solution (4.2.20) of the system (4.2.10), we obtain the solution of the differential equations of projectile flight (4.4.17) in the following form:

$$
\left\{
\begin{aligned}
&u=u_0 \cdot e^{-B \cdot \bar{x}} \\[2ex]
&\bar{p}=\bar{p}_0+\frac{g(e^{2B \cdot \bar{x}}-1)}{2Bu_0^2} \\[2ex]
&w=\bar{p} \cdot u=u_0 \left(\bar{p}_0+\frac{g(e^{2B \cdot \bar{x}}-1)}{2Bu_0^2}\right) \cdot e^{-B\bar{x}}, \\[2ex]
&\bar{y}=\bar{p}_0\bar{x}-\frac{g\bar{x}}{2Bu_0^2}+\frac{g(e^{2B \cdot \bar{x}}-1)}{4B^2u_0^2} \\[2ex]
&t=\frac{e^{B \cdot \bar{X}}-1}{Bu_0}
\end{aligned}
\right. \tag{4.4.20}
$$

where $B=1.212 \cdot 10^{-4} c \cdot h(\hat{y})$.

The above outcomes can be applied as well when the angle of sight E is zero. The projection angle in this case is equal to the launching angle, i.e., $A=\alpha_0$.

Elements of Trajectory

Projectile Speed

Geometrically, we can find some relationships for the elements of the projectile trajectory.

From the triangle of velocities (figure 2, section 3.4):

$$v^2 = u^2 + w^2 - 2uw \cdot \sin\alpha_0 \qquad (4.4.21)$$

Range
   The distance to the target:

$$\overline{D} = \overline{x} \frac{\cos(A+E)}{\cos E} \qquad (4.4.22)$$

Projection Angle
   Applying the sine theorem (figure 2, section 3.4), we find that

$$\sin A = \frac{\overline{y}}{\overline{D}} \cos(E+A). \qquad (4.4.23)$$

Since the projection angle A is narrow, we can consider

$$\sin A \approx A, \quad \cos(E+A) \approx \cos E,$$

and as a result, the aiming angle E in radian can be approximated using the relation

$$E = \frac{\overline{y}}{\overline{D}} \cos(A). \qquad (4.4.24)$$

Terminal Angle
   The angle α_T the projectile velocity forms with the x-axis at the point of impact is

$$\alpha_T = \alpha_0 - p_T \cos\alpha_0. \qquad (4.4.25)$$

Cartesian Coordinates of Projectile
   The Cartesian coordinates of projectile at any moment are

$$x = \overline{x}\cos\alpha_0, \quad y = \overline{x}\sin\alpha_0 - \overline{y}. \qquad (4.4.26)$$

Example 1. Use the above approach to solve the problem of exercise 2 in section 4.2:

The antitank cannon 82 mm fires a projectile of mass 2 kg with initial speed 160 m/s on a hovering helicopter located at a distance $D=80$ m from the cannon.

The coefficient of the projectile form is 0.56. The angle of sight (angle of elevation) is $E=60°$. Find the projection angle A.

Metrological conditions are standard. The cannon is at sea level.

Solution

Supposing as a first approach that the projection angle is $A=0$, we can write

$$c=\frac{id^2}{m}\cdot 1000=\frac{(0.56)(0.082)^2}{2}\cdot 1000=1.88272$$

$$h(\bar{y})=(\frac{289.08-0.006328y}{289.08})^{4.4}=(\frac{289.08-0.006328\cdot(34.64)}{289.08})^{4.4}=0.999667$$

since $\hat{y}=69.3(1/2)=34.64$.

Thus,

$$B=1.212\cdot 10^{-4}\,c\cdot h(\hat{y})=1.212\cdot 10^{-4}(1.88272)(0.99967)=2.2811\cdot 10^{-4}.$$

We employ (4.4.23) to find the projection angle. i.e.

$$\sin A=\frac{\bar{y}}{D}\cos(E+A),$$

where

$$\bar{y}=-\frac{g\bar{x}}{2Bv_0^2}+\frac{g(e^{2b\bar{x}}-1)}{4B^2v_0^2}$$

is the equation of projectile trajectory.

We consider $\bar{x}=\bar{D}=80$. Substituting in the trajectory equation, we find

$$\bar{y}=-\frac{g\bar{x}}{2Bu_0^2}+\frac{g(e^{2B\bar{x}}-1)}{4B^2u_0^2}=-\frac{g(80)}{2\cdot(0.00022811)(160)^2}+\frac{g(e^{2(0.00022811)\cdot(80)}-1)}{4\cdot(0.00022811)^2(160)^2}=1.241m.$$

The projection angle is

$$\sin A = \frac{\bar{y}}{D}\cos(E+A) \approx \frac{1.244}{80}\cos(60) = 0.0077755.$$

Hence,

$$A = \sin^{-1}(0.0077755) = 0.44436°.$$

Improving the Outcomes
The new value of $\bar{x}$ is

$$\bar{x} = \bar{D}\frac{\cos(E)}{\cos(A+E)} = 80\frac{\cos(60°)}{\cos(60.444362°)} = 81.0917m$$

,

$$\bar{y} = \frac{g\bar{x}}{2Bu_0^2} + \frac{g(e^{2B\bar{x}}-1)}{4B^2u_0^2} = \frac{g(81.0917)}{2\cdot(0.00022811)(160)^2} + \frac{g(e^{2(0.00022811)\cdot(81.0917)}-1)}{4\cdot(0.00022811)^2(160)^2} = 1.2752m,$$

$$\sin A = \frac{\bar{y}}{D}\cos(E+A) = \frac{1.2752}{81.0917}\cos(60.444362) = 0.007757.$$

Hence,

$$A = \sin^{-1}(0.007757) = 0.4444°.$$

The last outcome practically is the same as the result we obtained before.
Note: Employing the QBasic.Bas PC program **ANTIARC.BAS**, we find approximately the same result for the launching angle.
Using the QBasic PC program **ANTIARC.BAS**[38]

INPUT:
$x = D\cdot\cos\alpha_0 = 80\cdot\cos(60) = 40$, $y = D\cdot\cos\alpha_0 = 80\cdot\sin(60) = 69.28$
Launching Speed = 160, Ballistics Coefficient = 1.88272

---

[38] Request an electronic copy or a CD with all PC programs to the author at: *gklimi@pace.edu*, or iven24@aol.com.

RESULTS:
Launching Angle = 60.46875[Degree], Time of Flight = 0.512[Sec.]
Terminal Angle =59.55 [Degree], Speed = 152.81

The projection angle is

$$A=\alpha_0-E=60.46875°-60°=0.468575°.$$

4.5 Trajectory of the Projectile Flight for Any Launching Angle. Euler's Method

The differential equations of projectile flight in general are solved numerically using Runge-Kutta's methods, or other similar methods, because they contain the function of resistance and the density function that decreases with the projectile altitude.

In 1753 Leonard Euler solved the differential equations of projectile motion in presence of air resistance considering the drag function proportional to the square of projectile speed, $D(v)=Bv^2$, and the density of air unchanged with the altitude of projectile. Euler's brilliant solution improved fundamentally the accuracy of ballistics tables that were usually calculated using the parabolic theory.

When a projectile is launched with speed less than 256 m/s we can solve the differential equations of flying projectiles in quadratures using Euler's approach that considers constant the density function.

The following approach to the solution of differential equations of projectile flight when the projectile is launched with speed less than 256 m/s is based on the approximation method introduced by the great mathematician Leonard Euler.

Using a graphing calculator it is possible to integrate in quadratures the differential equations of projectile flight avoiding the use of Otto's tables.

Consider the system of differential equations of projectile flight (2.6.20) when the projectile speed is less than 256 m/s,

$$\begin{cases} \dfrac{dv_x}{dp}=\dfrac{c \cdot h(y)}{g}K_D(v)\dfrac{v_x}{(1+p^2)^{1/2}} \\ \dfrac{dt}{dp}=-v_x/g \\ \dfrac{dx}{dp}=-v_x^2/g \\ \dfrac{dy}{dp}=-pv_x^2/g \end{cases} , \qquad (4.5.1)$$

where

$$p=v_y/v_x=\tan\alpha, \quad v=v_x(1+p^2)^{1/2},$$

$$c=\dfrac{id^2}{m}\cdot 1000, \quad h(y)=(\dfrac{289.08-0.006328y}{289.08})^{4.4},$$

and

$$K_D(v)=1.212 \cdot 10^{-4}v^2. \qquad (4.5.2)$$

We assume the following:

- The initial speed of projectile is less than 256 m/s.
- The density of atmospheric air is uniformly distributed with height.
- The density function $h(y)$ is considered to be a constant that is determined by

$$h(y)=h(\bar{y}),$$

where

- $\bar{y}=(2/3)y_{max}$, for field artillery,
- $\bar{y}=(1/2)y_{max}$, for antiaircraft artillery (downhill or uphill shooting), and
- y_{max} is the maximum altitude of the trajectory.

For projectiles launched at relatively narrow angles, the density function $h(y)$ can be considered constant and equal to one, i.e., $h(y) \approx 1$.

Since the launching speed is less than 256 m/s and

$$v = v_x (1+p^2)^{1/2},$$

the Siacci function of resistance (4.5.2) can be written as

$$K_D(v) = 1.212 \cdot 10^{-4} v^2 = 1.212 \cdot 10^{-4} v_x^2 (1+p^2). \tag{4.5.3}$$

Substituting in the first equation of (4.5.1), we have

$$\frac{dv_x}{dp} = \frac{c \cdot h(\bar{y}) \cdot 1.212 \cdot 10^{-4}}{g} v_x^3 (1+p^2)^{1/2}. \tag{4.5.4}$$

Denoting

$$b = c \cdot 1.212 \cdot 10^{-4}/g, \quad B = bh(\bar{y}), \tag{4.5.5}$$

we can write (4.5.4) as

$$\frac{dv_x}{dp} = B v_x^3 (1+p^2)^{1/2}. \tag{4.5.6}$$

Thus, we obtain this system of differential equations

$$\begin{cases} \dfrac{dv_x}{dp} = B \cdot v_x^3 (1+p^2)^{1/2} \\[2mm] \dfrac{dt}{dp} = -v_x/g \\[2mm] \dfrac{dx}{dp} = -v_x^2/g \\[2mm] \dfrac{dy}{dp} = -p v_x^2/g \end{cases}, \tag{4.5.7}$$

which describes the motion of projectile in air. We solve the above system for the following initial conditions:

$t=0$, $x=0$, $y=0$, $v_{0x}=v_0\cos\alpha_0$, $v_{0y}=v_0\sin\alpha_0$, when $p=p_0=\tan\alpha_0$.

Integrating the first differential equation, we can write

$$\frac{1}{v_0^2\cos^2\alpha_0}-\frac{1}{v_x^2}=B[(p(1+p^2)^{1/2}+\ln(p+(1+p^2)^{1/2}))-(p_0(1+p_0^2)^{1/2}+\ln(p_0+(1+p_0^2)^{1/2}))]$$

Hence, for the horizontal component of the projectile velocity, we have

$$v_x^2=\frac{v_0^2\cos^2\alpha_0}{1-Bv_0^2\cos\alpha_0^2[p(1+p^2)^{1/2}+\ln|(+(1+p^2)^{1/2})-p_0(1+p_0^2)^{1/2}-\ln(p_0+(1+p_0^2)^{1/2})]}\cdot\quad(4.5.8)$$

We denote

$$f^2(p)=1-Bv_0^2\cos^2\alpha_0[p\sqrt{1+p^2}+\ln(p+\sqrt{1+p^2})-p_0\sqrt{1+p_0^2}-\ln(p_0+\sqrt{1+p_0^2})]\cdot\quad(4.5.9)$$

Substituting (4.5.9) in (4.5.8), we find the horizontal component of the projectile velocity

$$v_x=v_0\cos\alpha_0\cdot\frac{1}{f^2(p)}\cdot\qquad(4.5.10)$$

Employing the first equation of (4.5.2), we find the y-component of projectile velocity

$$v_y=v_0\cos\alpha_0\cdot\frac{p}{f^2(p)}\cdot\qquad(4.5.11)$$

Substituting (4.5.10) and (4.5.11) respectively in the third and the fourth equation of system (4.5.1), we have

$$\frac{dx}{dp}=-v_0^2\cos^2\alpha_0\frac{1}{g\cdot f^2(p)}\qquad(4.5.12)$$

and

$$\frac{dy}{dp} = -v_0^2 \cos^2 \alpha_0 \cdot \frac{p}{g \cdot f^2(p)} . \tag{4.5.13}$$

Integrating (4.5.12) and (4.5.13) and substituting

$$\cos^2 \alpha_0 = \frac{1}{(1+\tan^2 \alpha_0)} = \frac{1}{1+p_0^2} ,$$

we obtain the coordinates of the projectile as functions of p:

$$x = -\frac{v_0^2}{g(1+p_0^2)} \cdot \int_{p_0}^{p} \frac{1}{f^2(p)} dp , \tag{4.5.14}$$

$$y = -\frac{v_0^2}{g(1+p_0^2)} \cdot \int_{p_0}^{p} \frac{p}{f^2(p)} dp . \tag{4.5.15}$$

Substituting (4.5.10) in the second equation of (4.5.7) and then integrating, we find the time of flight of projectile to the target

$$t = -\frac{v_0}{g(1+p_0^2)^{1/2}} \cdot \int_{p_0}^{p} \frac{1}{f(p)} dp . \tag{4.5.16}$$

The integrals in (4.5.16) can be evaluated using numerical methods or using graphing calculators.

Note: The function $f(p)$ that appears in (4.5.16) is

$$f(p) = \{1 - Bv_0^2 \cos^2 \alpha_0 [p\sqrt{1+p^2} + \ln(p+\sqrt{1+p^2}) - p_0\sqrt{1+p_0^2} - \ln(p_0+\sqrt{1+p_0^2})]\}^{1/2} .$$

In example 1 is illustrated the estimation of projectile flight using a TI-83 Plus graphing calculator. In example 2, we illustrate the use of an Excel worksheet that helps to estimate the trajectory elements employing equations (4.5.1)–(4.5.3).[39]

---

[39]   Leonardo Volpi. *Xnumbers.xla v 5.3*, by Foxes Team, 2007.
        http://digilander.libero.it/foxes. [Web site]

Example 1. A projectile is fired from a 60 mm trench mortar at a speed of 89 m/s, at an angle of 60°.

Determine the point of impact of the projectile on the horizontal plane, time of flight, the velocity of impact, the terminal angle, and the maximum height of the flight.

Projectile mass is 1.31 kg, projectile caliber is 60.7 mm, and the coefficient of projectile form is 0.622. The atmospheric conditions are standard, and the trench mortar and the target are at sea level.

Solution

First we evaluate

$$p_0=\tan60°=1.73205081$$

and

$$c=\frac{id^2}{m}1000=\frac{(0.622)(0.0607)^2}{1.31}\cdot1000=1.74943.$$

Considering as a first approach $h(y)\approx1$, we find

$$B=bh(y)=c\cdot h(y)\cdot1.212\cdot10^{-4}/g=(1.74943)(1)\cdot1.212\cdot10^{-4}/(9.80665)=2.16211\cdot10^{-5}.$$

(a) Maximum Height of Projectile Trajectory

Maximum height of projectile trajectory corresponds to the vertex point on the trajectory where the derivative of y with respect to x is zero, i.e., at the point on the trajectory where the vertical component of the projectile velocity is zero, or

$$p=\tan0°=0.$$

Substituting

$$v_0=89m/s,\quad p_0=\tan60°=1.73205081,\quad B=2.16211\cdot10^{-5},$$
$$p_m=\tan0°=0$$

in the second equation of (4.5.2) and integrating using a TI-83 Plus calculator, we find the y-coordinate of the trajectory vertex

$$y_m = -\frac{(89)^2}{g(1+1.732^2)} \int_{1.732}^{0} \frac{p}{f^2(p)} dp = 280.61m$$

since

$$\int_{1.732}^{0} \frac{p}{f^2(p)} dp = 1.389662$$

Now we can reevaluate the value of B, employing the approximation for the density function

$$h(y) = h(\bar{y}),$$

where

$$\bar{y} = (2/3)y_{max}, \text{ for field artillery.}$$

Substituting the value of maximum height, we find

$$\bar{y} = (2/3)y_{max} = (2/3)(280.61) = 187.07.$$

Substituting, for the density function, we obtain

$$h(\bar{y}) = (\frac{289.08 - 0.006328y}{289.08})^{4.4} = (\frac{289.08 - 0.006328 \cdot (187.073)}{289.08})^{4.4} = 0.9821055.$$

For B, we find

$$B = bh(\bar{y}) = c \cdot h(\bar{y}) \cdot 1.212 \cdot 10^{-4}/g = (1.74943)(0.9821055) \cdot 1.212 \cdot 10^{-4}/(9.81) = 2.123424 \cdot 10^{-5}.$$

Substituting the above value in (4.5.2), we obtain the trajectory vertex

$$y_m = -\frac{(89)^2}{g(1+1.732^2)} \int_{1.732}^{0} \frac{p}{f^2(p)} dp = 281 \ m,$$

where

$$f^2(p) = 1.200744 - 0.041987 \cdot p(1+p^2)^{1/2} - 0.041987 \cdot \ln(p + (1+p^2)^{1/2}).$$

Using the first equation (4.5.2), we find that the abscissa of the trajectory vertex is

$$x_m = -\frac{v_0^2}{g(1+p_0^2)} \cdot \int_{p_0}^p \frac{1}{f^2(p)} dp = -\frac{v_0^2}{g(1+p_0^2)} \int_{1.732}^0 \frac{1}{f^2(p)} dp = 315 \ m.$$

(b) Horizontal Distance x_T

At the point of impact the y-coordinate is zero. Substituting zero for y in the second equation of (4.5.2), we obtain the following integral equation:

$$-\frac{v_0^2}{g(1+p_0^2)} \cdot \int_{1.732}^p \frac{p}{f^2(p)} dp = 0$$

or

$$\int_{1.732}^p \frac{p}{f^2(p)} dp = 0, \tag{1}$$

The above integral, equation (1), can be solved considering that at the point of impact on the horizontal plane, the terminal angle is negative and has an absolute value greater than the value of the initial angle $60°$.

Using the trial and error procedure, we find

$$p_T = \tan(\alpha_T) = -1.77486$$

Hence, for the angle of impact we have

$$\alpha_T = -62.602°.$$

Using the first equation of (4.5.2), we obtain the horizontal range

$$x = -\frac{v_0^2}{g(1+p_0^2)} \cdot \int_{1.732}^{-1.7749} \frac{1}{f^2(p)} dp = 615.14 \ m.$$

(c) Time of Flight

To find the time of flight of projectile to the target, we substitute

$$p_T=-1.77486, \quad v_0=89, \quad p_0=\tan 60°=1.732$$

in (4.5.16); then integrating using a graphing calculator, we find the time of flight

$$t_T=-\frac{(89)}{g(1+(-1.732)^2)^{1/2}}\cdot\int_{1.732}^{-1.77486}\frac{1}{f(p)}dp=15.14\text{sec}.$$

The estimated elements of the trajectory are presented in the following table together with the results obtained from the differential equations of flight using the PC software RangeC.Bas.

Table 1

	Launching angle	Range	Time of flight	Angle of impact	Abscissa of vertex	Ordinate of vertex
Example	$a_0=60°$	$x=615$	$t=15$	$a_T=-62.602°$	$x_m=315$	$y_m=281$
RangeC	$a_0=60°$	$x=615$	$t=15.14$	$a_T=-62.614°$	$x_m=315.20$	$y_m=281$

Example 2. A projectile of mass 22 kg is fired from a 122 mm trench mortar at a speed of 183 m/s at an angle of $a_0=40°$.

Determine the point of impact of the projectile on the horizontal plane, time of flight, the velocity of impact, the terminal angle, and the maximum height of the flight. The coefficient of projectile form is $i=0.645$. The atmospheric conditions are standard, and the trench mortar and the target are at sea level.

Solution

First, we evaluate

$$p_0=\tan 40°=0.8390996$$

and

$$c=\frac{id^2}{m}1000=\frac{(0.645)(0.122)^2}{22}\cdot1000=0.4363718182.$$

Considering as a first approach $h(y) \approx 1$, we find

$$B=bh(y)=c \cdot h(y) \cdot 1.212 \cdot 10^{-4}/g=(0.43637182)(1) \cdot 1.212 \cdot 10^{-4}/(9.80665)=5.39308429610^{-5}.$$

(a) Estimation of the Maximum Altitude of Projectile Trajectory

The maximum altitude of projectile trajectory corresponds to the vertex point on the trajectory where the derivative of y with respect to x is zero, i.e., at the point on the trajectory where the vertical component of the projectile velocity is zero, or

$$p = \tan 0° = 0.$$

Substituting

$$v_0 = 183 m/s, \quad p_0 = \tan 40° = 0.83909963, \quad B = 5.39308429610^{-5},$$
$$p_m = \tan 0° = 0$$

in the second equation of (4.5.2) for the y-coordinate of the trajectory vertex, we can write

$$y_m = -\frac{(183)^2}{g(1+0.8391^2)} \int_{0.8391}^{0} pf^{-2}(p)dp, \tag{1}$$

where

$$f^{-2}(p)=[1.196951-0.10598606p\sqrt{1+p^2}-0.10598606 \cdot \ln(p+\sqrt{1+p^2})]^{-1}.$$

Using a TI-83 Plus graphing calculator, we find that

$$\int_{-0.8.391}^{0} pf^{-2}(p)dp=-0.3294764.$$

Substituting in (1), we find the y-coordinate of the trajectory vertex

$$y_m = -\frac{(183)^2}{g(1+0.8391^2)} \int_{0.8391}^{0} pf^{-2}(p)dp=660.26m.$$

Now we can reevaluate the value of b, employing the approximation for the density function

$$h(y)=h(\bar{y}),$$

where

$$\bar{y}=(2/3)y_{max}, \text{ for field artillery.}$$

Substituting the value of maximum height, we find

$$\bar{y}=(2/3)y_{max}=(2/3)(660.172)=440.17m.$$

Substituting, for the density function, we obtain

$$h(\bar{y})=(\frac{289.06-0.006328y}{289.06})^{4.4}=(\frac{289.06-0.006328\cdot(440.17)}{289.06})^{4.4}=0.95829.$$

For B, we find

$$B=bh(\bar{y})=c\cdot h(\bar{y})\cdot 1.212\cdot 10^{-4}/g=(1.74943)(0.95829)\cdot 1.212\cdot 10^{-4}/(9.80665)=5.1681579\cdot 10^{-5}.$$

For the above value of B, we have

$$f^{-2}(p)=[1.1887366-0.10156454p\sqrt{1+p^2}-0.1015654\cdot\ln(p+\sqrt{1+p^2})]^{-1}$$

and

$$y_m=-\frac{(183)^2}{g(1+0.8391^2)}\int_{0.8391}^{0}pf^{-2}(p)dp=661.98m.$$

Using the first equation of (4.5.2), we find that the abscissa of the trajectory vertex is

$$x_m=-\frac{v_0^2}{g(1+p_0^2)}\cdot\int_{p_0}^{p}f^{-2}(p)dp=-\frac{(183)^2}{g(1+0.8391^2)}\int_{0.8391}^{0}f^{-2}(p)dp=1534.05m.$$

(b) Horizontal Distance x_T

At the point of impact, the y-coordinate is zero. Substituting zero for y in the second equation of (4.5.2), we obtain the following integral equation:

$$-\frac{v_0^2}{g(1+p_0^2)} \cdot \int_{0.8391}^{p} p \cdot f^{-2}(p)dp=0, \text{ or } \int_{0.8391}^{p} p \cdot f^{-2}(p)dp=0, \tag{2}$$

where

$$f^{-2}(p)=[1.1887366-0.10156454 p\sqrt{1+p^2}-0.1015654 \cdot \ln(p+\sqrt{1+p^2})]^{-1}.$$

The above integral is solved using the Excel worksheet with add-ins of Leonardo Volpi and the Fox team "Xnumbers.xla v 5.3" 2007, *http://digilander.libero.it/foxes*.

To simplify the calculations, we are estimating the integral (2) for the values less than $-40°$ since the angle of impact is in absolute value somewhat greater than the launching angle.

The following Excel worksheet (table 2) shows the method used to find the angle of impact. In column I, there are displayed the values of guessed impact angle from $-40°$ to $-44°$; in column II, are presented the values of the integral (2) calculated using the following Excel formula:

= integr_nc("x*(1.1887366361-0.1015654254*x*(1+x^2)^0.5 - 0.1015654254* ln(x+(1+x^2)^0.5))^-1", TAN(40 π /180), TAN(A1* π /180), 1000)

In cell A1 is located the value of angle $\alpha=-40°$ ". The formula can be generated automatically in column II to estimate the integral for all other angles.

The second column of the table shows that the value of angle α_T for which the equation (2) is satisfied is between $\alpha=-43.25°$ and $\alpha=-43.50°$. In column III, are shown some intermediate values between $\alpha=-43.25°$ and $\alpha=-43.50°$; while in column IV, are estimated the corresponding values of the integral that is on the left side of (2).

In the same way, there is constructed column III and the corresponding column IV of the values of integral (2).

Column IV shows that the value of impact angle α_T that satisfies equation (2) is between $\alpha=-43.075°$ and $\alpha=-43.10°$.

To get more accurate results, are constructed columns V and VI. In column VI, we see that the integral in (4.5.20) is approximately zero when the impact angle is $\alpha_T=-43.08°$.

Table 2. Estimation of impact angle α_T

$a_0=40$,		$p_0=\tan\alpha_0=\tan(40)=0.8391$,		$p_0=\tan\alpha$	
I	II	III	IV	V	VI
Impact angle α	$\int_{0.8391}^{p} p\,f^{-2}(p)dp$	Impact angle α	$\int_{0.8391}^{p} p\,f^{-2}(p)dp$	Impact angle α	$\int_{0.8391}^{p} p\,f^{-2}(p)dp$
-40	-0.06124	-43.000	-0.00172	-43.075	-0.0000906
-40.25	-0.05668	-43.025	-0.00118	-43.076	-0.0000689
-40.5	-0.05205	-43.050	-0.00063	-43.077	-0.0000471
-40.75	-0.04735	**-43.075**	**-0.00009**	-43.078	-0.0000253
-41	-0.04258	**-43.100**	**0.00045**	**-43.079**	**-0.0000035**
-41.25	-0.03774	-43.125	0.00100	-43.080	0.0000183
-41.5	-0.03282	-43.150	0.00155	-43.081	0.0000401
-41.75	-0.02783	-43.175	0.00210	-43.082	0.0000619
-42	-0.02277	-43.200	0.00264		
-42.25	-0.01762	-43.225	0.00319		
-42.5	-0.01240	-43.250	0.00374		
-42.75	-0.00710				
-43	**-0.00172**				
-43.25	**0.00374**				
-43.5	0.00929				
-43.75	0.01493				
-44	0.02065				

Substituting the above value $\alpha_T=-43.08°$ in the first equation of (4.5.20), we obtain the horizontal range

$$x_T=-\frac{v_0^2}{g(1+p_0^2)}\cdot\int_{0.8391}^{-0.935096} f^{-2}(p)dp=-\frac{183^2}{g(1+0.8391^2)}(-1.4919588)=2989.83m \; .$$

(c) Time of Flight

To find the time of flight of projectile to the target, we substitute

$$p_T=-0.935096 \; , \; v_0=183 \; , \; p_0=\tan 40°=0.8391$$

in (4.5.3), and then integrating using a graphing calculator, we find the time of flight

$$t_T=-\frac{(183)}{g(1+(-0.8391)^2)^{1/2}}\cdot\int_{0.8391}^{-1}f^{-1}(p)dp=23.23\text{sec}.$$

(d) Speed of Impact

Substituting

$$p_T=-0.935096 \; , \; v_0=183$$

and

$$f(p_T)=[1.1887366-0.1015654\cdot p_T(1+p_T^2)^{1/2}-0.1015654\cdot\ln(p_T+(1+p_T^2)^{1/2})]^{1/2}$$
$$=(1.403543)^{1/2}=1.1847122$$

in both equations of (1), we find respectively the x-component and the y-component of the impact velocity:

$$v_{x_T}=v_0\cos\alpha_0\cdot\frac{1}{f(p_T)}=183\cos(40)\frac{1}{1.1847122}=118.33m/s \; ,$$

$$v_y=v_0\cos\alpha_0\cdot\frac{p}{f(p)}=pv_x=(-0.935096)\cdot(118.33)=-110.65 \; .$$

The impact speed is

$$v_T=(v_{xT}^2+v_{yT}^2)^{1/2}=(118.33)^2+(110.64)^2=162m/s \; .$$

The estimated elements of the trajectory are presented in the following table.

In the same table are displayed the results obtained using the QBasic PC program RangeC.Bas, presented in the second row of table 3.

Table 3

Launch angle	Range in meters	Time of flight	Angle of impact	Impact Speed	Abscissa of vertex	Ordinate of vertex
α_0=40°	x_T=2990	t_T=23.23	α_T=−43.079°	v_T=162	x_m=1534	y_m=662
α_0=40°	x_T=2988	t_T=23.22	α_T=−43.055°	v_T=162	x_m=1534	y_m=661

5

The Siacci Method

Introduction

In this chapter, we will introduce the exceptional method of the Italian mathematician Colonel Francisco Siacci (1839–1907) that has been employed successfully since the 1880s to solve complicated ballistics problems.

Particularly, the Siacci method, as it is presented in this chapter, brings back to life the Siacci technique that nowadays is somewhat abandoned because of the large amount of pages filled with tabulated Siacci functions necessary to estimate the elements of projectile trajectory.

The modern approximate Siacci function of resistance, as it is presented in (2.3.2),

$$K_D(v)=\begin{cases} 1.212 \cdot 10^{-4} v^2 & for \quad v \leq 256 m/s \\ (v-240)/3 & for \quad v > 256 m/s \end{cases}, \tag{1}$$

allows us to solve analytically and relatively easy a large variety of ballistics problems with a very good accuracy and so avoid the huge amount of pages of Siacci's tabulated functions.

The methods we have used in chapter 3 and chapter 4 to solve the system of differential equations of projectile flight give approximate solutions for some particular cases, such as for the flight of projectiles for relatively narrow launching angles or when the time of flight is relatively short.

The Siacci method, as it was introduced initially more than a century ago (and as it is used nowadays), gives very good solutions for ballistics problems when the launching angle is not greater than 10–15°.[40]

The accuracy of the Siacci method is better for the elements of the trajectory at the point of impact (terminal point). The Siacci method gives less accurate results for intermediate points of the projectile trajectory.

Anyway, the Siacci method can be used to obtain approximate solutions of differential equations of projectile flight for launching angles greater than 15°, for uphill and downhill fire as well, especially for low-speed projectiles.

5.1 The Siacci Method. The Pseudo-speed

Consider the system of differential equations (2.7.20), i. e.,

$$
\begin{cases}
\dfrac{dv_x}{dp} = \dfrac{c \cdot h(y)}{g} K_D(v) \dfrac{v_x}{(1+p^2)^{1/2}} \\[2mm]
\dfrac{dt}{dp} = -v_x/g \\[2mm]
\dfrac{dx}{dp} = -v_x^2/g \\[2mm]
\dfrac{dy}{dp} = -pv_x^2/g
\end{cases}
\qquad (5.1.1)
$$

---

[40]   Herrmann, E. E. *Exterior Ballistics*, p. 48, p. 53, U.S. Naval Institute, The College Press, 1935.

where

$$p = v_y / v_x = \tan\alpha$$

and

$$K_D(v) = \begin{cases} 1.212 \cdot 10^{-4} v^2 & for \quad v \leq 256 \\ (v-240)/3 & for \quad v > 256 \end{cases}.$$ (5.1.2)

Initial conditions are

$x=0$, $y=0$, $v_x = v_0 \cos\alpha_0$, $v_Y = v_0 \sin\alpha_0$, $t=0$, when $p = p_0 = \tan\alpha_0$.

Siacci introduced the "pseudo-speed" u defined by the equation

$$u = v \frac{\cos\alpha}{\cos\alpha_0}.$$ (5.1.3)

Siacci considered that

$$h(y)K_D(v) = h(y)K_D(\frac{u \cdot \cos\alpha_0}{\cos\alpha}) \approx \beta \cdot K_D(u) \cdot \frac{\cos\alpha_0}{\cos\alpha}.$$

(5.1.4)

In the above approximation Siacci considered that

$$K_D(u \cdot \cos\alpha_0 / \cos\alpha) \approx K_D(u) \cdot \cos\alpha_0 / \cos\alpha,$$

which is true for departure angles till around 15 degrees and, at the same time, he introduced a constant factor β to compensate for the errors related with the variation of the density function with the altitude.

The estimation of the value of β, as it was introduced by Siacci and his followers, is a very difficult and time-consuming problem.

To compensate for the errors introduced in trajectory of flight ignoring the change of atmospheric density with the y-coordinate, we consider $h(y)$ as a constant that is determined as follows:[41]

$$h(y)=h(\bar{y}),$$
(5.1.5)

where

$$h(\bar{y})=(\frac{289.08-0.006328\cdot\bar{y}}{289.08})^{4.4}$$
(5.1.6)

and

- $\bar{y}=(2/3)y_{max}$, for field artillery, target on the ground,
- $\bar{y}=(1/2)y_{max}$, for antiaircraft artillery (uphill); target is on the ascending point of the trajectory; y_{max} is the maximum height of the projectile trajectory.

The value of the Siacci factor β can be approximated by[42]

$$\beta=\frac{h(\bar{y})}{\sqrt{\cos\alpha_0}} \quad \text{and} \quad \beta=h(\bar{y})/\cos\alpha_0,$$
(5.1.7)

respectively, for the field artillery and the antiaircraft shooting. Substituting (5.1.4) in the first equation of (5.1.1), we can write that equation in the form

$$\frac{dv_x}{dp} = \frac{c}{g}\beta\cdot K_D(u)\cdot\frac{\cos\alpha_0}{\cos\alpha}v_x\cos\alpha,$$

---

[41] Herrmann, E. E. *Exterior Ballistics*, p. 48, p. 50. U.S. Naval Institute, The College Press, 1935.
Okunev, B. H. *Fundamentals of Ballistics*, Vol.1, Book 2, p. 186. Moskva, 1943.
[42] McCoy, R. L. *Modern Exterior Ballistics*, p. 100. Schiffer Publishing ltd, 1999.
Herrmann, Ernest E. *Exterior Ballistics*, p. 52-53. U.S. Naval Institute, The College Press, 1935.
Shapiro, J. M. *Vneshnaja Balistika*, p. 94, Oborongiz 50'.

since $(1+p)^{-1/2} = \cos\alpha$.

From the above equation, we can write:

$$\frac{dv_x}{dp} = \frac{c}{g}\beta \cdot K_D(u) \cdot v_x \cos\alpha_0 \qquad (5.1.8)$$

From (5.1.3), we can write for v_x

$$v_x = u \cdot \cos\alpha_0 . \qquad (5.1.9)$$

Substituting (5.1.9) in (5.1.8), we have

$$\frac{du}{dp} = \frac{c}{g}\beta K_D(u) \cdot u \cdot \cos^2\alpha_0 . \qquad (5.1.10)$$

Then, substituting (5.1.9) in the three last differential equations of (5.1.1) and replacing the first equation of (5.1.1) with (5.1.10), system (5.1.1) can be written in the following Siacci form:

$$\left\{ \begin{array}{l} \dfrac{du}{dp}=\beta\dfrac{c}{g}K_D(u)u\cos^2\alpha_0 \\[2mm] \dfrac{dt}{dp}=-\dfrac{u}{g}\cos\alpha_0 \\[2mm] \dfrac{dx}{dp}=-\dfrac{u^2}{g}\cos^2\alpha_0 \\[2mm] \dfrac{dy}{dp}=-p\dfrac{u^2}{g}\cos^2\alpha_0 \end{array} \right. , \qquad (5.1.11)$$

where

$$K_D(u)=\begin{cases} 1.212\cdot10^{-4}u^2 & for \quad u\leq256m/s \\ (u-240)/3 & for \quad u>256m/s \end{cases}, \qquad (5.1.12)$$

$$c=\frac{id^2}{m}1000, \quad u=v\frac{\cos\alpha}{\cos\alpha_0}, \quad p=\tan\alpha, \tag{5.1.13}$$

and

$$\beta=h(\bar{y})/\cos\alpha_0, \tag{5.1.14}$$

for antiaircraft fire, or

$$\beta=\frac{h(\bar{y})}{\sqrt{\cos\alpha_0}}, \tag{5.1.15}$$

for field-artillery fire.

Dividing each of the last three differential equations of (5.1.11) with the first one and then flipping the first equation of (5.1.11), we obtain the system of differential equations of projectile flight where the variable of integration is the pseudo-speed u

$$\begin{cases} \dfrac{dp}{du}=-\dfrac{g}{\beta \cdot c K_D(u)\cdot u\cdot\cos^2\alpha_0} \\[2mm] \dfrac{dt}{du}=-\dfrac{1}{\beta \cdot c K_D(u)\cdot\cos\alpha_0} \\[2mm] \dfrac{dx}{du}=-\dfrac{u}{\beta \cdot c K_D(u)} \\[2mm] \dfrac{dy}{du}=-p\dfrac{u}{\beta \cdot c K_D(u)} \end{cases} \tag{5.1.16}$$

The x-coordinate, x_m, of the point of the maximum altitude of the trajectory can be found using the approximate empirical formula introduced by Vallier[43]

$$x_m=(0.5+0.0001\cdot v_0)\cdot x_T, \tag{5.1.17}$$

where x_T is the horizontal range of fire.

---

[43]   Okunev, B. H. *Fundamentals of Ballistics*, Vol.1, Book 2, p. 186. Moskva, 1943.

The y-coordinate of the maximum altitude y_m of the trajectory can be estimated using

$$y_m=0.25x_T(0.5+0.0001 \cdot v_0) \cdot \tan(\alpha_0)+(0.5-0.0001 \cdot v_0)|\tan\alpha_T|. \quad (5.1.18)$$

Example 1. The artillery cannon projectile caliber $d=0.107$ m, mass $m=17.04$ kg, fired with initial speed $v_0=562$ m/s, launched with an angle of $10.39°$ hits the ground at the horizontal distance of $x=6462$ m (see example 1, section 5.3). The impact angle is $-15.96°$. Find the maximum altitude and the abscissa of the point where the altitude is maximum.

Solution

Substituting in (5.1.18) for the maximum altitude, we obtain

$$y_m=0.25(6462)[(0.5+0.0001 \cdot (562) \cdot \tan(10.39°)+(0.5-0.0001 \cdot 562_0) |\tan(-15.96°)]=369.80m.$$

For the abscissa of the maximum altitude, employing (5.1.17), we find

$$x_m=(0.5+0.0001 \cdot v_0) \cdot x_T=(0.5+0.0001 \cdot 562) \cdot (6462)=3594m.$$

Note: The results obtained in example 1 of section 5.3 are respectively $y_{max}=371.m$ and $x_m=3564$. The results of numerical integration are respectively $x_m=3591$ and $y_{max}=374.m$.

5.2 The Siacci Method for High Speeds

We apply the Siacci method of solution of system (5.1.11) when the projectile speed during the flight to the target is $v \geq 256 m/s$. For such speeds, the Siacci function of resistance is

$$K_D(u)=(u-240)/3. \quad (5.2.1)$$

Substituting (5.2.1) in the first equation of (5.1.16), we obtain the system of differential equations

$$\begin{cases} \dfrac{dp}{du} = \dfrac{1}{B(u-240)\cdot u\cdot \cos^2\alpha_0} \\[2mm] \dfrac{dt}{du} = -\dfrac{1}{gB(u-240)\cdot \cos\alpha_0} \\[2mm] \dfrac{dx}{du} = -\dfrac{u}{gB(u-240)} \\[2mm] \dfrac{dy}{du} = -p\dfrac{u}{gB(u-240)} \end{cases} \qquad (5.2.2)$$

where

$$b=c/(3g),\ \ \beta=h(\bar{y})/\cos\alpha_0,\ \text{or } \beta=\frac{h(\bar{y})}{\sqrt{\cos\alpha_0}}),\ \ u=v\frac{\cos\alpha}{\cos\alpha_0},\ \ B=\beta\cdot b\,.(5.2.3)$$

Siacci Functions

From the first differential equation of system (5.2.2), we can write

$$dp=\frac{1}{B\cos^2\alpha_0}\cdot\frac{du}{u(u-240)}\,. \qquad (5.2.4)$$

Integrating, we obtain $p=\tan\alpha$ as a function of the pseudo-velocity u,

$$p=p_0+\frac{1}{240\,B\cos^2\alpha_0}[\ln(\frac{u-240}{u})-\ln(\frac{v_0-240}{v_0})]\,. \qquad (5.2.5)$$

The above equation can be written in the form

$$p=p_0+\frac{1}{240\,B\cos^2\alpha_0}[J(u)-J(v_0)]\,, \qquad (5.2.6)$$

where

$$J(u)=\ln(\frac{u-240}{u})\,,\quad J(v_0)=\ln(\frac{v_0-240}{v_0})\,. \qquad (5.2.7)$$

Integrating the second and the third differential equations of (5.2.2), we find that the time of flight t and the x-coordinate of projectile are respectively

$$t=-\frac{1}{Bg\cos\alpha_0}[\ln(u-240)-\ln(v_0-240)] \qquad (5.2.8)$$

and

$$x=-\frac{1}{Bg}[(u+240\cdot\ln(u-240))-(v_0+240\cdot\ln(v_0-240)]. \qquad (5.2.9)$$

The time of flight can be written as

$$t=-\frac{1}{Bg\cos\alpha_0}[T(u)-T(v_0)], \qquad (5.2.10)$$

where

$$T(u)=\ln(u-240), \quad \text{and} \quad T(v_0)=\ln(v_0-240). \qquad (5.2.11)$$

We write the x-coordinate of the projectile given in (5.2.9) in the following compact form:

$$x=-\frac{1}{Bg}[D(u)-D(v_0)], \qquad (5.2.12)$$

where

$$D(u)=[u+240\cdot\ln(u-240)], \quad \text{and} \quad D(v_0)=[v_0+240\cdot\ln(v_0-240)]. \qquad (5.2.13)$$

From the third and the fourth differential equations of (5.2.2), we obtain

$$dy=pdx, \qquad (5.2.14)$$

while from the third one, we have

$$dx=-\frac{1}{Bg}\cdot\frac{u}{(u-240)}du \qquad (5.2.15)$$

Substituting (5.2.6) in (5.2.14), we can write

$$dy=\{p_0+\frac{1}{240B\cos^2\alpha_0}[J(u)-J(v_0)]\}dx. \qquad (5.2.16)$$

Integrating, we have

$$y=p_0x+\frac{1}{240B\cos^2\alpha_0}[\int_0^x J(u)dx-J(v_0)x]. \qquad (5.2.17)$$

Substituting (5.2.15) in (5.2.17) yields

$$y=p_0x+\frac{1}{240B\cos^2\alpha_0}\cdot[\frac{1}{-Bg}\int_{v_0}^u\frac{u}{u-240}J(u)du-J(v_0)x]. \qquad (5.2.18)$$

We can write (5.2.18) as follows:

$$y=p_0x+\frac{x}{240B\cos^2\alpha_0}[\frac{1}{-Bgx}\int_0^u\frac{u}{u-240}J(u)du-J(v_0)x] \qquad (5.2.19)$$

or

$$y=p_0x+\frac{x}{240B\cos^2\alpha_0}\cdot\{\frac{1}{-Bgx}[A(u)-A(v_0)]-J(v_0)\}, \qquad (5.2.20)$$

where

$$A(u)-A(v_0)=\int_{v_0}^u\frac{u}{u-240}J(u)du. \qquad (5.2.21)$$

Considering (5.2.7), the above integral can be written in the form

$$A(u)-A(v_0)=\int_{v_0}^u\frac{u}{u-240}\ln\frac{u-240}{u}du. \qquad (5.2.22)$$

Employing equation (5.2.12), we find that the quantity $(-Bgx)$, on the right side of (5.2.20), can be expressed as follows:

$$-Bgx=[D(u)-D(v_0)]. \qquad (5.2.23)$$

Substituting (5.2.23) in equation (5.2.20), we write equation (5.2.20) in the following form:

$$y=p_0 x+\frac{x}{240 B \cos^2 \alpha_0}\cdot[\frac{A(u)-A(v_0)}{D(u)-D(v_0)}-J(v_0)]. \qquad (5.2.24)$$

Equation (5.2.24) is the equation of the projectile trajectory.

The Siacci function of resistance we are using in this material allows us to solve different ballistics problems without employing the tabulated functions of Siacci. Employing (5.2.6), (5.2.10), (5.2.12), (5.2.17), (5.2.18), or (5.2.24), we are able to find the elements of projectile trajectory and solve the main problems of exterior ballistics.

The functions $J(u)$, $T(u)$, $D(u)$, and $A(u)$, defined respectively by (5.2.7), (5.2.11), (5.2.13), and (5.2.22), are called Siacci functions. They are tabulated in different ballistics textbooks.

In appendix A, are tabulated the Siacci functions based on the approximate Siacci function of resistance (5.2.1).

The Trajectory Elements

The equations obtained above using the Siacci method express the elements of trajectory of a projectile as functions of the dimensionless pseudo-speed u. We can find elements of the projectile trajectory.

Launching Angle

At the point of impact, the y-coordinate is zero. Substituting $y=0$ into the left side of the trajectory equation (5.2.24), we can write

$$p_0 x+\frac{x}{240 B \cos^2 \alpha_0}\cdot[\frac{A(u)-A(v_0)}{D(u)-D(v_0)}-J(v_0)]=0.$$

For field artillery fire

$$p_0 = \tan\alpha_0, \quad B = \beta \cdot b, \quad \beta = \frac{h(\bar{y})}{\sqrt{\cos\alpha_0}},$$

we obtain the launching angle from the following equation:

$$\sin(2\alpha_0) = -\frac{1}{120B} \left[\frac{A(u) - A(v_0)}{D(u) - D(v_0)} - J(v_0) \right]. \tag{5.2.26}$$

Range, Time of Flight to the Target, and Angle of Impact

Using formulas (5.2.12), (5.2.10), and (5.2.6), i.e.,

$$x = -\frac{1}{Bg} [D(u) - D(v_0)], \tag{5.2.27}$$

$$t = -\frac{1}{Bg\cos\alpha_0} [T(u) - T(v_0)], \tag{5.2.28}$$

$$p = p_0 + \frac{1}{240B\cos^2\alpha_0} [J(u) - J(v_0)], \tag{5.2.29}$$

or their equivalent ones (5.2.9), (5.2.8), and (5.2.5), i.e.,

$$x = -\frac{1}{Bg} [(u + 240\cdot\ln(u - 240)) - (v_0 + 240\cdot\ln(v_0 - 240)], \tag{5.2.30}$$

$$t = -\frac{1}{Bg\cos\alpha_0} [\ln(u - 240) - \ln(v_0 - 240)], \tag{5.2.31}$$

$$p = p_0 + \frac{1}{240B\cos^2\alpha_0} [\ln(\frac{u - 240}{u}) - \ln(\frac{v_0 - 240}{v_0})], \tag{5.2.32}$$

we can find all trajectory elements of the point of impact of the projectile to the target.

The Elements of the Trajectory Vertex

The elements at the trajectory vertex can be found using the fact that the angle of flight at that point is zero.

Thus, substituting $p=0$ in (5.2.29) and solving the obtained equation

$$p_0+\frac{1}{240B\cos^2\alpha_0}[J(u)-J(v_0)]=0 \tag{5.2.33}$$

for the pseudo-speed u and then substituting the obtained value of u in (5.2.10), (5.2.12), and (5.2.24), we are able to find all the other elements of projectile trajectory.

Example 1. An artillery cannon projectile of mass $m=17.04$ kg is fired with initial speed $v_0=562$ m/s. The projectile form coefficient is $i=0.525$ and its caliber is $d=0.107$ m. Determine the launching angle α_0 needed to hit a target located on the ground at the horizontal distance of $x=6462$ m.

Solution

The ballistics coefficient is

$$c=\frac{id^2}{m}1000=\frac{(0.525)(0.107)^2}{17.04}1000=0.3527421$$
.

(I) As a first approach, we consider $h(\bar{y})=1$ and $\cos(\alpha_0)=\cos(0)=1$. The values of β and b are

$$\beta=1, \quad b=c/(3g)=0.011989894,$$

and so

$$B=\beta\cdot b=0.011989894.$$

Approximate Value of the Pseudo-speed at the Point of Impact

Using equation (5.2.10) we can write

$$D(u)=D(v_0)-Bgx ,$$
(1)

where

$$D(u)=[u+240 \cdot \ln(u-240)]$$

and

$$D(v_0)=[v_0+240 \cdot \ln(v_0-240)]=562+240 \cdot \ln(562-240)=1947.892371 .$$

Thus, substituting the above value in (1), we find that

$$D(u)=[u+240 \cdot \ln(u-240)]=$$
$$=(1947.892371)-(0.011989894)(9.80665)(6462)=1188.085926 .$$

Hence,

$$\ln(u-240)]=(1188.085926-u)/(240) .$$
(2)

The above equation can be solved numerically or using a TI-83 Plus graphing calculator.

First, using TI-83 Plus, we graph each of the following functions,

$$y_1=\ln(u-240) , \quad y_2=\frac{1}{240}(1188.085926-u) ,$$

which represent respectively the left side and the right side of equation (2).

Using a TI-83 Plus, we find the coordinates of intersection of the above functions:

$$u=283.36566 \text{ m/s}, \quad y_1=\ln(u-240)=3.7696678 .$$

Approximate Value of the Launching Angle

To find the launching angle, we use (5.2.26b),

$$\sin(2\alpha_0)=-\frac{1}{120B}[\frac{A(u)-A(v_0)}{D(u)-D(v_0)}-J(v_0)].\qquad(3)$$

First, we find that

$$D(u)-D(v_0)=-Bgx=-759.806446 .$$

Using (5.2.22) and a graphing calculator TI-83 Plus, we obtain

$$A(u)-A(v_0)=\int_{v_0}^{u}\frac{u}{u-240}\ln\frac{u-240}{u}du=\int_{562}^{283.4}\frac{u}{u-240}\cdot\ln\frac{u-240}{u}du=813.37812 .$$

For $J(v_0)$, we can write

$$J(v_0)=\ln(\frac{v_0-240}{v_0})=\ln\frac{562-240}{562}=-0.5569503043 .$$

Substituting in (5.2.26), we find

$$\sin(2\alpha_0)=-\frac{1}{120B}[\frac{A(u)-A(v_0)}{D(u)-D(v_0)}-J(v_0)]=$$

$$=-\frac{1}{120(0.011989894)}[\frac{813.37812}{-759.8064446}-(-0.5569503043)]=0.356937195 .$$

Hence, we obtain

$$2\alpha_0=\sin^{-1}(0.356937195)=20.912217° .$$

The launching angle is

$$\alpha_0=20.912217°/2=10.45610871° .$$

Coordinates of the Trajectory Vertex

We need to find the trajectory vertex to better estimate $h(\bar{y})$. At the point where the vertex is located, the projectile velocity is parallel to the x-axis. Thus,

$$p=\tan\alpha=0.$$

Using (5.2.33),

$$p_0+\frac{1}{240 B\cos^2\alpha_0}[J(u)-J(v_0)]=0,$$

we find

$$J(u)=J(v_0)-240 Bp_0\cos^2\alpha_0 \qquad (4)$$

Substituting in the above equation

$$J(u)=\ln(\frac{u-240}{u}), \quad J(v_0)=\ln(\frac{v_0-240}{v_0})=-0.5569503043,$$

$$p_0=\tan\alpha_0=\tan(10.4561087144°)=0.1845467956,$$

and

$$240 Bp_0\cos^2\alpha_0=240(0.011989894)(0.1845467956)\cos^2(10.4561087144°)=0.5135566968,$$

we have

$$\ln\frac{u-240}{u}=J(v_0)-240bp_0\cos^2\alpha_0=(-0.5569503043-0.5135566968=-1.070507.$$

Hence,

$$\frac{u-240}{u}=e^{-1.070507}=0.3428346558$$

Solving the last equation for u, we obtain the pseudo-speed at the point where the vertex of the trajectory is located:

$$u_m=365.20489 \text{ m/s.}$$

Substituting in (5.2.30), we find that the x-coordinate of the trajectory vertex is

$$x_m=-\frac{1}{Bg}[(u+240\cdot\ln(u-240))-(v_0+240\cdot\ln(v_0-240)]=$$

$$=-\frac{1}{0.1175806999}[1524.393225-1947.892]=3601.774324$$

.

We find as well that

$$D(u_m)-D(v_0)=-bgx_m=-(0.011989894)(9.80665)(3601.774324)=-423.4991247$$.

Then, substituting

$$x=3601.774324m,$$

$$J(v_0)=\ln(\frac{v_0-240}{v_0})=-0.5569503043,$$

$$A(u)-A(v_0)=\int_{v_0}^{u}\frac{u}{u-240}\ln\frac{u-240}{u}du$$,

$$=\int_{562}^{365.20}\frac{u}{u-240}\ln\frac{u-240}{u}=329.0952245$$

$$D(u)-D(v_0)=-423.4991247$$,

$$\alpha_0=10.45610871°$$,

$$p_0=\tan\alpha_0=\tan(10.45610871°)=0.1845467956$$,

and

$$B=0.011989894$$

into the trajectory equation (5.2.24), we find that the maximum height of the projectile trajectory is

$$y_{max}=p_0 x+\frac{x}{240Bcos^2\alpha_0}[\frac{A(u)-A(v_0)}{D(u)-D(v_0)}-J(v_0)]=379.7747654m$$

(II) Improving the Accuracy

The values of elements of trajectory were found supposing that $\beta=1$. We can find more accurate results for the trajectory elements considering that $y_{max}=379.7747654m$.

Using the above value of the maximum height, we find that

$$\bar{y}=(2/3)y_{max}=\frac{2}{3}(379.775)=253.18m$$

and

$$h(\bar{y})=(\frac{289.08-0.006328\bar{y}}{289.08})^{4.4}=(\frac{289.08-0.006328(253.18)}{289.08})^{4.4}=0.9758413288$$

Thus,

$$\beta=h(\bar{y})/\sqrt{cos\alpha_0}=(0.9738194)/\sqrt{cos10.45610871°}=0.98404596$$

and

$$B=\beta\cdot b=(0.984046)(0.0119898946)=0.01179861$$

For the above value of B, repeating the mathematical operations we performed in I, we find the following:

The New Value of the Pseudo-speed at the Point of Impact $x_T=6462$

$$u_T=285.156m/s$$

The New Launching Angle

$$\sin(2\alpha_0)=0.354777154$$

Hence, for the launching angle, we obtain

$$2\alpha_0=\sin^{-1}(0.354777154)=20.779787°$$

and

$$\alpha_0=20.779787°/2=10.3899°=10°23'.$$

The New Coordinates of the Trajectory Vertex

The pseudo-speed u at the point where the trajectory has the maximum height is

$$u_m=365.4 \text{ m/s.}$$

The speed of projectile v_m at the vertex is

$$v_m=u_m\frac{\cos\alpha_0}{\cos\alpha_m}=(365.20)\frac{\cos(10.3899°)}{\cos(0°)}=359.2m/s.$$

The x-coordinate and y-coordinate of the trajectory vertex are respectively

$$x_m=3564,\ y_{max}=371.m.$$

(III) The Other Elements of the Trajectory

Time of Flight to the Target

Using the new value of β,

$$\beta=h(\bar{y})/\sqrt{\cos\alpha_0}=(0.984499)/\sqrt{\cos10.3899°}=0.984499,$$

we have

$$B=\beta \cdot b = 0.011804 \, .$$

Substituting $u_T = 285.156 m/s$, we find that the projectile time of flight to the target is

$$t = -\frac{1}{Bg\cos\alpha_0}[\ln(u-240)-\ln(v_0-240)]=$$

$$= -\frac{1}{(0.011804)(9.80665)\cdot\cos(10.33899°)}[\ln(285.156-240)-\ln(562-240)] \, .$$

$$= 17.25 \text{ sec.}$$

Angle of Impact

For the point of impact, we can write

$$p_T = p_0 + \frac{1}{240 B\cos^2\alpha_0}[\ln(\frac{u_T-240}{u_T})-\ln(\frac{v_0-240}{v_0})]=$$

$$= \tan(10.3899°) + \frac{1}{2.739568}[\ln(\frac{285.156-240}{285.156})-\ln(\frac{562-240}{562})]=-0.28905 \, .$$

Hence, for the angle of impact, we obtain

$$\alpha_T = \tan^{-1}(0.28605) = -15.96326°$$

because $p_T = \tan\alpha_T$.

The Speed of Impact

Using the formula for the projectile pseudo-speed,

$$u = v\frac{\cos\alpha}{\cos\alpha_0} \, ,$$

we find that the speed of projectile at the impact point is

$$v_T = u_T \frac{\cos\alpha_0}{\cos\alpha_T} = (285.156)\frac{\cos(10.3899)}{\cos(-15.96)} = 291.77 m/s \,.$$

In the following table are given for comparison the solution of the problem using Siacci functions and the solution using the PC QBasic program **ANGLEC.BAS** (see appendix B).

Table 1

	Range (m)	Launching Angle	Time of Flight (s)	Impact Speed (m/s)	Angle of Impact
Siacci	6462	10.39°	17.25	292	- 15.96°
ANGLEC.BAS	6462	10.35°	17.22	292	- 15.86°

Note: Use of QBasic PC program **ANGLEC.BAS** (see appendix B)

Input Data: Range=6462, Initial Speed Ballistics=562 m/s, Coefficient=0.3527421

Results: Launching Angle = 10.35376°; Time of Flight = 17.21 s; Terminal Speed = 292 m/s; Terminal Angle = -15.86261°; Trajectory Vertex (3591, 374)

Example 2. An antiaircraft artillery projectile is fired with an initial speed of v_0=800 m/s at a launching angle of α_0=72°. The ballistics coefficient of the projectile is c=0.4433 (caliber is d=0.085 m).

Use the Siacci method to find the elements of the projectile trajectory at a point with abscissa x=1700m.

Solution
(I) First Approach
    Estimation of parameter b

As a first approach, we consider the y_{max}-coordinate approximately:

$$y_m = x\tan(\alpha_0) = (1701)\tan(72°) = 5235m$$

Thus,

$$\bar{y} = (1/2)y_{max} = (1/2)\cdot(5235) = 2618$$

and

$$h(\bar{y}) = (\frac{289.08-0.006328\bar{y}}{289.08})^{4.4} = (\frac{289.08-0.006328(2618)}{289.08})^{4.4} = 0.77133069 \, .$$

For b, we find

$$b = c/(3g) = (0.4433)\cdot/(3\cdot(9.80665)) = 0.0150680066 \, ,$$

$$\beta = h(\bar{y})/\cos\alpha_0 = (0.77133069)/\cos(72°) = 2.49607855 \, ,$$

and

$$B = \beta\cdot b = 0.03761093 \, , \quad Bg = 0.3688372071 \, .$$

Pseudo-speed at $x=1700m$
Substituting in (5.2.9),

$$x = -\frac{1}{Bg}[(u+240\cdot\ln(u-240))-(v_0+240\cdot\ln(v_0-240)] \, ,$$

we have

$$1070 = -\frac{1}{(0.36884)}[(u+240\cdot\ln(u-240))-(800+240\cdot\ln(800-240)] \, .$$

Solving the last equation, we find the pseudo-speed

$$u = 425.9136579 \, .$$

Thus,

$$A(u)-A(v_0)=\int_{v_0}^{u}\frac{u}{u-240}\ln\frac{u-240}{u}du$$

$$=\int_{800}^{425.914}\frac{u}{u-240}\ln\frac{u-240}{u}=347.639$$

Substituting in the trajectory equation, we obtain the y-coordinate

$$y=p_0x+\frac{x}{240B\cos^2\alpha_0}\cdot[\frac{A(u)-A(v_0)}{D(u)-D(v_0)}-J(v_0)]=4868.854066\,\text{m}.$$

For the time of flight and the speed, we obtain respectively

$$t=-\frac{1}{Bg\cos\alpha_0}[\ln(u-240)-\ln(v_0-240)]=9.504\text{sec}.$$

and

$$v=u\frac{\cos\alpha_0}{\cos\alpha}=425.91\frac{\cos(72)^\circ}{\cos(68.509)}=359.25m/s\,.$$

In the following table are given for comparison the solution of the problem using the Siacci method and the solution obtained using the QBasic PC program **RangeC.BAS** (see appendix C).

Table 1

	x-coordinate	y-coordinate	Launching Angle	Time of Flight	Speed	Terminal Angle
Siacci	1700 m	4,869	72°	9.50 s	360 m/s	68.51°
RangeC.bas	1700 m	4,853	72°	9.64 s	371 m/s	68.40°

Note: The above table shows that the Siacci method gives very approximate results for launching angles greater than $15°$, though the accuracy is not so satisfying.

Employing QBasic PC Program RangeC.BAS:

Input Data
Initial x-coordinate = 0, Initial y-coordinate =0
Launching Angle = 72, Initial Speed = 800, Ballistics Coefficient
   =0.4433
Initial Time = 0, Integration Step h0 = 1
Input: x-coordinate of a Point on Trajectory = 1700

Displayed Results
Range = 9,774; Error in y-coordinate = -1.405; Terminal Speed = 312
Terminal Angle = -78.78; Coordinates of Vertex (5213, 9532)

Abscissa of the Given Point on the Trajectory = 1700,
Corresponding y-coordinate of x = 1700 is 4853; Speed = 371;
Time = 9.6.3; Angle = 68.40

Example 3. Evaluate the ballistics coefficient and the coefficient of
form of a bullet of a Russian semiautomatic rifle of caliber $d=7.62mm$
using the following data obtained from the range table of the rifle:
mass $m=7.9g$, initial speed of bullet $v_0=735m/s$, launching angle
$\alpha_0=0.084°$, range $x=100m$; the speed of impact and time of flight are
respectively $v=640m/s$ and $t=0.14s$.

Solution
    Substituting $\beta=1$, $u=640m/s$, $x=100m$ in (4.7.17),

$$x=-\frac{1}{bg}[(u+240\cdot\ln(u-240))-(v_0+240\cdot\ln(v_0-240)]$$,

we have

$$100=-\frac{1}{b(9.80665)}[(640+240\cdot\ln(640-240))-(735+240\cdot\ln(735-240)]$$.

Hence, we find that

$$b=\frac{14.90237}{100}=0.1490237.$$

Substituting the above value in $b=ch(y)/(3g)$ and solving the obtained equation for c, we find the value of the ballistics coefficient $c=4.384271151$.

Substituting the value of c in

$$c=i\frac{d^2}{m}\cdot 1000,$$

we have

$$4.384271151=i\frac{\cdot(0.00762)^2}{(0.0079)}\cdot 1000.$$

Hence, we find that the form coefficient is

$$i=0.5965056.$$

Note 1: This result ($i=0.5965056$) is the same as the outcome obtained in example 3, section 2.6.

Note 2: The methods used in this example and the method used in example 3, section 2.6, as well as the method used in example 2 of section 2.6, give the same results.

Example 4. (Refer to example 3 of section 3.5) A 7.9 g bullet is fired with initial speed 735 m/s. Determine the projection angle needed to hit a target located on a hill at a slant distance of 300 m from a 7.62 mm Russian rifle. The elevation angle is 65°.

Use the data obtained in example 2 of section 3.3. (The ballistics coefficient is $c=4.219797$.)

Solution

(I) As a first approach, we consider the launching angle $\alpha_0 = E \approx 65°$. So

$$\cos(\alpha_0) = \cos(E+A) \approx \cos(65°) = 0.4226182617 \ .$$

The x-coordinate and the y-coordinate at the slant distance $D=300m$ are respectively

$$x_T = D \cdot \cos(E) = (300)\cos(65°) = 126.7854785 \ ,$$

$$y_T = D \cdot \sin(E) = (300) \cdot \sin(65°) = 271.8923361 \ .$$

We find

$$\bar{y} = (1/2) y_{max} = (1/2) \cdot (271.89) = 135.95m$$

and

$$h(\bar{y}) = (\frac{289.08 - 0.006328\bar{y}}{289.08})^{4.4} = 0.9869713155 \ .$$

The values of the parameters b and β are respectively

$$b = c/(3g) = 0.1434331805$$

and

$$\beta = h(\bar{y})/\cos\alpha_0 = (0.9869713155/\cos(65°)) = 2.33537309 \ ,$$

while

$$B = \beta \cdot b = 0.3349699899 \ .$$

Approximate Value of the Pseudo-speed at the Point of Impact

Using equation (5.2.12), we can write

$$D(u_T) = D(v_0) - Bgx \tag{1}$$

where

$$D(u_T)=u_T+240\ln(u_T-240),$$

$$D(v_0)=[v_0+240\cdot\ln(v_0-240)]$$
$$=735+240\cdot\ln(735-240)=2224.0939,$$

and

$$Bgx=-416.4818596.$$

Thus, substituting the above values in (1), we find that

$$D(u_T)=u_T+240\ln(u_T-240)=1807.612.$$

Hence,

$$\ln(u-240)]=(1807.612-u)/(240).\tag{2}$$

The above equation can be solved numerically or using a TI-83 Plus graphing calculator.

First, using TI-83 Plus, we graph each of the following functions,

$$y_1=\ln(u-240),\quad\text{and}\quad y_2=\frac{1}{240}(1765.181-u),$$

which represent respectively the left side and the right side of equation (2).

Using a TI-83 Plus, we find the coordinates of intersection of the above functions:

$$u=486.1683\text{ m/s},\quad y_1=\ln(u-240)=5.5060154.$$

Approximate Value of the Launching Angle

To find the launching angle, we use the equation of projectile trajectory (5.2.24),

$$y_{max}=p_0 x+\frac{x}{240 B \cos^2 \alpha_0}\cdot[\frac{A(u)-A(v_0)}{D(u)-D(v_0)}-J(v_0)]=411.8885m \quad . \qquad (3)$$

First, we find that

$$D(u)-D(v_0)=-bgx=-416.4818595 \quad .$$

Using (5.2.22) and a graphing calculator TI-83 Plus, we obtain

$$A(u)-A(v_0)=\int_{v_0}^{u}\frac{u}{u-240}\ln\frac{u-240}{u}du=\int_{735}^{486.1683}\frac{u}{u-240}\cdot\ln\frac{u-240}{u}du=215.89478 \quad .$$

For $J(v_0)$, we can write

$$J(v_0)=\ln(\frac{v_0-240}{v_0})=\ln\frac{562-240}{562}=-0.3953127366 \quad .$$

Substituting in (3), we can write

$$271.89=(126.7855)p_0+\frac{(126.7855)(1+p^2)^{0.5}}{240h(\bar{y})\cdot b}[\frac{215.89478}{-416.48186}+0.39531274))] \quad .$$

Hence, we obtain the following equation

$$126.7855p_0-0.45924\cdot(1+p_0^2)^{0.5}-271.892=0 \quad .$$

The solution obtained using a TI-83 Plus graphing calculator is

$$p_0=\tan(\alpha_0)=2.1531059 \quad .$$

Hence, we find the launching angle

$$\alpha_0=\tan^{-1}(2.1531059)=65.08770769 \quad .$$

The angle of sight is

$$A=\alpha_0-E=65.08770769°-65°=0.08771°$$

or

$$A=\alpha_0-E=5.262'$$.

Comment: The solution obtained using the Siacci method is the same as the solution obtained in example 2 of section 3.5 using the curvilinear coordinates.

Example 5. A projectile of mass $m=27.3kg$ is fired from 122 mm artillery cannon with initial speed $v_0=885m/s$ m/s. Determine the horizontal range if the launching angle is $\alpha_0=14.483°$. The ballistics coefficient is $c=0.251$ ($i=0.46038$).

Solution
I.   As a first approach, we consider $\beta=1$. The value b is

$$b=c/(3g)=(0.251)/(3 \cdot g)=0.0085316256$$.

Thus,

$$B=\beta \cdot b=0.0085316256$$.

II. Coordinates of the Trajectory Vertex

We need to find the trajectory vertex to better estimate $h(\bar{y})$ and the correction coefficient β. At the point where the vertex is located, the projectile velocity is parallel to the x-axis. Thus,

$$p=\tan\alpha=0$$.

Using (5.2.33),

$$p_0+\frac{1}{240B\cos^2\alpha_0}[J(u)-J(v_0)]=0$$,

we find

$$J(u)=J(v_0)-240Bp_0\cos^2\alpha_0.$$

Substituting in the above equation

$$J(u)=\ln(\frac{u-240}{u}),$$

$$J(v_0)=\ln(\frac{v_0-240}{v_0})=\ln(\frac{885-240}{885})=-0.3163373282,$$

$$p_0=\tan\alpha_0=\tan(14.48333°)$$

and

$$240Bp_0\cos^2\alpha_0=240(0.0085316256)\tan(14.48333°)\cos^2(14.48333°)=0.495824677,$$

we obtain

$$\ln\frac{u-240}{u}=J(v_0)-240Bp_0\cos^2\alpha_0=(-0.3163373283-0.4916558971=-0.812162.$$

Hence,

$$\frac{u-240}{u}=e^{-0.812162}=0.44389732.$$

Solving the last equation for u, we obtain the pseudo-speed at the point where the vertex of the trajectory is located:

$$u_m=431.575m/s.$$

Substituting the above value of the pseudo-speed in (5.2.30), we find that the x-coordinate of the trajectory vertex is

$$x_m=-\frac{1}{Bg}[(u+240\cdot\ln(u-240))-(v_0+240\cdot\ln(v_0-240)]=8901.73.$$

We find as well

$$D(u_m)-D(v_0)=-Bgx_m=-744.77808.$$

Then, substituting

$$x_m=8901.73,$$

$$J(v_0)=J(885)=\ln(\frac{885-240}{885})=-0.3163373282,$$

$$A(u_m)-A(v_0)=\int_{v_0}^{u_m}\frac{u}{u-240}\ln\frac{u-240}{u}du$$

$$=\int_{885}^{426.56}\frac{u}{u-240}\ln\frac{u-240}{u}=376.71025,$$

$$D(u_m)-D(v_0)=-744.77808,$$

$$\alpha_0=14.48333°,$$

$$p_0=\tan\alpha_0=\tan(14.48333°),$$

and

$$B=0.0085316256$$

in the trajectory equation (5.2.24), we find that the maximum height of the projectile trajectory is

$$y_{max}=p_0x+\frac{x}{240B\cos^2\alpha_0}\cdot[\frac{A(u)-A(v_0)}{D(u)-D(v_0)}-J(v_0)]=1420.74m.$$

We can find more accurate results for the trajectory elements considering that $y_{max}=1420.74m$.

Using the above value of the maximum height, we find that

$$\bar{y}=(2/3)y_{max}=\frac{2}{3}(1420.74)=947.160747m$$

and

$$h(\bar{y})=(\frac{289.08-0.006328\bar{y}}{289.08})^{4.4}=(\frac{289.08-0.006328(947.1607471)}{289.08})^{4.4}=0.91192936 \; .$$

Thus,

$$\beta=h(\bar{y})/\sqrt{\cos\alpha_0}=(0.91192936)/\sqrt{\cos(14.4833333°)}=0.9267745$$

and

$$B=\beta{\cdot}b=(0.9267745)(0.0085316256)=0.0079068931 \; .$$

III. Repeating the same mathematical operations we performed in II using the above value of B, we find that the new value of the pseudo-speed that corresponds to the vertex point and the coordinates of the point where the vertex is located are respectively

$$u_m=444.6997766m/s \; , \qquad x_m=9230.70m \; , \qquad y_m=1461.50m \; .$$

IV. Horizontal Range

Consider the equation of trajectory (5.2.24),

$$y=p_0x+\frac{x}{240B\cos^2\alpha_0}[\frac{A(u)-A(v_0)}{D(u)-D(v_0)}-J(v_0)] \; . \qquad (1)$$

At the point of impact, the y-coordinate is zero. Substituting in (1), we write

$$\tan(14.48333)+\frac{1}{240(0.0079068931)\cos^2(14.48333)}[\frac{A(u)-A(885)}{D(u)-D(885)}-J(885)]=0, \qquad (2)$$

where

$$J(885)=-0.3163373282, \quad D(v_0)=2437.620076$$

$$A(u)-A(885)=\int_{885}^{u}\frac{u}{u-240}\ln(\frac{u-240}{u})du,$$

$$D(u)=[u+240\cdot\ln(u-240)].$$

Hence, we obtain the following equation:

$$\int_{885}^{u}\frac{u}{u-240}\ln(\frac{u-240}{u})du=1881.488661-0.7758549957\cdot[u+240\cdot\ln(u-240)].$$

Using the trial and error and a TI-83 Plus graphing calculator, we find the value of the pseudo-speed at the point of impact

$$u_T=291.16m/s.$$

The range of fire is

$$x_T=-\frac{1}{Bg}[(u+240\cdot\ln(u-240))-(v_0+240\cdot\ln(v_0-240)]=15,502.55m.$$

Angle of Impact
Substituting in (5.2.6),

$$p_T=p_0+\frac{1}{240B\cos^2\alpha_0}[J(u_T)-J(v_0)],$$

where

$$J(u)=\ln(\frac{u-240}{u}), \quad J(v_0)=\ln(\frac{v_0-240}{v_0}),$$

THE SIACCI METHOD

$$p_T=\tan(\alpha_T)=-0.541361961.$$

Hence, we find that the impact angle is

$$\alpha_T=\tan^{-1}(-0.541361961)=28.43°.$$

Speed of Impact

$$v_T=u_T\frac{\cos\alpha_0}{\cos\alpha_T}=291.16\frac{\cos(14.48333°)}{\cos(-28.43°)}=320.57.$$

Time of Flight
   The time of flight to the target is

$$t_T=-\frac{1}{Bg\cos\alpha_0}[\ln(u_T-240)-\ln(v_0-240)]=33.76s.$$

Note: The table below compares the elements of the trajectory obtained theoretically using the Siacci method, the PC QBasic program RANGEC.BAS (see appendix C), and the corresponding elements of the trajectory given in the range table of artillery cannon 122 mm.

Table 1

	Range	Height	Impact angle	Time of flight	Impact speed
Siacci Method	15,503	1,462.32	- 28.43°	33.8	320.57
RangeC.Bas	15, 397	1,437.00	- 28.09°	33.6	314.77
Range Table	15,400	1,450.00	- 28.00°	34.0	314.00

Example 6. Find an appropriate ballistics coefficient and the corresponding form coefficient for a 12.8 g bullet 8x57 S fired from a Mauser K98k considering the following experimental data: initial

speed $v_0=755m/s$ and terminal speed at the range of 100 m
$v_T=706$ m/s. [44] Consider standard atmosphere on the ground (at $y=0$).

Solution

$$h(y)=h(0)=1$$

Substituting $\beta=1$, $u=706m/s$, $x=100m$, $B=\beta \cdot b=b$ in

$$x=-\frac{1}{bg}[(u+240 \cdot \ln(u-240))-(v_0+240 \cdot \ln(v_0-240)]$$,

we have

$$100=-\frac{1}{b(9.80665)}[(706+240 \cdot \ln(706-240))-(755+240 \cdot \ln(755-240)]$$.

Hence, we find that

$$b=0.074435$$.

Substituting the above value in $b=ch(y)/(3g)$ and solving the
obtained equation for c we find the value of the ballistics coefficient

$$c=2.189874$$.

Note 1: This result ($c=2.189874$) is the same as the outcome we
obtained using (2.6.26).
 Indeed, substituting $x=100m$ and $v_T=706$ m/s in (2.6.26), we find

$$c=3[(v_1-v_2)+240 \cdot \ln\frac{v_1-240}{v_2-240}]+[h(y)(x_2-x_1)]=3[(755-706)+240\ln\frac{755-240}{706-240}]+(100)=2.189865$$

---

[44]   Mori, E. *Balistica Teorica e Pratica.*
      http://www.earmi.it/balistica/coefball.htm [Web site]

Example 7. Use the Siacci method to find an appropriate ballistics coefficient for a 12.8 g bullet 8x57 S fired from a Mauser K98k considering the following experimental data:

initial speed $v_0=755m/s$ and terminal speed at the range of 500 m $v_T=534m/s$.[45]

On the ground, the atmospheric air has standard values.

Use the obtained ballistics coefficient to find the time of flight, trajectory vertex, and the angle of impact.

Note: The coefficient of form we are going to determine plays the role of an "adjusting" coefficient (see chapter 7).

Solution

Substituting $\beta=1$, $u=v_T=534m/s$, $x=500m$, $B=\beta \cdot b=b$ in (5.2.30),

$$x=-\frac{1}{bg}[(u+240 \cdot \ln(u-240))-(v_0+240 \cdot \ln(v_0-240)]$$,

we have

$$500=-\frac{1}{b(9.80665)}[(534+240 \cdot \ln(534-240))-(755+240 \cdot \ln(755-240)]$$.

Hence, we find that

$$b=\frac{36.25508}{500}=0.0725101665$$.

Substituting the above value in $b=ch(y)/(3g)$ and solving the obtained equation for c, we find the value of the ballistics coefficient $c=2.133245$.

---

[45]   Edoardo Mori. *Balistica Teorica e Pratica.*
http://www.earmi.it/balistica/coefball.htm [Web site]

Time of Flight

$$t=-\frac{1}{bg\cos\alpha_0}[T(u)-T(v_0)]=\frac{1}{0.07251g\cos(17.167')}[\ln(534-240)-\ln(755-240)]=0.788$$

Coordinates of Trajectory Vertex

At the point where the vertex is located, the projectile velocity is parallel to the x-axis,

$$p=\tan\alpha=0$$.

Using (5.2.33),

$$p_0+\frac{1}{240B\cos^2\alpha_0}[J(u)-J(v_0)]=0$$,

we find

$$J(u)=J(v_0)-240Bp_0\cos^2\alpha_0$$. \hfill (4)

Substituting in the above equation

$$J(u)=\ln(\frac{u-240}{u}),\ J(v_0)=\ln(\frac{v_0-240}{v_0})=-0.38255085$$

and

$$p_0=\tan\alpha_0=\tan(17.167°)=0.004993622$$,

we have

$$\ln\frac{u-240}{u}=J(v_0)-240bp_0\cos^2\alpha_0=-0.469449896$$.

Hence,

$$\frac{u-240}{u}=e^{-0.49449896}$$.

Solving the last equation for u, we obtain the pseudo-speed at the point where the vertex of the trajectory is located:

$$u_m = 640.59 m/s.$$

Substituting in (5.2.30), we find that the x-coordinate of the trajectory vertex is

$$x_m = -\frac{1}{Bg}[(u+240 \cdot \ln(u-240)) - (v_0 + 240 \cdot \ln(v_0 - 240)] = 245.70m.$$

We find as well

$$D(u_m) - D(v_0) = -bgx_m = -174.703.$$

Then, substituting

$$x = 245.70m,$$

$$J(v_0) = \ln(\frac{v_0 - 240}{v_0}) = -0.38255085,$$

$$A(u) - A(v_0) = \int_{v_0}^{u} \frac{u}{u-240} \ln\frac{u-240}{u} du$$

$$= \int_{755}^{640.59} \frac{u}{u-240} \ln\frac{u-240}{u} = 74.008957,$$

$$\alpha_0 = 17.167',$$

and

$$B = 0.0725102$$

in the trajectory equation (5.2.24), we find that the maximum height of the projectile trajectory is

$$y_{max}=p_0 x+\frac{x}{240 B\cos^2\alpha_0}\cdot[\frac{A(u)-A(v_0)}{D(u)-D(v_0)}-J(v_0)]=0.65m$$

Impact Angle

$$p=p_0+\frac{1}{240 B\cos^2\alpha_0}[J(u)-J(v_0)]=-0.00732$$

Hence, for the impact angle we find

$$\alpha_T=-25.16'.$$

The following table compares the data obtained in this example with the data given by E. Mori. (The last row of table 1 corresponds to the data given by E. Mori's table.)

Table 1

	Range	Launching Angle	Impact angle	Vertex Coordinate x	y	Time to target	Speed of impact
Example 7	500 m	17'10"	- 25'10"	246 m	0.65 m	0.79 m	534 m/s
E. Mori	500 m	17'10"	- 19'40"	259 m	0.7 m	0.79 m	534 m/s

5.3 The Siacci Method for Low Speeds

Consider again the system of differential equation (5.1.16),

$$\begin{cases} \dfrac{dp}{du}=-\dfrac{g}{\beta\cdot cK_D(u)\cdot u\cdot\cos^2\alpha_0} \\ \dfrac{dt}{du}=-\dfrac{1}{\beta\cdot cK_D(u)\cdot\cos\alpha_0} \\ \dfrac{dx}{du}=-\dfrac{u}{\beta\cdot cK_D(u)} \\ \dfrac{dy}{du}=-p\dfrac{u}{\beta\cdot cK_D(u)} \end{cases} \qquad (5.3.1)$$

where

$$c=\frac{id^2}{m}1000 \ , \quad u=v\frac{\cos\alpha}{\cos\alpha_0} \ , \quad p=\tan\alpha \ ,$$

$$\beta=h(\bar{y})\,/\cos\alpha_0 \text{ (antiaircraft artillery), or } \beta=\frac{h(\bar{y})}{\sqrt{\cos\alpha_0}} \text{ (field artillery),}$$

and

$$K_D(u)=\begin{cases}1.212{\cdot}10^{-4}\,u^2 & for \quad u{\le}256m/s \\ (u-240)/3 & for \quad u{>}256m/s \end{cases} \tag{5.3.2}$$

When the projectile speed is less than 256 m/s, we can write (5.3.1) as follows:

$$\begin{cases}\dfrac{dp}{du}=\dfrac{1}{Bu^3{\cdot}\cos^2\alpha_0} \\[2mm] \dfrac{dt}{du}=-\dfrac{1}{Bgu^2\cos\alpha_0} \\[2mm] \dfrac{dx}{du}=-\dfrac{1}{Bgu} \\[2mm] \dfrac{dy}{du}=-p\dfrac{1}{Bgu}\end{cases} \tag{5.3.3}$$

where

$$B=1.212{\cdot}10^{-4}\,\beta\frac{c}{g} \ , \quad c=\frac{id^2}{m}1000 \ , \quad u=v\frac{\cos\alpha}{\cos\alpha_0} \ , \quad p=\tan\alpha \ , \tag{5.3.4}$$

and

$$\beta=h(\bar{y})/\cos\alpha_0 \ , \tag{5.3.5}$$

for antiaircraft shooting, or

$$\beta=\frac{h(\bar{y})}{\sqrt{\cos\alpha_0}} \ ,$$ (5.3.6)

for field artillery shooting.

The first differential equation of (5.3.3) can be written as

$$dp=\frac{1}{Bu^3\cos^2\alpha_0}du \ .$$ (5.3.7)

Integrating, we find p as a function of the pseudo-speed u

$$p=p_0-\frac{1}{2B\cos^2\alpha_0}(\frac{1}{u^2}-\frac{1}{v_0^2}) \ .$$ (5.3.8)

Solving the other two differential equations of system (5.3.3), we obtain respectively

$$t=\frac{1}{Bg\cos\alpha_0}(\frac{1}{u}-\frac{1}{v_0})$$ (5.3.9)

and

$$x=-\frac{1}{Bg}(\ln u-\ln v_0) \ .$$ (5.3.10)

From the third and the fourth differential equations of (5.3.3), we have

$$dy=pdx \ ,$$ (5.3.11)

while from the third one, we have

$$dx=-\frac{g}{Bu}du \ .$$ (5.3.12)

Integrating the differential equation (5.3.11) and considering (5.3.12), (5.3.8), and (5.3.10), we obtain the trajectory equation

$$y=p_0x+\frac{x}{2B\cos^2\alpha_0}[\frac{1}{2}(\frac{1}{u^2}-\frac{1}{v_0^2})\cdot(\frac{1}{\ln u-\ln v_0})+\frac{1}{v_0^2}]$$ (5.3.13)

The above four equations—(5.3.8), (5.3.9), (5.3.10), and (5.3.13)—describe the projectile flight in standard atmosphere when the projectile speed all over the trajectory is less than 256 m/s.

Example 1. (See example 1 of section 4.5). A projectile is fired from a 60 mm trench mortar with a speed of 89 *m/s* under an angle of 45°. Determine the point of impact of the projectile on the horizontal plane, time of flight, the velocity of impact, the terminal angle, and the maximum height of the flight. Projectile mass is 1.31 kg, projectile caliber is 60.7 mm, and the coefficient of projectile form is 0.622. The atmospheric conditions are normal. The trench mortar and the target are at sea level.

Solution
    First, we evaluate

$$p_0=\tan45°=1$$

and

$$c=\frac{id^2}{m}1000=\frac{(0.622)(0.0607)^2}{1.31}\cdot1000=1.749$$

Considering as a first approach $h(\bar{y})\approx1$, we find

$$\beta=\frac{h(\bar{y})}{\sqrt{\cos\alpha_0}}=\frac{1}{\sqrt{\cos45°}}=1.189207115$$

and

$$B=1.212\cdot10^{-4}\beta\frac{c}{g}=1.212\cdot10^{-4}(1.189207115)(1.479)/(9.80665)=2.57057\cdot10^{-5}$$

Maximum Value of y-coordinate of the Projectile

Maximum height of the projectile is located at the point where $p=\tan\alpha=\tan(0°)=0$.

Substituting the above value in (5.3.8), we have

$$p_0 - \frac{1}{2B\cos^2\alpha_0}(\frac{1}{u^2}-\frac{1}{v_0^2})=0$$
.

Substituting all known values, we have

$$1 - \frac{1}{2(2.57057\cdot10^{-5})\cos^2(45°)}(\frac{1}{u^2}-\frac{1}{89^2})=0$$.

From the above equation, we find that the pseudo-speed at the vertex of the trajectory is $u=81.12342 m/s$. Substituting in (5.3.10), we find the abscissa

$$x=-\frac{1}{Bg}(\ln u-\ln v_0)=-\frac{1}{(2.57057\cdot10^{-5})(9.80665)}[\ln(81.1234)-\ln(89)]=367.59m$$

of the trajectory vertex.

Substituting in the equation of the trajectory (5.3.11), we find that the y-coordinate of the vertex is

$$y=p_0x+\frac{x}{2B\cos^2\alpha_0}[\frac{1}{2}(\frac{1}{u^2}-\frac{1}{v_0^2})\cdot(\frac{1}{\ln u-\ln v_0})+\frac{1}{v_0^2}]=$$

$$=367.59+\frac{367.59}{2(2.57057\cdot10^{-5})\cos(45°)^2}[\frac{1}{2}(\frac{1}{81.1234^2}-\frac{1}{89^2})\cdot\frac{1}{\ln 81.1234-\ln 89}+\frac{1}{89^2}]=189.468m$$

We find that

$$\bar{y}=(2/3)y_{max}=(2/3)(189.468)=126.31$$

and

$$h(\bar{y})=(\frac{289.08-0.006328\bar{y}}{289.08})^{4.4}=(\frac{289.08-0.006328\cdot(126.31)}{289.08})^{4.4}=0.98788358$$
.

The New Value of β

The new value is

$$\beta = \frac{h(\bar{y})}{\sqrt{\cos\alpha_0}} = \frac{0.98788358}{\sqrt{\cos 45°}} = 1.1747982 ,$$

while

$$B = 1.212 \cdot 10^{-4} \, \beta \frac{c}{g} = 2.5394228 \cdot 10^{-5} .$$

The Range and the Time of Flight

The range x is at the point where y-coordinate is zero. Substituting $y=0$, in the equation of trajectory (5.3.13), we have

$$1 + \frac{1}{2B\cos^2\alpha_0} [\frac{1}{2}(\frac{1}{u^2} - \frac{1}{v_0^2}) \cdot (\frac{1}{\ln u - \ln v_0}) + \frac{1}{v_0^2}] = 0 .$$

Substituting the other values, we get the following equation

$$1 + \frac{1}{2(2.5394 \cdot 10^{-5})\cos^2(45°)} [\frac{1}{2}(\frac{1}{u^2} - \frac{1}{(89)^2}) \cdot (\frac{1}{\ln u - \ln(89)}) + \frac{1}{(89)^2}] = 0 .$$

Solving the last equation for the pseudo-speed u using a TI-83 Plus graphing calculator, we find

$$u = 74.487707 m/s .$$

Substituting the value of pseudo-speed in equations (5.3.9) and (5.3.10), we find respectively the time and the range of projectile flight to the target

$$t_T = \frac{1}{(2.5394 \cdot 10^{-5})(9.80665)\cos(45°)} (\frac{1}{74.49} - \frac{1}{89}) = 12.43 sec.$$

and

$$x_T=-\frac{1}{Bg}(\ln u-\ln v_0)=-\frac{1}{(2.5394\cdot10^{-5})(9.80665)}(\ln74.49-\ln89)=714.78m$$

The Impact Angle

The angle of projectile at the point of impact can be found substituting in (5.3.8)

$$p=p_0\frac{1}{2B\cos^2\alpha_0}(\frac{1}{u^2}-\frac{1}{v_0^2})=1-\frac{1}{2(2.5394\cdot10^{-5})\cos^2(45°)}(\frac{1}{74.49^2}-\frac{1}{89^2})=-1.1258709.$$

Since $p=\tan\alpha_T$, we find that the impact angle is

$$\alpha_T=\tan^{-1}(-1.12602525)=-48.38847657°=-48°23'$$

Impact Speed

Employing the formula (5.3.4) for the pseudo-speed we can write

$$v=u\frac{\cos\alpha_0}{\cos\alpha}=74.47\frac{\cos(45°)}{\cos(-48.39°)}=79.3m/s.$$

In the following table are given, for comparison, the data obtained in example 1, sec. 4.4 and the results obtained using the Siacci method, as well as the values of the same elements as they are shown in the range table of the 60 mm trench mortar.

Table 1

	Launching Angle	Angle of impact	Range	Time of flight	Max Height
Siacci method	$\alpha_0=45$	$\alpha=-48\ 23'$	$x=714.78$	$t=12.43$	$y=189.5$
Example 1 Sec. 4.4	$\alpha_0=45$	$\alpha=-48\ 10'$	$x=713.78$	$t=12.40$	$y=189$
Range Table	$\alpha_0=45$	$\alpha=-48\ 15'$	$x=716$	$t=12$	$y=186$

Comment: Comparing the outcomes obtained by the Siacci method and the outcomes obtained using method of section 4.4, we see that the results show that both methods give quite the same values for the corresponding elements of their trajectories.

This also shows that the Siacci method gives satisfying results for launching angles greater than 15° for speeds less than 256 m/s.

Example 2. (Given: form coefficient "i", launching speed "v_0", range x_T. Find Launching angle "α_0"). A projectile is fired from a 120mm cannon with a speed of $183 \, m/s$. Find the launching angle needed to hit a target located at a horizontal distance of 2974m. Projectile mass is 22 kg and the coefficient of projectile form is 0.668. The atmospheric conditions are normal. The cannon and the target are at the sea level.

Determine as well all the other elements of trajectory at the point of impact and the maximum height of the flight trajectory.

Solution

First we evaluate

$$c=\frac{id^2}{m}1000=\frac{(0.668)(0.120)^2}{22}\cdot1000=0.43724$$

I.   Approximate Solution

Since the maximum height of the trajectory and the launching angle are unknowns we consider as a first approach $h(\bar{y})=1$, and $\cos\alpha_0=1$. Thus

$$\beta=h(\bar{y})/\sqrt{\cos\alpha_0}=1$$

and

$$B=1.212\cdot10^{-4}\,\beta\frac{c}{g}=1.212\cdot10^{-4}(0.43724)/(9.80665)=5.4037869\cdot10^{-6}$$

The following estimated values are approximate.

Pseudo-speed at the point of impact
Solving (5.3.10) for the pseudo-speed "u" we find

$$u = e^{(\ln v_0 - Bgx)} = e^{\ln(183) - 0.0000054 \cdot 9.80665 \cdot 2972)} = 156.32 m/s$$

Launching angle
At the point of impact the y-coordinate is zero. Substituting $y = 0$ on the left side of (5.3.11) as well as all known values we have

$$\tan(\alpha_0) + \frac{1}{2(5.403787 \cdot 10^{-6}) \cos^2(\alpha_0)} [\frac{1}{2}(\frac{1}{156.32^2} - \frac{1}{183^2}) \cdot \frac{1}{\ln(156.32) - \ln(183)} + \frac{1}{183^2}] = 0$$

Since

$$\cos^2 \alpha_0 = (1 + \tan^2 \alpha_0)^{-1} = (1 + p_0^2)^{-1},$$

we can write the above equation as follows

$$p_0 - 0.484967(1 + p_0^2) = 0$$

Hence we find two values for $p_0 = \tan \alpha_0$

$$p_0 = \tan \alpha_0 = 0.780053276, \quad p_0 = \tan \alpha_0 = 1.28111762$$

For the corresponding launching angle we obtain respectively

$$\alpha_0 = \tan^{-1}(0.7800876) = 37.95735°$$

and

$$\alpha_0 = \tan^{-1}(1.2819073) = 52.04265°$$

Let's solve the problem for the small launching angle $\alpha_0 = 37.95735°$. In a similar way we can solve the problem for the other angle.

The coordinates of the trajectory vertex

Considering $h(\bar{y}) \approx 1$. We estimate

$$\beta = \frac{h(\bar{y})}{\sqrt{\cos \alpha_0}} = \frac{1}{\sqrt{\cos(37.95735°)}} = 1.12620$$

and

$$B = 1.212 \cdot 10^{-4} \, \beta \frac{c}{g} = 1.212 \cdot 10^{-4} (1.12620)(0.43724)/(9.80665) = 6.0856 \cdot 10^{-6}$$

Maximum height of the projectile is located at the point where $p = \tan\alpha = \tan(0°) = 0$.

Using (5.3.8) we have

$$p_0 - \frac{1}{2B\cos^2\alpha_0}(\frac{1}{u^2} - \frac{1}{v_0^2}) = 0$$

Substituting we can write

$$p_0 - \frac{1}{2B\cos^2\alpha_0}(\frac{1}{u^2} - \frac{1}{v_0^2}) = \tan(37.97) - \frac{1}{2(6.08629 \cdot 10^{-6})\cos^2(37.97°)}(\frac{1}{u^2} - \frac{1}{183^2}) = 0$$

From the above equation we find that the pseudo-speed at the maximum point of the trajectory is

$$u_m = 167.22 m/s$$

The x-coordinate of the trajectory vertex is

$$x_m = -\frac{1}{Bg}(\ln u_m - \ln v_0) = -\frac{1}{(6.0856 \cdot 10^{-6})(9.80665)}[\ln(167.22) - \ln(183)] = 1511 m$$

Employing (5.3.13) for the y-coordinate of the trajectory vertex we have

$$y_m = (1511) \cdot \tan(37.95735°) + \frac{1511}{2(6.0856 \cdot 10^{-6})\cos^2(37.95735°)}[\frac{1}{2}(\frac{1}{167.22^2} - \frac{1}{183^2})\frac{1}{\ln(167.22) - \ln(183)} +$$

$$+ \frac{1}{183^2}] = 607.17 m$$

We find that

$$\bar{y}=(2/3)y_{max}=(2/3)(607.17)=405.4m$$

and

$$h(\bar{y})=(\frac{289.08-0.006328\bar{y}}{289.08})^{4.4}=(\frac{289.08-0.006328\cdot(405)}{289.08})^{4.4}=0.96151$$

II. Improving results

The new value of β is

$$\beta=\frac{h(\bar{y})}{\sqrt{\cos\alpha_0}}=\frac{0.961651}{\sqrt{\cos(37.95735°)}}=1.08293$$

while

$$B=1.212\cdot10^{-4}\quad\beta\frac{c}{g}=5.852\cdot10^{-6}$$

The new value of pseudo-speed
Repeating again the calculations performed in (I) we find more accurate values of the pseudo-speed at the point of impact

$$u = e^{(\ln v_0-Bgx)} = e^{\ln(183)-0.0000058541\cdot\ 9.80665\cdot\ 2974)} = 154.28m/s$$

The new value of the Launching angle
At the point of impact the y-coordinate is zero. Substituting $y = 0$ on the right side of (5.3.13) as well as all known values we have

$$\tan(\alpha_0)+\frac{1}{2(5.8541\cdot10^{-6})\cos^2(\alpha_0)}[\frac{1}{2}(\frac{1}{154.28^2}-\frac{1}{183^2})\frac{1}{\ln(154.28)-\ln(183)}+\frac{1}{183^2}]=0$$

Since

$$\cos^2\alpha_0 = (1+\tan^2\alpha_0)^{-1} = (1+p_0^2)^{-1},$$

we can write the above equation as follows

$$p_0 - 0.489667(1 + p_0^2) = 0$$

Hence we find two different values for $p_0 = \tan \alpha_0$

$$p_0 = \tan \alpha_0 = 0.8145832 \quad p_0 = \tan \alpha_0 = 1.22762$$

For the corresponding launching angle we obtain respectively

$$\alpha_0 = \tan^{-1}(0.81458) = 39.165682°$$

and

$$\alpha_0 = \tan^{-1}(1.2298728) = 50.8857°$$

Let's solve the problem for the small launching angle $\alpha_0 = 39.165682°$.

III. **Note**: We can still continue to improve the accuracy of our calculation repeating all steps we did at (II). In a similar way as in (II) we obtain:

$u_m = 167.51m/s$, $x_m = 1528.38m$, $y_m = 647m$, $\bar{y} = 427m$, $h(\bar{y}) = 0.959117$,

$$B = 5.88661 \cdot 10^{-6} \quad u_T = 154.13m/s \quad \alpha_0 = 39.2219°$$

IV. Estimation of the Other Elements of the Trajectory

Using the data obtained in (III) we estimate the other elements of the trajectory.

The Impact Angle
Substituting in (5.3.8) we find the value of p at the point of impact of projectile to the target

$$p_T = p_0 - \frac{1}{2B\cos^2\alpha_0}(\frac{1}{u^2} - \frac{1}{v_0^2}) =$$

$$= \tan(39.165682°) - \frac{1}{2(5.852 \cdot 10^{-6})\cos^2(39.165682°)}(\frac{1}{154.13^2} - \frac{1}{183^2}) = -0.91258$$

Hence, since $p_T = \tan\alpha_T$, we find that the impact angle is

$$\alpha_T = \tan^{-1}(-0.91258) = -42.38295°$$

Impact Speed

Employing the pseudo-speed formula, i.e. the third formula of (5.3.4), we can find

$$v_T = u_T\frac{\cos\alpha_0}{\cos\alpha_T} = 154.13\frac{\cos(39.2219°)}{\cos(-42.3829°)} = -161.65 m/s$$

Time of Flight

Using (5.3.9) we find that the time of flight is

$$t = \frac{1}{Bg\cos\alpha_0}(\frac{1}{u} - \frac{1}{v_0}) = \frac{1}{(5.852 \cdot 10^{-6})(9.80665)\cos(39.2219°)}(\frac{1}{154.13} - \frac{1}{183}) = 23s$$

In the following table are given for comparison the data obtained above, with the data obtained by numerical integration of differential equations of projectile flight.

Table 2

Solution Method	Launching Angle [Degree]	Angle of impact	Impact speed [m/s]	Time of flight [sec.]	Maximum Height [meter]
Siacci method	$\alpha_0 = 39.22$	$\alpha_0 = -42.38$	$v_T = 162$	$t = 23$	$y_m = 647$
Numerical Integration	$\alpha_0 = 39.16$	$\alpha = -42.19$	$v_T = 162$	$t = 23$	$y_m = 639$

5.4 The Ingalls Functions of Resistance and the Siacci Method

The system of differential equations of projectile flight (5.1.16) in English units can be written as

$$\begin{cases} \dfrac{dp}{du} = \dfrac{g}{\beta \cdot c K_D(u) \cdot u \cdot \cos^2 \alpha_0} \\[2mm] \dfrac{dt}{du} = -\dfrac{1}{\beta \cdot c K_D(u) \cdot \cos \alpha_0} \\[2mm] \dfrac{dx}{du} = -\dfrac{u}{\beta \cdot c K_D(u)} \\[2mm] \dfrac{dy}{du} = -p \dfrac{u}{\beta \cdot c K_D(u)} \end{cases} , \qquad (5.4.1)$$

where

$$c = \frac{1.422334331}{C} , \qquad C = \frac{m}{id^2} \text{ (the U.S. Army ballistics coefficient)}, (5.4.2)$$

$$u = v \frac{\cos \alpha}{\cos \alpha_0} \text{ (Siacci's pseudo-speed)}, \quad p = \tan \alpha , \qquad (5.4.3)$$

and

$$\beta = h(\bar{y})/\cos \alpha_0 , \qquad (5.4.4)$$

for antiaircraft fire, or

$$\beta = \frac{h(\bar{y})}{\sqrt{\cos \alpha_0}} , \qquad (5.4.5)$$

for field artillery fire, while

$$h(y) = e^{-0.00003159145 y} \qquad (5.4.6)$$

and (see section 2.8 and section 2.9)

$$K_D(v) = \begin{cases} 1.1260 \cdot 10^{-5} v^2 & \text{for} \quad v \leq 840 \, ft./s \\ (v - 787.4016)/3 & \text{for} \quad v > 840 \, ft./s \end{cases}$$ (5.4.7)

For projectile speeds $v \geq 840 \, ft./s$, system (5.41) can be written

$$\begin{cases} \dfrac{dp}{du} = \dfrac{1}{B(u - 787.4) \cdot u \cdot \cos^2 \alpha_0} \\[2ex] \dfrac{dt}{du} = -\dfrac{1}{Bg(u - 787.4) \cdot \cos \alpha_0} \\[2ex] \dfrac{dx}{du} = -\dfrac{u}{Bg(u - 787.4)} \\[2ex] \dfrac{dy}{du} = -p \dfrac{u}{Bg(u - 787.4)} \end{cases},$$ (5.4.8)

where

$$b = c/(3g), \quad c = \dfrac{1.422334331}{C}, \quad C = \dfrac{m}{id^2}, \quad g = 32.17405 \, ft./s.$$ (5.4.9)

and

$$\beta = h(\bar{y})/\cos\alpha_0, \quad \text{or} \quad \beta = \dfrac{h(\bar{y})}{\sqrt{\cos\alpha_0}}, \quad \text{and} \quad B = \beta \cdot b.$$ (5.4.10)

Solving system (5.4.8) in the same way as system (5.2.11), we obtain the following elements of the projectile trajectory as functions of the pseudo-speed u,

The Parameter $p = \tan\alpha$

$$p = p_0 + \dfrac{1}{787.4 \, B \cos^2 \alpha_0} [\ln(\dfrac{u - 787.4}{u}) - \ln(\dfrac{v_0 - 787.4}{v_0})].$$ (5.4.11)

The Coordinates (x, y) of the Projectile Trajectory

$$x=-\frac{1}{Bg}[(u+787.4\cdot\ln(u-787.4))-(v_0+787.4\cdot\ln(v_0-787.4)],\qquad (5.4.12)$$

$$y=p_0x+\frac{1}{787.4\,B\cos^2\alpha_0}[-\frac{1}{Bg}\int_0^u\frac{u}{u-787.4}\ln(\frac{u-787.4}{u})du-J(v_0)x].\qquad (5.4.13)$$

The Trajectory Equation

$$y=p_0x+\frac{1}{240\,B\cos^2\alpha_0}[\int_0^x\ln(\frac{u-787.4}{u})dx-J(v_0)x]$$

$$(5.4.14)$$

The Time of Flight to the Point (x, y)

$$t=-\frac{1}{Bg\cos\alpha_0}[\ln(u-787.4)-\ln(v_0-787.4)]\qquad (5.4.15)$$

The above equations can be written respectively in the following compact form:

The Parameter $p=\tan\alpha$

$$p=p_0+\frac{1}{787.4\,B\cos^2\alpha_0}[J(u)-J(v_0)]\qquad (5.4.11.1)$$

The Coordinates (x, y) of the Projectile

$$x=-\frac{1}{Bg}[D(u)-D(v_0)],\qquad (5.4.12.1)$$

$$y=p_0x+\frac{x}{787.4\,B\cos^2\alpha_0}\cdot\{-\frac{1}{Bgx}[A(u)-A(v_0)]-J(v_0)\},\qquad (5.4.13.1)$$

The Trajectory Equation

$$y = p_0 x + \frac{x}{787.4\,B\cos^2\alpha_0}\cdot[\frac{A(u)-A(v_0)}{D(u)-D(v_0)}-J(v_0)] \qquad (5.4.14.1)$$

The Time of Flight to the Point (x, y)

$$t = -\frac{1}{Bg\cos\alpha_0}[T(u)-T(v_0)], \qquad (5.4.15.1)$$

The Siacci Functions

$$J(u)=\ln(\frac{u-787.4}{u}), \text{ where } J(v_0)=\ln(\frac{v_0-787.4}{v_0}) \qquad (5.4.16)$$

$$D(u)=[u+2787.4\cdot\ln(u-787.4)], \text{ where}$$
$$D(v_0)=[v_0+787.4\cdot\ln(v_0-787.4)] \qquad (5.4.17)$$

$$A(u)-A(v_0)=\int_{v_0}^{u}\frac{u}{u-787.4}J(u)du, \text{ or}$$

$$A(u)-A(v_0)=\int_{v_0}^{u}\frac{u}{u-787.4}\ln\frac{u-787.4}{u}du \qquad (5.4.18)$$

$$T(u)=\ln(u-787.4), \text{ and } T(v_0)=\ln(v_0-787.4) \qquad (5.4.19)$$

where

$$A(u)=\int_{3300}^{u}\frac{u}{u-787.4}\ln\frac{u-787.4}{u}du \qquad (5.4.20)$$

Launching Angle α_0 :

For field artillery fire

$$\sin(2\alpha_0)=-\frac{1}{393.7B}[\frac{A(u)-A(v_0)}{D(u)-D(v_0)}-J(v_0)] \qquad (5.4.21)$$

Using the above equations, we can find the elements of the projectile trajectory with a satisfactory approximation.

Example 1. A projectile of mass $m=2100lb.$ of an artillery cannon $d=16"$ is fired with an initial speed $v_0=2600\,ft./s$. The projectile form coefficient is $i=0.61$. Determine the launching angle needed to hit a target located on the ground at the horizontal distance of $x=30,000\,ft$.

Solution
The ballistics coefficients are:

$$C=\frac{m}{id^2}=\frac{2100}{(0.61)\cdot(16)^2}=13.4477459$$

and

$$c=\frac{1.422334331}{C}=\frac{1.422334331}{1.447}=0.10576749 \quad .$$

I.   As a first approach, we consider $h(\bar{y})=1$ and $\cos(\alpha_0)=\cos(0)=1$. The values of β and b are

$$\beta=1_,\qquad b=c/(3g)=(0.10577)/[(3)\cdot(32.17405)]=0.0010958_,$$

and so

$$B=\beta\cdot b=0.0010958 \quad .$$

Approximate Value of the Pseudo-speed at the Point of Impact

Using equation (5.4.12.1), we can write

$$D(u)=D(v_0)-Bgx_, \tag{1}$$

where

$D(u)=[u+787.4\cdot\ln(u-787.4)]$ and

$D(v_0)=[v_0+787.4\cdot\ln(v_0-787.4)]$.

We find that

$D(2600)=(2600)+787.4\cdot\ln[(2600)-787.4)]=8507.482325$.

Thus, substituting the above values in (1), we find that

$[u+787.4\cdot\ln(u-787.4)]=8507.48-(0.001096)\cdot(32.17405)\cdot(30,000)=7449.807$.

Hence,

$$\ln(u-787.4)=(7449.8-u)/787.4 .\qquad(2)$$

The solution of (2) obtained using a TI-83 Plus graphing calculator is

$$u_T=1915.63\,ft./s ,\quad y_1=\ln(u-787.4)=7.0284 .$$

Approximate Value of the Launching Angle

To find the launching angle we use (5.4.22),

$$\sin(2\alpha_0)=-\frac{1}{393.7B}\cdot[\frac{A(u)-A(v_0)}{D(u)-D(v_0)}-J(v_0)] .\qquad(3)$$

First, for $u=1915.63\,ft./s$, we find that

$$D(u)-D(v_0)=D(1915.63)-D(2600)=-Bgx=-1,057.675 .$$

Using (5.2.18) and a graphing calculator TI-83 Plus, we obtain

$$A(u)-A(v_0)=\int_{v_0}^{u}\frac{u}{u-787.4}\ln\frac{u-787.4}{u}du ,$$

$$A(u)-A(v_0)=\int_{v_0}^{u}\frac{u}{u-787.4}\ln\frac{u-787.4}{u}du=\int_{2600}^{1915}\frac{u}{u-787.4}\cdot\ln\frac{u-787.4}{u}du=461.69$$

For $J(v_0)$ we can write

$$J(v_0)=\ln(\frac{v_0-787.4}{v_0})=\ln(\frac{2600-787.4}{2600})=-0.3607492 \ .$$

Substituting in(3), we find

$$\sin(2\alpha_0)=-\frac{1}{393.7\,B}\cdot[\frac{A(u)-A(v_0)}{D(u)-D(v_0)}-J(v_0)]=$$

$$=-\frac{1}{393.7(0.0010958)}[\frac{461.69}{-1,057.675}-(-0.3607492)]=0.17562129 \ .$$

Hence, we obtain

$$2\alpha_0=\sin^{-1}(0.117562)=10.11481515° \ .$$

The launching angle is

$$\alpha_0=10.187144°/2=5.0574076° \ .$$

Coordinates of the Trajectory Vertex

We need to find the altitude of the trajectory vertex to better estimate $h(\bar{y})$. At the point where the vertex is located, the projectile velocity is parallel to the x-axis. Thus,

$$p=\tan\alpha=0 \ .$$

Using (5.4.11.1),

$$p_0+\frac{1}{787.4\,B\cos^2\alpha_0}[J(u)-J(v_0)]=0 \ ,$$

we find

$$J(u)=J(v_0)-787.4\,Bp_0\cos^2\alpha_0 \ . \tag{4}$$

Substituting in the above equation

$$J(u)=\ln(\frac{u-787.4}{u}), \quad J(v_0)=\ln(\frac{v_0-787.4}{v_0})=-0.3607492,$$

$$p_0=\tan\alpha_0=\tan(5.05758°)=0.0884984,$$

and

$$787.4\,Bp_0\cos^2\alpha_0=787.4(0.00109578)(0.0884985)\cos^2(5.057408°)=0.075765,$$

we have

$$\ln\frac{u-787.4}{u}=J(v_0)-787.4bp_0\cos^2\alpha_0=(-0.3607492-0.075765)=-0.436514.$$

Hence,

$$\frac{u-787.4}{u}=e^{-0.437050}=0.645939.$$

Solving the last equation for u, we obtain the pseudo-speed of the projectile at the point where the vertex of the trajectory is located:

$$u_m=2226.09\,ft/s.$$

Substituting in (5.4.12.1), we find that the x-coordinate of the trajectory vertex is

$$x_m=-\frac{1}{Bg}[(u+787.4\cdot\ln(u-787.4))-(v_0+787.4\cdot\ln(v_0-787.4)]=$$
$$=15,765.47\,ft.$$

We find as well

$$D(u_m)-D(v_0)=-bgx_m=-(0.00109578)(32.17405)(15,765.47)=-555.82.$$

Then, substituting

$$x=15,765.47\,ft$$,

$$J(v_0)=\ln(\frac{v_0-787.4}{v_0})=-0.3607492$$,

$$A(u)-A(v_0)=\int_{v_0}^{u}\frac{u}{u-787.4}\ln\frac{u-787.4}{u}du$$

$$=\int_{2600}^{2226}\frac{u}{u-240}\ln\frac{u-240}{u}=220.47933$$,

$$D(u)-D(v_0)=-555.82$$,

$$\alpha_0=5.0574°$$,

$$p_0=\tan\alpha_0=\tan(5.0574°)=0.088498$$,

and

$$B=0.00109578$$

in the trajectory equation (5.4.14.1), we find that the maximum height of the projectile trajectory is

$$y_{max}=p_0x+\frac{x}{787.4\,Bcos^2\,\alpha_0}\cdot[\frac{A(u)-A(v_0)}{D(u)-D(v_0)}-J(v_0)]=733.65\,ft$$.

II. Improving the Accuracy

The values of elements of trajectory were found assuming that $\beta=1$. We can find more accurate results for the trajectory elements considering that $y_{max}=763.65\,ft$.

Using the above value of the maximum height, we find that

$$\bar{y}=(2/3)y_{max}=\frac{2}{3}(763.65)=509.1\,ft.$$

Substituting in (6), we have

$$h(\bar{y})=e^{-0.00003159145\bar{y}}=e^{-0.00003159145(489)}=0.984045437.$$

Thus,

$$\beta=h(\bar{y})/\sqrt{\cos\alpha_0}=(0.984667)/\sqrt{\cos(5.0574°)}=0.9859665$$

and

$$B=\beta b=(0.98596655)\cdot(0.0010957846)=0.001080422.$$

For the above value of B, repeating the mathematical operations we performed in I, we find the following:

The New Value of Pseudo-speed at the Point of Impact $x_T=30,000\,ft.$
   is $u_T=1,924.9\,ft./s$

The New Launching Angle

$$\alpha_0=5.0358034°$$

The New Coordinates of the Trajectory Vertex
   The pseudo-speed u at the point where the trajectory has the maximum height is

$$u_m=2231.71\,ft/s.$$

The speed of projectile v_m at the vertex is

$$v_m=u_m\frac{\cos\alpha_0}{\cos\alpha_m}=(2231.08)\frac{\cos(5.0343°)}{\cos(0°)}=2223.10\,ft/s./$$

The x-coordinate and y-coordinate of the trajectory vertex are respectively

$$x_m=15{,}753\,ft. \qquad y_{max}=729.50\,ft.$$

(III).   The Other Elements of the Trajectory

Time of Flight to the Target:

Substituting $u_T=1{,}924.9\,ft./s$, we find that the projectile time of flight to the target is

$$t=-\frac{1}{Bg\cos\alpha_0}[\ln(u-787.4)-\ln(v_0-787.4)]=$$

$$=-\frac{1}{(0.0010804)(32.14705)\cdot\cos(5.035803°)}[\ln(1924.9-787.4)-\ln(2600-787.4)].$$

$$=13.47 \text{ sec.}$$

Angle of Impact
    For the point of impact, we can write

$$p_T=p_0+\frac{1}{787.4\,B\cos^2\alpha_0}[\ln(\frac{u_T-787.4}{u_T})-\ln(\frac{v_0-787.4}{v_0})]=$$

$$=\tan(5.035803°)+\frac{1}{0.84417}[\ln(\frac{1924.9-787.4}{1924.9})-\ln(\frac{2600-787.4}{2600})]=-0.1076859.$$

Hence, for the angle of impact, we obtain

$$\alpha_T=\tan^{-1}(-0.1076859)=-6.14626218°,$$

since $p_T=\tan\alpha_T$.

The Speed of Impact
    Using the formula for the projectile pseudo-speed,

$$u=v\frac{\cos\alpha}{\cos\alpha_0},$$

we find that the speed of projectile at the impact point is

$$v_T=u_T\frac{\cos\alpha_0}{\cos\alpha_T}=(1924.9)\frac{\cos(5.035803)}{\cos(-6.146262)}=1928.55\,ft./s \text{ ec.}$$

In table 1, given (for comparison) are the solution of the problem using the Siacci functions and the solution of the same problem using the numerical integration, i.e., the QBasic PC program **INGAAN.BAS** (see appendix D), which is compiled in English units.

Table 1. Projectile caliber $d=16''$, $m=2100lb$.
Given: Range $x=30,000\,ft$,
Launching Speed $v_0=2600\,ft./s$; Form Coefficient $i=0.61$

	Range (ft)	Launching Angle	Time of Flight (s)	Impact Speed (ft)	Angle of Impact
Ingalls	30,000	5.0358°	13.47	1928.60	- 6.1463°
INGAAN.BAS	30,000	5.0451°	13.47	1926.73	- 6.1574°

The coordinates of the trajectory vertex are respectively

- Example 1:

$$(x_m=15,753\,ft., \quad y_{max}=729.50\,ft)$$

- Numerical Integration (using PC program INGAAN.BAS):

$$(x_m=15,760\,ft., \quad y_{max}=730.70\,ft).$$

Instructions on using INGAAN.BAS (see appendix D)

Input:

Range $x = 30,000\,ft$, Initial Speed = 2,600, Ballistics
Coefficient = 13.44775, $n=10$

Output:

Launching Angle = 5.0451°, Time = 13.47 s, Terminal Speed
= 1,926.73,

Terminal Angle = - 6.15744°

5.5 The Siacci Method for Any Projectile Speed

The use of the Siacci methods we studied in the first section of this chapter is valid when the speed v of projectile along its entire trajectory is greater than $256m/s$. On the other hand, the Siacci method for relatively low speeds is appropriate when the projectile speed is always less than $256m/s$.

In many problems of exterior ballistics, we encounter problems where the projectile speed during the flight changes from greater than $256m/s$ to speeds less than $256m/s$.

To solve such problems, we will use both methods:

- The first Siacci method (section 5.1) to describe the projectile flight from launching point to the point on the trajectory where the speed of projectile becomes less than $256m/s$
- The second Siacci method to describe the flight of projectile from the point where the projectile speed just turns to be less than $256m/s$

It is obvious that the point where the projectile speed is $v_c = 256m/s$ can be considered as the point of "impact" for the projectile launched with initial speed greater than $v_c = 256m/s$, and at the same time, it can be considered as the launching point for the projectile thrown from that point with initial speed $v_c = 256m/s$.

The following example illustrates the use of the Siacci method for the general case.

Example 1. A projectile is fired from a rifle with initial speed $v_0=875m/s$. Find the range if the launching angle is $\alpha_0=19°$ and the ballistics coefficient is $c=3.6$.

Solution

Initial Part of Trajectory

Since the projectile is launched with speed greater than $v_c=256m/s$, we use the Siacci method for speeds greater than $256m/s$ to describe the trajectory from the launching point till the critical point with (x_c, y_c).

(I) Critical Point

To find the critical point, we use equation (5.2.5), i.e.,

$$p_c=p_0+\frac{1}{240B\cos^2\alpha_0}[\ln(\frac{u_c-240}{u_c})-\ln(\frac{v_0-240}{v_0})] \qquad (5.5.1)$$

where the values of β and b are respectively

$$b=c/(3g)=(3.6)/(3g)=0.1223659456 \,,$$

$$\beta=\frac{h(\bar{y})}{\sqrt{\cos\alpha_0}}\approx\frac{1}{\sqrt{\cos(19°)}}=1.028407 \,,$$

$$B=\beta\cdot b=0.1258419786 \,,$$

while the pseudo-speed is

$$u_c=v_c\frac{\cos\alpha_c}{\cos\alpha_0}=(256)(1+p_c^2)^{-1/2}/\cos(19°)=270.75(1+p_c^2)^{-1/2}. \qquad (5.5.2)$$

As a first approach, we have to consider $h(\bar{y})=1$. Substituting in (5.5.1) for the critical value of $p_c=\tan\alpha_c$, we can write

$$p_c=(0.3443276133)+0.03703592[\ln(\frac{270.75(1+p_c^2)^{-1/2}-240}{270.75(1+p_c^2)^{1/2}})+0.3205989)] .$$

Solving the last equation with a TI-83 Plus graphing calculator, we find

$$p_c=\tan\alpha_c=0.26410364 .$$

Hence, we find the angle the velocity of projectile makes with the x-axis at the critical point.

$$\alpha_c=\tan^{-1}(0.26410364)=14.79423°$$

The Pseudo-speed
The value of pseudo-speed at critical point is

$$u_c=v_c\frac{\cos\alpha_c}{\cos\alpha_0}=(256)\frac{\cos(14.79°)}{\cos(19°)}=261.7752609m/s .$$

Substituting in (5.2.9) and (5.2.24), we find the abscissa of the critical point

$$x_c=-\frac{1}{Bg}[(u_c+240\cdot\ln(u_c-240))-(v_0+240\cdot\ln(v_0-240)]=1152.84 \; m .$$

Substituting in (5.2.24), we find the ordinate of the maximum point,

$$y_c=p_0x_c+\frac{x_c}{240B\cos^2\alpha_0}\cdot[\frac{A(u_c)-A(v_0)}{D(u_c)-D(v_0)}-J(v_0)]=368.0182$$

since

$$A(u_c)-A(v_0)=\int_{v875}^{261.775}\frac{u}{u-240}\ln\frac{u-240}{u}du=1420.324079$$

and

$$[D(u_c)-D(v_0)]=-1422.706543$$

(II) Improving Outcomes

The values of elements of trajectory were found supposing that

$$\beta=\frac{h(\bar{y})}{\sqrt{\cos\alpha_0}}=\frac{1}{\sqrt{\cos(19°)}}=1.028407$$

and that the trajectory will reach the target on the ground before decelerating to the critical speed.

We need to modify that approach.

Considering $y_{max}=368.0182m$, we do find more accurate results for the trajectory elements at the critical point. Since the critical point is at the ascending part of projectile trajectory, we have

$$\bar{y}=(1/2)y_{max}=(1/2)\cdot(368.01817)=184.001m$$

and

$$h(\bar{y})=(\frac{289.08-0.006328\bar{y}}{289.08})^{4.4}=0.9823966320$$

The values of the b and β are respectively

$$b=c/(3g)=0.1223659456$$

and

$$\beta=h(\bar{y})/\cos\alpha_0=(1.03902995)/\cos(19°)=1.039003$$

while

$$B=\beta\cdot b=0.1271386$$

The New Critical Angle

Using the above values and substituting in (5.5.1), we find

$$p_c=(0.3443276133)+0.0366582056[\ln(\frac{270.75(1+p_c^2)^{-1/2}-240}{270.75(1+p_c^2)^{1/2}})+0.3205989)]=0.26484744 \ .$$

Hence, for the projectile angle at the critical point, we have

$$\alpha_c=\tan^{-1}(0.26484744)=14.8341° \ .$$

The New Pseudo-speed

$$u_c=v_c\frac{\cos\alpha_c}{\cos\alpha_0}=(256)\frac{\cos(14.834°)}{\cos(19°)}=261.7271377m/s$$

The New Coordinates of Critical Point

For the abscissas of the critical point, substituting $u_c=261.7271377m/s$, we obtain

$$x_c=-\frac{1}{Bg}[(u_c+240\cdot\ln(u_c-240))-(v_0+240\cdot\ln(v_0-240)]=1141.549458m \ .$$

Substituting in (5.2.24), we find that the new ordinate of the critical point is

$$y_c=p_0x_c+\frac{x_c}{240B\cos^2\alpha_0}[\frac{A(u_c)-A(v_0)}{D(u_c)-D(v_0)}-J(v_0)]=364.681 \ ,$$

where

$$A(u_c)-A(v_0)=\int_{875}^{261.727}\frac{u}{u-240}\ln\frac{u-240}{u}du=1421.7647$$

and

$$[D(u_c)-D(v_0)]=-1423.28 \ .$$

Time of Flight to Critical Point

Substituting in (5.2.8), we find the time of flight to the critical point

$$t_c = -\frac{1}{Bg\cos\alpha_0}[\ln(u_c - 240) - \ln(v_0 - 240)] = 2.86295s.$$

Second Part of the Trajectory

II. Since the speed of the trajectory behind the critical point is less than $256m/s$, we use the Siacci method for speeds less than $256m/s$ to describe the trajectory of the projectile "fired" at the "launching point" that corresponds to the critical point,

$$(x_c = 1141.549458, y_c = 364.681),$$

with "initial speed" $v_c = 256m/s$ until it hits the ground at $y=0$.

The origin of a new coordinate system will be on the x-axis, at the point with coordinates $(x_c = 1141.549458, \ 0)$.

The initial conditions are

$$(x_0 = 0, \ y_0 = y_c = 364.681), \quad v_0 = 256m/s, \quad \alpha_0 = 14.8341°. \tag{3}$$

As a first approximation, we consider

$$\bar{y} = (2/3)y_{max} = (2/3)\cdot(364.681) = 243.1207m.$$

Substituting in $h(\bar{y})$, we have

$$h(\bar{y}) = (\frac{289.08 - 0.006328\bar{y}}{289.08})^{4.4} = 0.9767928116$$

and

$$\beta = \frac{h(\bar{y})}{\sqrt{\cos\alpha_0}} = \frac{(0.9767928)}{\sqrt{\cos(14.83141°)}} = 0.9934828352.$$

We find

$$B=1.212 \cdot 10^{-4} \beta \frac{c}{g}=4.42022944 \cdot 10^{-5} . \tag{4}$$

Substituting $y=0$ into the equation of trajectory (5.3.11), we can write

$$y_0 + p_0 x_T + \frac{x_T}{2 B \cos^2 \alpha_0} [\frac{1}{2}(\frac{1}{u_T^2} - \frac{1}{v_0^2}) \cdot (\frac{1}{\ln u_T - \ln v_0}) + \frac{1}{v_0^2}] = 0 \tag{5}$$

Substituting in (5) the equation (5.3.10), i.e., substituting

$$x_T = -\frac{1}{Bg}(\ln u_T - \ln v_0)$$

in equation (5), we find that

$$y_0 + p_0 [\frac{-1}{Bg}(\ln u_T - \ln v_0)] - \frac{1}{2 B^2 g \cos^2 \alpha_0}(\ln u_T - \ln v_0)[\frac{1}{2}(\frac{1}{u_T^2} - \frac{1}{v_0^2}) \cdot (\frac{1}{\ln u_T - \ln v_0}) + \frac{1}{v_0^2}] = 0 . \tag{6}$$

Substituting in (6) the values given in (3) and (4) and then solving the obtained equation, we find that the pseudo-speed at the point of the impact is

$$u_T = 92.42895119 m/s .$$

The x-coordinate of the point of impact is

$$x_T = -\frac{1}{Bg}(\ln u_T - \ln v_0) = 2355m .$$

Trajectory Vertex
Trajectory vertex is located at the point where $p=0$. Substituting in equation (5.3.8), we have

$$p_0 - \frac{1}{2 B \cos^2 \alpha_0}(\frac{1}{u_m^2} - \frac{1}{v_0^2}) = 0 \tag{7}$$

Substituting in (7) the known values and solving, we get the pseudo-speed of vertex

$$u_m=164.0936152 m/s.$$

For the coordinates of the vertex that corresponds to the pseudo-speed $u_m=164.0936152 m/s$, we find

$$x_m=-\frac{1}{Bg}(\ln u_m-\ln v_0)=-\frac{1}{4..3348\cdot10^{-4}}(\ln164.09-\ln256)=1031.13$$

and

$$y_m=p_0x+\frac{x}{2Bcos^2\alpha_0}[\frac{1}{2}(\frac{1}{u_m^2}-\frac{1}{v_0^2})\cdot(\frac{1}{\ln u_m-\ln v_0})+\frac{1}{v_0^2}]=520.55563684.$$

III. Improving Accuracy

Because the altitude changes, we calculate an average altitude maximum height as follows:

$$h(\bar{y})=\frac{h(y_1)+h(y_2)}{2}=\frac{0.97678467+0.9670361364}{2}=0.9719104.$$

In the same way as in II, we find

$$B=4.313122\cdot10^{-5}.$$

For this new value of B, we find that the pseudo-speed at the point of impact is

$$u_T=92.709313 m/s.$$

The abscissa of point of impact in the new system of coordinates is

$$x_T=-\frac{1}{Bg}(\ln u_T-\ln v_0)=2354.93m.$$

The $p_T=\tan\alpha_T$ at the point of impact is

$$p=p_0-\frac{1}{2Bcos^2\alpha_0}(\frac{1}{u_T^2}-\frac{1}{v_0^2})=-0.964966432$$

Hence, for the angle of projectile at impact point, we obtain

$$\alpha_T = -43.99°.$$

Time of Flight
   The time of flight is

$$t_T = \frac{1}{Bg\cos\alpha_0}(\frac{1}{u_T} - \frac{1}{v_0}) = 16.5\text{sec}.$$

V. Summary Data

   Combining results of both parts of the trajectory, for the elements of the trajectory, we obtain the following:

Range: $x_T = 1141.5 + 2355 = 3496m$
Angle of Impact: $\alpha_T = -43.99°$
Impact Speed: $v_T = u_T\frac{\cos\alpha_0}{\cos\alpha_T} = (92.7093)\frac{\cos(19°)}{\cos(43.99°)} = 121.92m/s$
Time of Flight: $t_T = 2.86 + 16.5 = 19.40\text{sec}.$

Trajectory Vertex: $x_m = 1031.13 + 1141.55 = 2172.70m$ $y_m = 520.56$

Comment: The results obtained using the general Siacci methods are approximate since the Siacci ballistics coefficient $c = 3.6$ is not appropriate to be used for a large range of projectile velocities.

Note: Using QBasic PC program RangeC.Bas (See appendix C), we obtain the following data:

Range: $x_T = 3552m$
Angle of Impact: $\alpha_T = -44°$
Impact Speed: $v_T = 120m/s$
Time of Flight: $t_T = 19.60\text{sec}.$
Trajectory Vertex: $x_m = 2205m$ $y_m = 526$.

5.6 The Tabulated Siacci Functions

The analytical solutions obtained in preceding sections using the Siacci method can be programmed using different PC programming languages.

Another alternative way to find the elements of the projectile trajectory that is traditionally used in exterior ballistics and is familiar to ballisticians is based on the tables of Siacci functions (see appendix A).

Siacci function tables can save time and are suitable for ballisticians that prefer to estimate the elements of projectile trajectory using the formulas obtained in preceding sections and the tabulated values of Siacci's functions.

We are going to construct the tables of Siacci functions only for projectile speeds greater than 256 m/s.

Consider the equations (5.2.27)–(5.2.29) that present the elements of projectile trajectory as functions of pseudo-speed u, i.e.,

Coordinates of Projectile:

Abscissa

$$x=-\frac{1}{Bg}[D(u)-D(v_0)],$$
(5.6.1)

Ordinate

$$y=p_0x+\frac{x}{240Bcos^2\alpha_0}[\frac{A(u)-A(v_0)}{D(u)-D(v_0)}-J(v_0)]$$
(5.6.2)

Time of Flight

$$t=-\frac{1}{Bgcos\alpha_0}[T(u)-T(v_0)],$$
(5.6.3)

Angle of Flight

$$p=p_0+\frac{1}{240B\cos^2\alpha_0}[J(u)-J(v_0)]$$

$$\text{(5.6.4)}$$

Launching angle:

For field artillery fire

$$\sin(2\alpha_0)=-\frac{1}{120B}\cdot[\frac{A(u)-A(v_0)}{D(u)-D(v_0)}-J(v_0)]$$

$$\text{(5.6.5)}$$

In equations (5.6.1)–(5.6.6), are given the following functions of pseudo-speed, known as Siacci functions:

$$D(u)=[u+240\cdot\ln(u-240)]\text{, } T(u)=\ln(u-240)\text{, } J(u)=\ln(\frac{u-240}{u}) \text{ (5.6.6)}$$

and

$$A(u)-A(v_0)=\int_{v_0}^{u}\frac{u}{u-240}\ln\frac{u-240}{u}du$$

$$\text{(5.6.7)}$$

Siacci function $A(u)-A(v_0)$ depends on the initial speed of projectile v_0. It can be written in the following form:

$$A(u)-A(v_0)=\int_{v_0}^{u}\frac{u}{u-240}\ln\frac{u-240}{u}du=\int_{256}^{u}\frac{u}{u-240}\ln\frac{u-240}{u}du-\int_{256}^{v_0}\frac{u}{u-240}\ln\frac{u-240}{u}du \text{ .}$$

Denoting

$$A_1(u)= \int_{256}^{u} \frac{u}{u-240} \ln\frac{u-240}{u} du \tag{5.6.8}$$

we can write (5.6.9)

$$A(u)-A(v_0)=A_1(u)-A_1(v_0) \tag{5.6.9}$$

To simplify the calculation of trajectory elements and solve the different ballistics problem using the Siacci method, we will use the table of the Siacci functions $D(u)$, $T(u)$, $J(u)$, and $A_1(u)$.[46]

Note: In Appendix A, Siacci's Functions, there is a disorder in the listing of the pseudo-speed and the corresponding values of the Siacci's functions.

Example 1. A projectile of mass 27.3 kg is launched at a speed of $v_0=885m/s$ and an angle of $\alpha_0=5.60°$ from a field artillery cannon 122 mm. Find the elements of the trajectory using the tables of Siacci's functions included in appendix A. Ballistics coefficient is $c=0.238845$.

Solution

I. As a first approach, we consider $\beta=1$. The value b is

$$b=c/(3g)=(0.238845)/(3 \cdot g)=0.00811847$$

Thus,

$$B=b\beta=0.00811847$$

---

[46] The Siacci tables included in Appendix A are constructed using Excel and the free software "XNUMBERS 5.3 - Multi Precision Floating Point Computing and Numerical Methods for EXCEL", prepared by Leonardo Volpi & Fox team. http://digilander.libero.it/foxes/ [Web site]

II. Coordinates of the Trajectory Vertex

Estimate $h(\bar{y})$ and the correction coefficient β. At the point where the vertex is located, the projectile velocity is parallel to the x-axis, i.e.,

$$p=\tan\alpha=0$$.

Using (5.2.33),

$$p_0+\frac{1}{240B\cos^2\alpha_0}[J(u)-J(v_0)]=0$$,

we find

$$J(u)=J(v_0)-240Bp_0\cos^2\alpha_0 \qquad (1)$$

In the table displayed in appendix A, we find that

$$J(v_0)=J(885)=-0.31634$$.

Substituting

$$p_0=\tan\alpha_0=\tan(5.60°)$$

in (1) and

$$240Bp_0\cos^2\alpha_0=240(0.00811847)\tan(5.60°)\cos^2(5.60°)=0.1892263$$,

we obtain

$$J(u)=J(v_0)-240Bp_0\cos^2\alpha_0=(-0.31634-0.189226305)=-0.5055663$$.

Using the Siacci function table (appendix A), we find the pseudo-speed at the point where the vertex of the trajectory is located:

$$u_m=605m/s$$.

Coordinates of the Trajectory Vertex
  Using the table included in appendix A, we find

$$x_m = -\frac{1}{Bg}[D(u_m) - D(v_0)] = -\frac{1}{0.0823337}[D(605) - D(885)] =$$

$$-\frac{1}{0.0079615}(2020.9754 - 2437.6201) = 5233.24m$$

  We find as well

$$D(u_m) - D(v_0) = -Bgx_m = -416.6447 \ .$$

  Then, substituting

$$x_m = 5233.24m \ ,$$

$$J(v_0) = J(885) = -0.31634 \ ,$$

and (see table in appendix A)

$$A(u_m) - A(v_0) = A_1(u_m) - A_1(v_0) = A_1(605) - A_1(885)$$
$$= (-1468.14096) - (-1634.42276) = 166.2818 \qquad ,$$

$$p_0 = \tan\alpha_0 = \tan(5.60°) = 0.09805086 \ ,$$

$$B = 0.00811847$$

in the trajectory equation (5.2.24), we find that the maximum height of the projectile trajectory is

$$y_{max} = p_0 x + \frac{x}{240 B \cos^2 \alpha_0} \cdot [\frac{A(u_m) - A(v_0)}{D(u_m) - D(v_0)} - J(v_0)] = 288.70m \ .$$

  We find more accurate results for the trajectory elements considering that $y_{max} = 288.70m$.

Using the above value of the maximum altitude, we find that

$$\bar{y}=(2/3)y_{max}=\frac{2}{3}(288.70)=192.50m$$

and

$$h(\bar{y})=(\frac{289.08-0.006328\bar{y}}{289.08})^{4.4}=(\frac{289.08-0.006328(191.046)}{289.08})^{4.4}=0.98159$$.

Thus,

$$\beta=h(\bar{y})/\sqrt{cos\alpha_0}=(0.98159)/\sqrt{cos(5.60°)}=0.983944$$,

and

$$B=\beta\cdot b=(0.98159)(0.00811847)=0.00798812$$.

III. Repeating the same mathematical operations we performed in II using the above value of B, we find that the new value of the pseudo-speed that corresponds to the vertex and the coordinates of the vertex are respectively

$$u_m=607.5m/s, \quad x_m=5265.80m, \quad y_m=292.53m$$.

IV. The Horizontal Range

Consider the equation of the projectile trajectory (5.2.24),

$$y=p_0x+\frac{x}{240Bcos^2\alpha_0}[\frac{A(u)-A(v_0)}{D(u)-D(v_0)}-J(v_0)]$$.   (2)

At the point of impact, the y-coordinate is zero. Substituting in (2), we write

$$tan(5.60°)+\frac{1}{240(0.00798812)cos^2(5.60°)}[\frac{A(u_T)-A(885)}{D(u_T)-D(885)}-J(885)]=0$$, (3)

where

$$J(885)=-0.31634, \quad D(885)=2437.6201,$$

$$A(u_T)-A(885)=A_1(u_T)-A_1(885)=A_1(u_T)-(-1634.42276).$$

Substituting in (3), we obtain the following equation:

$$A(u_T)+1634.42276)=-0.502528 \cdot [D(u_T)-2437.6201].$$

Hence, we have

$$A(u_T)=-1634.42276-0.502528 \cdot [D(u_T)-2437.6201]$$

Using the table included in appendix A and the trial-and-error procedure, we find that the value of the pseudo-speed at the point of impact is

$$u_T=435.5m/s.$$

The range is

$$x_T=-\frac{1}{Bg}[D(u_T)-D(v_0)]=-\frac{1}{0.0783367}[D(435)-D(885)]$$

$$=-\frac{1}{0.0783367}(1701.255-2437.6201)=9400m.$$

Angle of Impact
Substituting in (5.2.6) and using the table of appendix A, we have

$$p_T=p_0+\frac{1}{240Bcos^2\alpha_0}[J(u_T)-J(v_0)]=\tan(5.60°)+\frac{1}{1.9632028}[J(435.5)-J(885)]$$

$$=\tan(5.60°)+\frac{1}{1.89889}[(-0.800935)-(-0.31634)]=-0.1571478.$$

Thus,

$$p_T=\tan(\alpha_T)=-0.1571478.$$

Hence, we find that the impact angle is

$$\alpha_T = \tan^{-1}(-0.1571478) = -8.930866°\ .$$

Speed of Impact

$$v_T = u_T \frac{\cos\alpha_0}{\cos\alpha_T} = 435.5 \frac{\cos(5.60°°)}{\cos(-8.9309°)} = 438.74 m/s$$

Time of Flight
   The time of flight to the target is

$$t_T = -\frac{1}{Bg\cos\alpha_0}[T(u_T) - T(v_0)] = -\frac{1}{0.0779628}[(5.27500) - (6.46925)] = 15.30 \text{sec.}$$

Note: The table below compares the elements of the trajectory obtained theoretically using the Siacci method (tabulated Siacci functions) and the corresponding elements of the trajectory given in the range table of artillery cannon 122 mm.

Table 1

	Range	Height	Impact angle	Time of flight	Impact speed
Example 1	9,400 m	288.70 m	- 8.93°	15.30 sec.	438.74 m/s
Range Table	9,400 m	287.00 m	- 9.02°	15.00 sec.	436 m/s

Example 2. Use the tabulated Siacci function to construct the trajectory of flight for the projectile of example 1: projectile mass = 27.3 kg, caliber = 122 mm; initial speed of projectile $v_0 = 885 m/s$; launching angle $\alpha_0 = 5.60°$; ballistics coefficient $c = 0.2469$.

Solution
   To construct the graph, we estimate the coordinates of the trajectory for the given projectile employing the parametric formulas:

$$x = -\frac{1}{Bg}[D(u) - D(v_0)]$$

,

$$y=p_0 x+\frac{x}{240B\cos^2\alpha_0}\cdot[\frac{A(u)-A(v_0)}{D(u)-D(v_0)}-J(v_0)]$$,

which give respectively the x-coordinate and y-coordinate of projectile trajectory as functions of pseudo-speed u.

From the table of Siacci functions included in appendix A, we find the values of $D(u)$, and $A_1(u)$, and then we determine the coordinates of the projectile using the above formulas.

The results of the calculations are presented in table 1.

Table 1. Coordinates of Projectile 122 mm

u	D(u)	A₁(u)	x-coordinate	y-coordinate
885.0	2437.62	-1634.42	0.00	0.00
855.0	2396.19	-1621.04	511.56	48.40
825.0	2354.19	-1606.89	1030.17	93.85
795.0	2311.55	-1591.90	1556.59	135.98
765.0	2268.22	-1575.95	2091.68	174.36
735.0	2224.09	-1558.92	2636.46	208.45
705.0	2179.09	-1540.66	3192.15	237.62
675.0	2133.08	-1520.98	3760.19	261.08
645.0	2085.93	-1499.65	4342.37	277.84
615.0	2037.46	-1476.39	4940.85	286.65
585.0	1987.45	-1450.82	5558.35	285.89
555.0	1935.62	-1422.47	6198.35	273.45
525.0	1881.60	-1390.68	6865.35	246.44
495.0	1824.90	-1354.59	7565.36	200.86
465.0	1764.86	-1312.92	8306.68	130.93
435.0	1700.52	-1263.80	9101.16	27.93
428.5	1685.88	-1251.91	9282.00	0.00

The graph of the projectile trajectory is presented in the following chart (figure 12).

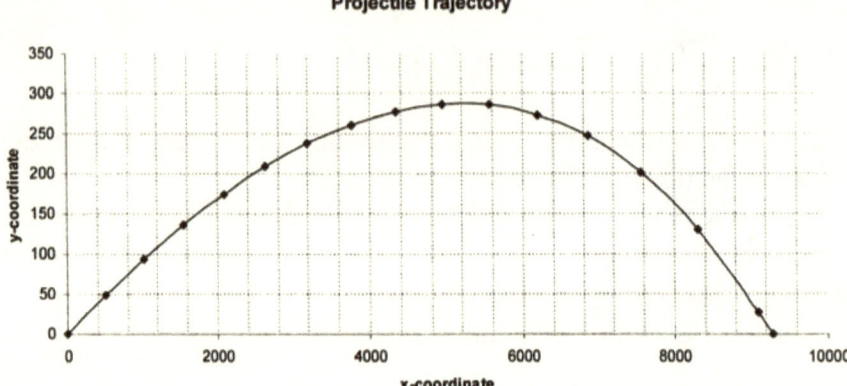

Figure 12

Note: We can use the QBasic PC program RangeC.BAS (see appendix C) to find the coordinates of the points that are on the trajectory.

RangeC.Bas Results

Input: Launching Angle = 5.60°, Initial Speed = 885, Ballistics Coefficient=0.2469, x-coordinate of a point on trajectory = 1,030 m

Output: y-coordinate =93.96.

The following tables are prepared by Prof. James Lewis
Trajectory
22 Caliber, CCI long Rifle, 40-grain Projectile
$b = 2.206727928$, $p_0 = 2.900381.10^{-3}$ *

Elevation Angle 0°

Method of Computation: Section 3.3 (See Example 6, section 3.5)

Author Prof. James Lewis

Horizontal Range (yards)	Distance (meters)	Drop (cm)	Elevation (Inches)	P (MOA)
0	0	0	-1.2	9.971
5	4.572	0.0728	-0.707	8.872
10	9.144	0.2931	-0.271	7.753
13.4302	12.281	0.5178	0	6.972
15	13.716	0.6637	0.105	6.611
20	18.288	1.1877	0.421	5.448
30	27.432	2.7070	0.867	3.055
42.2465	38.631	5.4534	1.064	0
50	45.721	7.7159	0.983	-2.006
60	54.865	11.2547	0.634	-4.677
65	59.437	13.2938	0.353	-6.049
70	64.009	15.5170	0	-7.445
75	68.581	17.9275	-0.427	-8.866
80	73.153	20.5286	-0.929	-10.312

Elevation Angle 30°
B=0.241024348, $\alpha_0 = 30.166179124°$
Method of Computation: Siacci

Distance (yards)	D(u)	A(u)	Drop (cm)	Elevation, y' (inches)
0	1556.2303	-1133.42940	0	-1.2
5	1545.4236	-1122.36154	0.0979	-0.541
9.213	1536.3171	-1112.87863	0.3333	0
10	1534.6169	-1111.09204	0.3931	0.101
15	1523.8103	-1099.62897	0.8777	0.546
20	1513.0036	-1087.92995	1.5932	0.962
30	1491.3902	-1063.90961	3.6330	1.554
40	1469.7769	-1038.99480	6.5481	1.800
50	1448.1635	-1013.15292	10.3702	1.690
60	1426.5501	-986.35387	15.1288	1.211
65	1415.7434	-972.56568	17.8847	0.823
70	1404.9368	-958.53367	20.8868	0.339
72.907	1398.6530	-950.25854	22.7434	0
75	1394.1301	-944.24262	24.13844	-0.244
80	1383.3234	-929.68926	27.6468	-0.928

* b established via a muzzle velocity of 1235ft/s and a second of 1036ft/s at 100yards.

6

Elements of the Theory of Corrections

Introduction

In preceding chapters, we studied the problems of exterior ballistics for a given projectile assuming that the projectile characteristics are standard and that it is launched in a standard atmosphere.

We ignored a set of other factors that influence the projectile flight. We assumed the following:

- The initial speed v_0 and the ballistics coefficient c for a projectile of a given mass and caliber are standard values.
- The atmospheric conditions are standard.
- The only forces that act on the projectile are the gravity and the air resistance.

In practice, the projectile characteristics as well as the meteorological factors—such as temperature, humidity, density, pressure, etc.—usually are different from the accepted standard ones.

Some other factor that influence the projectile flight that need to be considered are the wind, the change of gravity with altitude, the Coriolis and Magnus effects, etc.

Thus, for example, hereafter are shown some factors that influence the projectile flight:

(a) The initial speed of the projectile might be different from the standard values because of the consumption of the barrel as result of previous fires or as result of changes in temperature or in mass of the black powder the projectile shell contains, etc.

(b) The ballistics coefficient c might be different from the standard value, for example, as result of the variation in the mass of projectile.

(c) The projectile is launched usually in a windy atmosphere.

It is obvious that the ballistics problems for nonstandard values can be solved using methods of chapters 3 or chapter 4, making appropriate changes in the system of differential equations of projectile flights considering the variation of the flight parameters from the standard ones, for example, considering a different initial speed of the projectile, or a different projectile mass.

Another method of study of the projectile flight in nonstandard conditions is the "theory of corrections." The theory of corrections makes adjustments, called "corrections," to the elements of the projectile flight already obtained for standard conditions in order to take into account some relatively small variations of factors from the standard ones, such as the relatively small variations of temperature, launching speed, etc.

Using the theory of correction, we calculate the data for firearm shootings in different atmospheric conditions, for shooting uphill or downhill, by employing the range tables prepared for shooting in standard conditions.

In this chapter, we study the influence of some of the abovementioned factors to the projectile flight and reflect them in the elements of flight as "corrections."

Note: Some new elements of the Correction Theory are introduced in my book Exterior Ballistics: A New Approach, Xlibris, 2010.

6.1 The Correction Coefficients for High Speeds

The elements of projectile trajectory are functions of the projectile parameters and the parameters that characterize the atmospheric air.

Thus, the horizontal range of a projectile is a function of initial speed, the projectile mass, ballistics coefficient, initial launching angle, air temperature, atmospheric pressure, temperature of the projectile propellant, and so on:

$$x = f(v_0, \ c, \ \alpha_0, m \ T, H, T_c \ldots) \qquad (6.1.1)$$

Consider some relatively small changes dv_0, dc, $d\alpha_0$, dm, dT_0, dp_0, dT_c respectively corresponding to the values of initial speed v_0, ballistics coefficient c, launching angle α_0, the projectile mass m, the air temperature T_0, atmospheric pressure p_0, and the temperature of projectile propellant T_c. The linear change Δx of the horizontal range (obtained in standard conditions) can be considered approximately equal to the total differential of the horizontal range x,

$$\Delta x = \frac{\partial x}{\partial v_0} dv_0 + \frac{\partial x}{\partial c} dc + \frac{\partial x}{\partial \alpha_0} d\alpha_0 + \frac{\partial x}{\partial m} dm + \frac{\partial x}{\partial T_0} dT_0 + \frac{\partial x}{\partial p_0} dp_0 + \frac{\partial x}{\partial T_c} dT_c. \quad (6.1.2)$$

The partial derivatives

$$\frac{\partial x}{\partial v_0}, \ \frac{\partial x}{\partial c}, \ \frac{\partial x}{\partial \alpha_0}, \ \frac{\partial x}{\partial m}, \ \frac{\partial x}{\partial T_0}, \ \frac{\partial x}{\partial p_0}, \ \frac{\partial x}{\partial T_c}$$

are called "correction coefficients." They correspond respectively to the following individual changes of the respective parameters dv_0, dc, $d\alpha_0$, dm, dT_0, dp_0, dT_c.

There are different methods to estimate the correction coefficients needed to evaluate the linear change of horizontal distance or other trajectory elements.

Hereafter are presented the Siacci correction coefficients that are determined using the Siacci method of solution of ballistics problems when the projectile speed v all over the trajectory is greater than $256m/s$, i.e., $v>256m/s$. For such projectiles, the drag function is

$$K_D(v)=(v-240)/3.$$

The correction coefficients presented hereafter are not valid for projectiles flying with speeds $v<256m/s$.

In exterior ballistics are estimated the following correction coefficients:[47]

- The correction coefficient that corresponds to the change dv_0 of initial speed is

$$\frac{\partial x}{\partial v_0}=\frac{v_0}{\beta c K_D(v_0)}(1+\frac{\tan\alpha_0}{\tan|\alpha_T|}-\frac{gx}{v_0^2\cos^2\alpha_0\tan\alpha_T}). \qquad (6.1.3)$$

- The correction coefficient that corresponds to the change dc of ballistics coefficient is

$$\frac{\partial x}{\partial c}=-(1-\frac{\tan\alpha_0}{\tan|\alpha_T|})\frac{x}{c}. \qquad (6.1.4)$$

- The correction coefficient that corresponds to the change $d\alpha_0$ of launching angle is

$$\frac{\partial x}{\partial\alpha_0}=\frac{x\cos(2\alpha_0)}{\cos^2\alpha_0\tan|\alpha_T|}. \qquad (6.1.5)$$

- The correction coefficient that corresponds to the change dm of projectile mass is

$$\frac{\partial x}{\partial m}=(1-\frac{\tan\alpha_0}{\tan|\alpha_T|}-0.4\frac{v_0}{x}\frac{\partial x}{\partial v_0})\frac{x}{m}. \qquad (6.1.6)$$

---

[47]   Shapiro, J. M. *Vneshnaja Balistika*, p. 188. Oborongiz, 50'.

- The correction coefficient that corresponds to the change dT_0 of air temperature is

$$\frac{\partial x}{\partial T_0}=(1-\frac{v_0}{2x}\frac{\partial x}{\partial v_0})\frac{x}{T_0}.$$
(6.1.7)

- The correction coefficient that corresponds to the change dp_0 of atmospheric pressure is

$$\frac{\partial x}{\partial p_0}=-(1-\frac{\tan\alpha_0}{\tan|\alpha_T|})\frac{x}{p_0}.$$
(6.1.8)

- The correction coefficient that corresponds to the change dT_c of the temperature of explosive load of propellant is

$$\frac{\partial x}{\partial T_c}=0.001\cdot v_0\frac{\partial x}{\partial v_0},$$
(6.1.9)

where α_T is the angle of impact of projectile to the target,

$$K_D(v_0)=(v_0-240)/3,$$
(6.1.10)

and,

$$\beta=h(\bar{y})/\cos\alpha_0,\beta=\frac{h(\bar{y})}{\sqrt{\cos\alpha_0}},$$
(6.1.11)

respectively, for the antiaircraft fire and for field artillery fire.

The correction coefficients (6.1.3), (6.1.4), and (6.1.5) are basic coefficients, while the other correction coefficients determined in (6.1.6), (6.1.7), (6.1.8), and (6.1.9) are secondary, derived correction coefficients.

We note that any change in the temperature of the propellant load alters the initial speed and, as result, the horizontal range and other elements of the trajectory.

For example, the increase of the temperature of propellant when the loaded projectile is exposed to sunlight increases the thrust force and the projectile initial speed.

We will illustrate the application of correction formulas in the following examples:

Example 1. A 7.9 g bullet is fired from a Russian rifle caliber 7.62 mm with an initial speed of 735 m/s under a launching angle of $\alpha_0 = 0.216°$ in standard atmosphere (air density is $\rho_0 = 1.205 kg/m^3$, air temperature is 288.15 Kelvin the pressure $p_0 = 750 mm$ Hg, and humidity 50%.

The tableau values of the projectile are the projectile range $x = 300$ m and the angle of impact to the ground $\alpha_T = 0.276°$.

The ballistics coefficient is $c = 4.219797$ (see example 3 in section 3.3).

I.   Find the correction coefficients that correspond to the small changes in initial speed, pressure, air temperature, and the temperature of black powder propellant.

II. Use approximation (6.1.2) to determine the change in horizontal range when the change in initial speed is $dv_0 = 10 m/s$, the temperature change is $dT_0 = 10$ K, the pressure change is $dH_0 = 10 mm$ Hg, and the change in propellant temperature is $dT_c = 10$ K.

Solution

We can consider $\beta = 1$.

(a) Substituting in (6.1.3), we find the correction coefficient corresponding to the change in initial speed of projectile

$$\frac{\partial x}{\partial v_0} = \frac{v_0}{\beta \cdot c K_D(v_0)}(1 + \frac{\tan \alpha_0}{\tan|\alpha_T|} - \frac{gx}{v_0^2 \cos^2 \alpha_0 \tan \alpha_T})$$

$$= \frac{735}{(4.219797) \cdot (735 - 240)/3}(1 + \frac{\tan(0.216°)}{\tan(0.276°)} - \frac{9.80665 \cdot (300)}{735^2 \cdot \cos^2(0.216°)\tan(0.276°)})$$

$$= 1.055630272 \cdot (0.65207) = 0.688347$$

The change in the horizontal distance, which corresponds only to a change $dv_0=10m/s$ in launching speed (assuming that all other characteristics are standard), according to (6.1.2), is

$$\Delta x_1 = \frac{\partial x}{\partial v_0} dv_0 = (0.688347)(10) = 6.88347m .$$

(b) Change in air temperature T_0

Substituting in (6.1.7), we find the temperature correction coefficient

$$\frac{\partial x}{\partial T_0} = (1 - \frac{v_0}{2x} \frac{\partial x}{\partial v_0}) \frac{x}{T_0} = (1 - \frac{(735)}{2(300)} \cdot (0.688347)) \frac{300}{299.08} = 0.15104 .$$

The change in the horizontal distance that corresponds only to the change $dT_0=10Kelvin$, estimated using (6.1.2), is

$$\Delta x_2 = \frac{\partial x}{\partial T_0} dT_0 = (0.15104)(10) = 1.51m .$$

(c) Change in propellant (black powder load) temperature T_c
    Substituting in (6.1.9), we find

$$\frac{\partial x}{\partial T_c} = 0.001 \cdot v_0 \frac{\partial x}{\partial v_0} = 0.001 \cdot (735)(0.688347) = 0.5059 .$$

The change in the horizontal distance that corresponds only to the change in propellant temperature $dT_c=10Kelvin$ is

$$\Delta x_3 = \frac{\partial x}{\partial T_c} dT_c = (0.5059)(10) = 5.059m .$$

(d) Change in the atmospheric pressure H_0
    Substituting in (6.1.8), we find that

$$\frac{\partial x}{\partial p_0}=-(1-\frac{\tan\alpha_0}{\tan|\alpha_T|})\frac{x}{p_0}=-(1-\frac{\tan(0.216°)}{\tan(0.276)})\frac{300}{750}=-0.086957$$

The change in the horizontal distance that corresponds only to the change $dH_0=10mm$ Hg, is

$$\Delta x_4=\frac{\partial x}{\partial p_0}dp_0=-(0.086957)(10)=-0.86947m$$

The total change in distance as result of respective changes in initial speed, air temperature, propellant temperature, and atmospheric pressure is

$$\Delta x = 6.88347 + 1.5104 + 5.059 - 0.86947 = 12.6m.$$

Thus, the actual range of fire is

$$x=x_0+\Delta x=300+12.64=312.64m$$

The change in vertical direction (figure 1) is

$$\Delta y_1=\Delta x_1\cdot\tan\alpha_T=(12.64)\cdot(\tan0.276°)=0.061m$$

The bullet, at the distance $x_0=300m$, will pass 6.1 cm over the horizontal line.

Note: As it is shown in the following table, the results obtained above using the Siacci correction method match perfectly with the correction data given by the range table of the 7.62 mm Russian rifle.

Table 1. (Distance 300 m)

Data obtained from	Change in temperature of air & propellant $dT_0=10$, $dT_c=10$	Change in atmospheric pressure $dp_0=10mm$	Change in initial speed $dv_0=10m/s$
Estimated Change of Range Δx	$\Delta x=6.63$ m	$\Delta x=-0.87$ m	$\Delta x=6.88$ m
Change of Δx According to Range Table of 7.62 mm rifle.	$\Delta x=7$ m	$\Delta x=-1$ m	$\Delta x=7$

Example 2. Consider the projectile of the field artillery cannon 122 mm (Russian made). Use the following data (taken from the range table of the 122 mm cannon):

Initial Speed: $v_0=885m/s$; Launching Angle $\alpha_0=6.20°$; Impact Angle $\alpha_T=10°$;

Horizontal Range $x=10,000m$; Maximum Height of the Trajectory: $y_m=345m$;

Time of Flight: $t=17sec$.

to find the correction coefficients that correspond to the following small changes:

$dv_0=8.85m/s$, $d\alpha=0.06°=0.0010472\ radian$, $dT_0=10°$ C,

$dp_0=10mm\ Hg$, and $dT_c=10°$ C.

The ballistics coefficient is $c=0.2388$.

Solution
First, we estimate the coefficient β. We find

$$\bar{y} = (2/3)y_{max} = (2/3)\cdot(345) = 230m,$$

$$h(\bar{y})=(\frac{289.08-0.006328\bar{y}}{289.08})^{4.4}=(\frac{289.08-0.006328\cdot(230)}{289.08})^{4.4}=0.978034 ,$$

$$\beta=\frac{h(\bar{y})}{\sqrt{\cos\alpha_0}}=\frac{0.978034}{\sqrt{\cos(6.20°)}}=0.9809074$$

Substituting in (6.1.3), (6.1.4), (6.1.5), (6.1.6), (6.1.7), (6.1.8), and (6.1.9), we find respectively the correction coefficients and corresponding changes in range:

(a) Change in Initial Speed

$$\frac{\partial x}{\partial v_0}=\frac{v_0}{\beta\cdot cK_D(v_0)}\cdot(1+\frac{\tan\alpha_0}{\tan|\alpha_T|}-\frac{gx}{v_0^2\cos^2\alpha_0\tan\alpha_T})=$$

$$=\frac{885}{(0.98091)\cdot(0.2388)\cdot(885-240)/3}\cdot[1+\frac{\tan 6.20°}{\tan 10°}-\frac{9.80665\cdot(10,000)}{(885)^2\cdot\cos^2(6.20°)\cdot\tan(10°)}]= \cdot$$

$$= 17.5508122(0.897625) = 15.75405$$

Change in the horizontal range that corresponds to the change in the initial speed $dv_0=8.85m/s$ is

$$\Delta x_1=\frac{\partial x}{\partial v_0}dv_0=(15.75405)(8.85)=139.42m .$$

(c) Change in temperature of air.
   Substituting in (6.1.7) we find that

$$\frac{\partial x}{\partial T_0}=(1-\frac{v_0}{x}\frac{\partial x}{\partial v_0})\frac{x}{T_0}=(1-\frac{(885)}{2(10,000)}\cdot(15.75405))\frac{10,000}{288.15}=10.5113 .$$

Change of the horizontal range that corresponds to the change of the air temperature $dT_0=10°$ is

$$\Delta x_3=\frac{\partial x}{\partial T_0}dT_0=(10.5113)(10)=105.113m .$$

(d) Change in propellant (black powder) temperature (standard temperature of propellant is $T_c=15°$) is

$$\frac{\partial x}{\partial T_c}=0.001 \cdot v_0 \frac{\partial x}{\partial v_0}=0.001 \cdot (885)(15.75405)=13.9423$$.

Change of the horizontal range that corresponds to the change of the propellant temperature ($dT_c=10°$) is

$$\Delta x_4=\frac{\partial x}{\partial T_c}dT_c=(13.9423)(10)=139.9423m$$.

(e) Change in the atmospheric pressure p_0 (standard value $p_0=750mm\ Hg$).

Substituting in (6.1.8), we find that

$$\frac{\partial x}{\partial p_0}=-(1-\frac{\tan\alpha_0}{\tan|\alpha_T|})\frac{x}{p_0}=-(1-\frac{\tan(6.20°)}{\tan(10°)})\frac{10,000}{750}=-5.1187$$.

The change in the horizontal distance that corresponds only to the change $dp_0=10mm\ Hg$ is

$$\Delta x_4=\frac{\partial x}{\partial p_0}dp_0=-(5.1187)(10)=-51.19m$$.

Correction of the Horizontal Range
Employing (6.1.2), we find the total change in distance as a result of respective changes we estimated above

$$\Delta x=139.42+58.69+105.113+139.9423-51.19=391.98m$$.

Thus, the actual range of fire is

$$x=x_0+\Delta x=10000+391.98=10,391.98m$$.

The change in vertical direction is approximately

$$\Delta y = \Delta x \cdot \tan \alpha_T = (391.98) \cdot (\tan 10°) = 69.12 m .$$

The projectile, at the distance $x_0 = 10,000$ m, will pass $\Delta y = 69.12 m$ over the target.

Note: To hit the target located at the horizontal range 10,000 m from the cannon, we need to modify the launching angle $\alpha_0 = 6.20°$ with a quantity $d\alpha_0$ that can be determined by the equation

$$\Delta x = \frac{\partial x}{\partial \alpha_0} d\alpha_0 , \tag{1}$$

where

$$\Delta x = -391.98 m$$

and

$$\frac{\partial x}{\partial \alpha_0} = \frac{x \cos(2\alpha_0)}{\cos^2 \alpha_0 \tan|\alpha_T|} = \frac{10,000 \cdot \cos(2 \cdot 6.20)}{\cos^2 (6.20°) \cdot \tan(10°)} = 56,043.52 .$$

Substituting in (1), we have

$$(-391.98) = (56,043.52) d\alpha_0 .$$

Hence, we find

$$d\alpha_0 = -0.00699421 \ radian = -0.4007° .$$

The actual launching angle should be

$$\alpha_0' = (\alpha_0 + d\alpha_0 = (6.20°) - (0.4007°) = 5.80° .$$

Note: Using the QBasic PC program ANGMET.BAS (Request an electronic copy or a CD with all PC programs to the author at: *gklimi@pace.edu*, or *iven24@aol.com*), we find the same value for the launching angle.

Using ANGMET.BAS:

Input

Range: $x=10,000m$; Launching Speed: $v_0=885+8.85=893.85m/s$; Temperature of Air $T_0=25°$; Atmospheric Pressure: $p_0=760mm$ Hg; Pressure of Vapor (standard): $e=6.35mm$ Hg; Projectile Mass: $m=27.30kg$; Change in Projectile Mass: $dm=0$; Propellant Temperature: $T_c=25°$; Range Wind Speed: $w=0m/s$.

Results:

Launching Angle: $\alpha_0'=5.823242°$; Time of Flight: $t=16.16s$; Terminal Speed: $v_T=439m/s$; Terminal Angle $\alpha_T=-9.416066°$; Vertex of Trajectory is located at the point ($x=5610$, $y=323$).

Example 3. Solve the problem of example 2 if there is no change in propellant's temperature.

Solution

In example 3 (d), we found that change in range due to the change in propellants temperature is

$$\Delta x_4 = \frac{\partial x}{\partial T_c} dT_c = (13.9423)(10) = 139.9423m$$

Thus (refer to example 2), the total correction in range is

$$\Delta x = 391.98 - 139.42 = 252.04 \ m.$$

Repeating the procedure shown in the note of the above example, we find that change in the launching angle is

$$d\alpha_0 = -0.0044971783 \ radian = -0.2577°$$

For the launching angle, we have

$$\alpha_0' = \alpha_0 + d\alpha_0 = 6.20° - 0.2577° = 5.9423°$$

Comment. Using ANGMET.BAS, we get the following value for the launching angle needed to hit the target in the conditions given in example 3:

$$\alpha_0 = 5.958862°$$

6.2 Converting the Experimental Data of Shooting into Standard Data

In practice, to construct the range tables of a given firearm, the ballisticians use the theoretical approach combined with experimental data. The standard conditions, presented in section 2.2, are difficult to achieve, so the firearm tests to construct the range tables are performed in regular weather conditions.

Since the experiments are normally carried out in nonstandard conditions, the horizontal range needs to be "converted" or to be "brought" in standard conditions.

In other words, using the horizontal range obtained experimentally in nonstandard conditions, we determine what would have been the range (or other trajectory elements) if the firearm tests were performed in standard conditions.

To bring the experimental data into standard conditions, exterior ballistics uses the theory of correction.

The following example illustrates one of the methods we can use to convert the experimental shooting data into standard ones, and then use the differential equations of projectile flight to find all elements of the trajectory.

The illustration presents a method used in exterior ballistics to construct the range tables from the ballistics tables that contains the theoretical results of the solution of the differential equations of the projectile flight.

Example 1. The theoretical results obtained using the Siacci method for a projectile of a 122 mm artillery cannon (ballistics coefficient $c=0.242668$) fired in standard atmospheric conditions with speed $v_0=885 m/s$ under a launching angle of $\alpha_0=3.516667°$ are shown hereafter:

Theoretical Results
   Horizontal Range $x=7000m$; Impact Angle $\alpha_T=-5.144394°$; Maximum Height of the Trajectory: $y_m=131m$; Time of Flight: $t=10.34\text{sec}$

To verify the theoretical results and to construct the range tables, there were organized experiments firing some 122 mm projectiles of Russian field cannon in a nonstandard atmosphere at sea level.

The Experimental Data
   Initial Speed $v_0=880 m/s$, Launching Angle $\alpha_0=3.516667°$, Horizontal Range $x=7180$ m, Atmospheric Pressure $p_0=755mm$ Hg, Temperature $T=28°C$, Propellant Explosive Load Temperature $T_c=25°C$.
   The effect of ballistics wind is not considered since the direction of fire is perpendicular to the direction of ballistics wind (see section 6.5).

Solution
   To compare the theoretical outcomes obtained for a standard projectile in standard conditions, we can assume that the change in x,

$$\Delta x=7180-7000=180m,$$

is result of small changes of projectile parameters and atmospheric conditions from the standard ones.

Is that change $\Delta x=180m$ indeed the result of the fact that the experiments were performed in nonstandard conditions, or it is a result of the approximate value of the theoretical result?

In other words, if the experiment would be done in standard conditions, would the range have been $x=7000m$?

The Coefficient β

We find

$$\bar{y}=(2/3)y_{max}=(2/3)\cdot(131)=87.333m,$$

$$h(\bar{y})=(\frac{289.08-0.006328\bar{y}}{289.08})^{4.4}=(\frac{289.08-0.006328\cdot(87.33)}{289.08})^{4.4}=0.9916151,$$

$$\beta=\frac{h(\bar{y})}{\sqrt{\cos\alpha_0}}=\frac{0.9916151}{\sqrt{\cos(3.51666°)}}=0.98255.$$

(a) Change in initial speed

$$\frac{\partial x}{\partial v_0}=\frac{v_0}{\beta c K_D(v_0)}(1+\frac{\tan\alpha_0}{\tan|\alpha_T|}\frac{gx}{v_0^2\cos^2\alpha_0\tan\alpha_T})$$

$$=\frac{885}{(0.99255)(0.242668)(885-240)/3}(1+\frac{\tan3.51666°}{\tan5.144394°}\frac{9.80665\cdot(7,000)}{(885)^2\cos^2(3.51666°)\tan(5.144394°)})=$$

$$=12.055$$

Change in the horizontal range that corresponds to the change of the initial speed $dv_0=880-8.85=-5$ is

$$\Delta x_1=\frac{\partial x}{\partial v_0}dv_0=(12.055)(-5)=-60.276m.$$

(c) Change in air temperature

Substituting in (6.1.7), we find that

$$\frac{\partial x}{\partial T_0}=(1-\frac{v_0}{2x}\frac{\partial x}{\partial v_0})\frac{x}{T_0}=[1-\frac{885}{2(7000)}\cdot12.055]\frac{7000}{289.06}=5.762341.$$

Change of the horizontal range corresponds to the change of the air temperature $dT_0=28°-15°=13°$ is

$$\Delta x_2 = \frac{\partial x}{\partial T_0}dT_0=(5.762341)(13)=74.91m$$

(d) Change in propellant temperature

$$\frac{\partial x}{\partial T_c}=0.001 \cdot v_0 \frac{\partial x}{\partial v_0}=0.001 \cdot (885)(12.055)=10.66868$$

Change in the horizontal range that corresponds to the change of the propellant temperature $dT_c=10°$ is

$$\Delta x_3 = \frac{\partial x}{\partial T_c}dT_c=(10.66868)(10)=106.69m$$

(e) Change in atmospheric pressure p_0
Substituting in (6.1.8), we find that

$$\frac{\partial x}{\partial p_0}=-(1-\frac{\tan \alpha_0}{\tan|\alpha_T|})\frac{x}{p_0}=-(1-\frac{\tan(3.51666°)}{\tan(5.144394)})\frac{7000}{750}=-2.9623$$

The change in the horizontal distance that corresponds only to the change $dp_0=5mm$ Hg is

$$\Delta x_4 = \frac{\partial x}{\partial p_0}dp_0=-(2.53535)(5)=-14.81m$$

Correction of the Horizontal Range
Employing (6.1.2), we find the total change in distance as result of respective changes we estimated above is

$$\Delta x=-60.27+74.91+106.69-14.81=106.52m.$$

Thus, the experimental range brought into standard conditions is

$$x_s = x_{exp} - \Delta x = 7180 - 106.52 = 7073.5m$$
.

The result shows that the theoretical horizontal range $x=7000m$ in standard conditions is different from the experimental value $x_s=7073m$ brought in standard conditions.

In other words, if the shooting test will be done in standard conditions, the projectile range would have been $x_s=7073.5m$.

Comment. There is a discrepancy between the theoretical results and experimental results. The theoretical range is $\Delta x=73.5m$ shorter than the experimental result.

The theoretical result need to be corrected to match the experimental outcome.

6.3 Wind Deflection. Ballistics Wind. Ballistics Temperature

An important factor that influences the projectile flight is the wind. The wind is a very complicated motion of an enormous mass of turbulent air. The velocity of wind changes in magnitude and direction and depends by location and the time of wind measurements, altitude, and the characteristics of terrain (field, forest, valley, town, hills, mountains, etc).

One important characteristic of the wind blow is the fact that the speed of a horizontal wind as it approaches the ground decreases and is zero for the layer of wind that touches the surface of the terrain.

The trajectory of a projectile flight in a windy weather is different from the trajectory in the absence of wind. Wind deflects the trajectory of the projectile, and as result, the projectile will miss the target if the launching angle is set up to hit the target in absence of wind.

To hit the target in presence of wind, exterior ballistics makes adjustments, called "wind corrections."

Exterior ballistics introduces the concept of the ballistics wind.

The "ballistics wind" is a hypothetical average uniform wind, blowing with constant velocity $\vec{w}_B$, whose effect on the trajectory of projectile flight is the same as that of the real wind the projectile encounters in flight.

To estimate the "ballistics wind," exterior ballistics employs a simplified model of wind:

- The projectile trajectory is divided horizontally (in y-direction) in a series of "n" layers (parallel to the xz-plane). The width of each layer is such that the speed and the direction of wind in each of them can be considered constant. The vertical size of each layer is between 200 m and 800 m.
- The value of the wind velocity in each layer $\vec{w}_i$ ($i=1,2,3,...,n$) is obtained using the average wind velocity measured by meteorological stations (for each layer).

The ballistics wind can be considered as composition of the "range wind," which blows in the direction of fire, and the "crosswind," which blows perpendicular to the trajectory. The component of the wind in vertical direction (y-axis) is practically very slight and can be neglected.

To simplify the calculation of the wind effect on the projectile flight the velocity of ballistics wind $\vec{w}_B$ and the average velocity in each layer $\vec{w}_i$ ($i=1,2,3,...,n$) are projected on the direction of the x- and z-axes, i.e., we consider the components of the wind velocity w_{xi} and w_{zi}, respectively, along the direction of fire (range wind is along the x-axis) and perpendicular to the firing plane (crosswind is along the z-axis).

First, we calculate the range effect of the wind that blows along the x-axis, and then, in a similar way, we estimate the crosswind effect.

We denote Δx_i the change in the range of shooting due to a change of one unit (1 m/s) in the wind speed only in the layer included between the horizontal planes $y=y_i$ and $y=y_{i+1}$. The change in range for the wind that blows with speed of w_{xi} is $w_{xi} \cdot \Delta x_i$. The total change in range δx due to the wind that blows in n layer is

$$\delta x = \sum_{i=1}^{n} w_{xi} \cdot \Delta x_i \tag{6.3.1}$$

The same change in range δx will be obtained if we consider a fictive uniform wind that blows along the x-axis with the same speed w_{Bx} in each layer, such that

$$\delta x = \sum_{i=1}^{n} w_{Bx} \cdot \Delta x_i \tag{6.3.2}$$

or

$$\delta x = w_{Bx} \sum_{i=1}^{n} \Delta x_i \tag{6.3.3}$$

We denote

$$\Delta x = \sum_{i=1}^{n} \Delta x_i \tag{6.3.4}$$

From (6.3.1) and (6.3.3), we obtain

$$w_{Bx} \cdot \Delta x = \sum_{i=1}^{n} w_{Bx} \cdot \Delta x_i \tag{6.3.5}$$

Hence, we find that the speed of the ballistics wind along the x-axis wind is

$$w_{Bx} = (\sum_{i=1}^{n} w_{xi} \cdot \Delta x_i) / \Delta x = \sum_{i=1}^{n} w_{xi} \cdot (\frac{\Delta x_i}{\Delta x}) \tag{6.3.6}$$

The quotient $\Delta x_i / \Delta x$, inside the sign of summation, is called the "weighting factor"[48] of the ith layer and is denoted as "q_i". Thus,

---

[48] The number of layers and the relative weight of each layer depend on the maximum height of the projectile trajectory and on the time the projectile "spends" in each layer.

$$q_i = \Delta x_i / \Delta x \tag{6.3.7}$$

Substituting (6.3.7) in (6.3.6), we write the x-component of the velocity of the ballistics wind

$$w_{Bx} = \sum_{i=1}^{n} w_{xi} q_i \tag{6.3.8}$$

In a similar way, we find the z-component of the velocity of the ballistics wind is

$$w_{Bz} = \sum_{i=1}^{n} w_{z_i} q_i \tag{6.3.9}$$

Using (6.3.8) and (6.3.9) for the velocity of the ballistics wind, we can write for the ballistics wind

$$\vec{w}_B = w_{bx}\vec{i} + w_{Bz}\vec{k} = \sum_{i=1}^{n} \vec{i} \cdot w_{xi} q_i + \sum_{i=1}^{n} \vec{k} \cdot w_{zi} q_i = \sum_{i=1}^{n} \vec{w}_i \cdot q_i \tag{6.3.10}$$

The ballistics meteorological station measures the layers average velocities $\vec{w}_i$ ($i=1,2,3,...,n$), computes the velocity of the ballistics wind $\vec{w}_B$ and then transmits that to the artillery battery.

The hypothetical "ballistics wind" simplifies the estimation of wind corrections. Hereafter, we study the "range wind correction" and the "crosswind correction."

Ballistics Temperature
The same procedure resulting in a formula similar to (6.3.9),

$$T_B = \sum_{i=1}^{n} T_{xi} q_i \tag{6.3.12}$$

The reader can find that information in the aeronautics or meteorological books. Approximately the weight of each layer can be found using the ideal model of projectile.

is used to find the ballistics temperature T_B. The ballistics wind and the ballistics temperature are given by the meteorological stations.

6.4 Projectile Motion in Presence of Range Wind

The influence of wind velocity $\vec{w}$ in projectile flight can be studied using the vector differential equation in a three-dimensional Cartesian coordinate system,

$$\frac{d\vec{V}}{dt} = \vec{g} - c \cdot h(y)K_D(V)\frac{\vec{V}}{V},$$

where V, given by the equation

$$V^2 = (\vec{v} - \vec{w})^2,$$

is the relative speed of the projectile with respect to air.

To simplify the solution of the vector differential equation and to use the results already obtained for the projectile flight in absence of wind, we use the traditional method that considers the components of wind: range wind $\vec{w}_x$ and crosswind $\vec{w}_z$.

We will introduce another system of coordinates, moving with the velocity of wind, and change formally the notation letters of the relative speed and the absolute speed.

Consider a projectile that is launched with initial velocity $\vec{V}_0$ with respect to a Cartesian coordinate system x, y, z (figure 13) and a range wind (tailwind or headwind) that blows in the direction of the x-axis with constant velocity $\vec{w}_x = \vec{w}$. The drag force is a function of the projectile velocity relative to the atmospheric air.

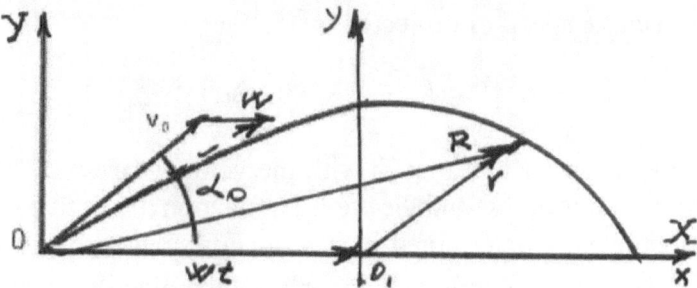

Figure 13

The relationship between position vectors $\vec{R}(t)$ and $\vec{r}(t)$ of the projectile in flight at time t is

$$\vec{R}(t)=\vec{r}(t)+\vec{w}\cdot t \ . \tag{6.4.1}$$

For the coordinates of the projectile with respect to the fixed system of coordinates at time t, we have

$$X(t)=x(t)+w\cdot t \ , \quad Y(t)=y(t) \ , \quad Z(t)=z(t)=0 \ .$$

Differentiating (6.4.1), we get the relation between the velocities of the projectile in both coordinate systems

$$\vec{V}(t)=\vec{v}(t)+\vec{w} \ ,$$

where $\vec{v}(t)$ is the relative velocity of projectile, i.e., the velocity with respect to the coordinate system that moves with the velocity of the wind. Hence, for the relative velocity of the given projectile, we have

$$\vec{v}(t)=\vec{V}(t)-\vec{w} \ . \tag{6.4.2}$$

The components of the relative velocity are

$$v_x(t)=V_X(t)-w \ , \quad v_y(t)=V_y(t) \ , \quad v_z(t)=V_z(t)=0 \ , \tag{6.4.3}$$

where the speed of wind "w" is positive for a tailwind and negative for a headwind.

The relative speed of projectile is

$$v=[(\vec{V}-\vec{w})^2]^{1/2}=(V^2+w^2-2Vw\cos\alpha)^{1/2} .$$

(6.4.4)

For the observer that moves with the velocity of wind "$\vec{w}$", the atmospheric air is at rest and the drag is a function of relative speed of the projectile $v(t)$ determined in (6.4.4). For such an observer, the vector differential equation that describes the projectile flight can be written as

$$\frac{d\vec{v}}{dt}=\vec{g}-c\cdot h(y)K_D(v)\frac{\vec{v}}{v} ,$$

(6.4.5)

considering that the relative speed (6.4.4) must be substituted in the Siacci function $K_D(v)$.

From (6.4.5), we can write the following system of differential equations in the coordinate system moving with the velocity of the range wind

$$\begin{cases} \dfrac{dv_x}{dt}=-ch(y)K_D(v)\dfrac{v_x}{(v_x^2+v_y^2)^{1/2}} \\[2mm] \dfrac{dp}{dt}=-\dfrac{g}{v_x} \\[2mm] \dfrac{dx}{dt}=v_x \\[2mm] \dfrac{dy}{dt}=v_y \end{cases}$$

(6.4.6)

where

$$p=\frac{v_y}{v_x}=\tan\alpha_1 ,$$

(6.4.7)

$$c=i\frac{\cdot d^2}{m}\cdot 1000 , \qquad h(y)=(\frac{289.08-0.006328y}{289.08})^{4.4} .$$

and

$$v=[(\vec{V}-\vec{w})^2]^{1/2}=(V^2+w^2-2Vw\cos\alpha)^{1/2}.$$

The initial conditions are

$$x_0=0, \quad y_0=0, \quad v_{x0}=(V_0\cos\alpha_0-w), \quad v_{y0}=V_0\sin\alpha_0,$$

$$p_0=\frac{v_{y0}}{(v_{x0}-w)}=\frac{V_0\sin\alpha_0}{V_0\cos\alpha_0-w_0}, \quad \text{for } t=0 \qquad (6.4.8)$$

The system of equations (6.4.6) can be solved for x using the integration methods presented in chapter 3 or chapter 4, the Siacci methods or numerical integration. Then we use equation (6.4.1) to find the solution of the projectile flight in the fixed system of coordinates, i.e.,

$$X=x+w\cdot t, \quad Y=y \qquad (6.4.9)$$

After the integration of (6.4.6), the actual abscissa X_w in presence of wind can be found using (6.4.9), i.e.,

$$X_w=x+w\cdot t. \qquad (6.4.10)$$

If we denote X_{nw} the projectile range in absence of wind, then the change in range is

$$\Delta X=X_w-X_{nw}=x+w\cdot t-X_{nw}. \qquad (6.4.11)$$

In fact we will use the correction method to consider the influence of wind.

Example 1. For the 122 mm projectile cannon: Initial Speed $v_0=885 m/s$; Launching angle $\alpha_0=6.20°$; Ballistics Coefficient $c=0.2388$. Use the solution of the system of differential equations of projectile (6.4.6) to find the range in presence of a range wind of $w=10 m/s$ blowing in the positive direction of the x-axis (tailwind).

Solution

We will solve the differential equations (6.4.6) using the PC program RANGEC.BAS (see appendix C) for the following initial conditions:

$$v_0=[(\vec{V}_0-\vec{w}_0)^2]^{1/2}=(V_0^2+w^2-2V_0 w\cdot\cos\alpha)^{1/2}$$
$$=[885^2+10^2-2\cdot(885)\cdot(10)\cdot\cos(6.2°)]=875m/s$$

and

$$p_0=\frac{v_{y0}}{(v_{x0}-w)}=\frac{V_0\sin\alpha_0}{V_0\cos\alpha_0-w_0}=\frac{885\sin(6.2°)}{885\cos(6.2°)-10}=0.10988364$$

Hence,

$$\alpha_0=6.2707°$$

Then we use the PC program RANGEC.BAS to find the range and the time of flight in the relative coordinate system moving with wind velocity:

PC program RANGEC.BAS

Input
Launching Speed $v_0=875m/s$
Launching angle $\alpha_0=6.2707°$
Ballistics Coefficient $c=0.2388$

Output
Range $x=9912m$
Time of Flight $t=16.76s$

Substituting in (6.4.10), the output values we find that the actual range (in absolute system of coordinates) is

$$X_w=x+w\cdot t=9912+(10)\cdot(16.76)=10079.6m$$

Note: The range in absence of wind is $X=10000m$.

6.5 **Range Wind Corrections**

Another method we use in exterior ballistics to find the range in presence of range wind is the "correction" method: to the range of the projectile in absence of wind, we add the "correction" due to the range wind.

We assume that there is no crosswind during the projectile flight. Thus, for the velocity of the range wind that moves along the x-axis, we have $\vec{w}_x = \vec{w}$.

The relative velocity of projectile (see [6.4.3]) is

$$\vec{v}(t) = \vec{V}(t) - \vec{w}.\tag{6.5.1}$$

The initial speed and the launching angle of the projectile with respect to the coordinate system in motion with the velocity of wind are respectively

$$v_0 = [(\vec{V}_0)^2 + (\vec{w})^2 - 2\vec{V}_0\vec{w})]^{1/2}$$

$$= (V_0^2 + w^2 - 2V_0 w \cdot \cos\alpha_0)^{1/2} = (1 - 2\frac{w}{V_0}\cos\alpha_0) + (w/V_0)^2)^{1/2}\tag{6.5.2}$$

and

$$\tan\alpha_{10} = \frac{V_0\sin\alpha_0}{V_0\cos\alpha_0 - w} = \frac{\tan\alpha_0}{1 - w/(V_0\cos\alpha_0)}.\tag{6.5.3}$$

We can consider that the quotient w/V_0 in (2) is less than one, i.e., $w/V_0 \ll 1$.

Expanding (6.5.2) in binomial series and neglecting the terms that contain $(w/v_0)^2$, we obtain

$$v_0 = V_0(1 - w \cdot \cos\alpha_0/V_0) = V_0 - w\cos\alpha_0.\tag{6.5.4}$$

Hence, we find that the change in speed is

$$dV_0 = v_0 - V_0 = -w\cos\alpha_0.\tag{6.5.5}$$

Expanding (6.5.3), we have

$$\tan\alpha_{10}=\tan\alpha_0\left(1+\frac{w}{V_0\cos\alpha_0}\right)=\tan\alpha_0+\frac{w}{V_0}\frac{\sin\alpha_0}{\cos^2\alpha_0}. \tag{6.5.6}$$

Hence, we can write

$$d(\tan\alpha_0)=\tan\alpha_{10}-\tan\alpha_0=\frac{w}{V_0}\frac{\sin\alpha_0}{\cos^2\alpha_0} \tag{6.5.7}$$

or

$$d(\tan\alpha_0)=\frac{d}{d\alpha_0}(\tan\alpha_0)\cdot d\alpha_0=\frac{d\alpha_0}{\cos^2\alpha_0}. \tag{6.5.8}$$

Substituting (6.5.8) on the left side of (6.5.7), we find

$$d\alpha_0=\frac{w}{V_0}\sin\alpha_0. \tag{6.5.9}$$

Consider the change in range obtained from (6.4.11),

$$\Delta X_T=X_w-X_{nw}=x_T+w{\cdot}t-X_{Tnw}. \tag{6.5.10}$$

The range x_T corresponds to the initial conditions (6.5.2) and (6.5.3), while X_{Tnw} correspond to the initial conditions $V(0)=V_0$ and $p(0)=p_0=\tan\alpha_0$.

The small change in range ΔX_T is a function of the initial speed, initial angle, and time of flight. It can be considered approximately equal to the differential of the right side of (6.5.10),

$$\Delta X_T\approx d(x+w{\cdot}t-X_n)=\frac{\partial(X_T)}{\partial V_0}dV_0+\frac{\partial(X_T)}{\partial\alpha_0}d\alpha_0+w \tag{6.5.11}$$

where

$$\frac{\partial X_T}{\partial v_0}=\frac{\partial(x-X_w)}{\partial v_0}, \qquad \frac{\partial X_T}{\partial \alpha_0}=\frac{\partial(x-X_w)}{\partial \alpha_0} \qquad (6.5.12)$$

are the "correction coefficients" related respectively to the change in initial speed and to the small change in the launching angle, and estimated by (6.1.3) and (6.1.5),

$$\frac{\partial X_T}{\partial v_0}=\frac{v_0}{\beta c K_D(v_0)}(1+\frac{\tan\alpha_0}{\tan|\alpha_T|}-\frac{gX_T}{v_0^2\cos^2\alpha_0\tan|\alpha_T|}) \qquad (6.5.13)$$

and

$$\frac{\partial X_T}{\partial \alpha_0}=\frac{X_T\cos(2\alpha_0)}{\cos^2\alpha_0\tan|\alpha_T|}. \qquad (6.5.14)$$

Substituting (6.5.5) and (6.5.9) on the right side of (6.5.11), we find that the change of horizontal range due to the small changes in the initial speed and the launching angle is

$$\Delta X_T=w(t-\frac{\partial(X_T)}{\partial v_0}\cdot\cos\alpha_0+\frac{\partial(X_T)\cdot\sin\alpha_0}{\partial \alpha_0}\frac{}{v_0}). \qquad (6.5.15)$$

Example 1. Use the data and the results obtained in example 2 of section 6.1 to find the change in horizontal range,

$$\Delta X==x_T+w\cdot t_T-X_{nw},$$

that corresponds to a range ballistics wind of $w=10$ m/s .

Initial Speed $v_0=885m/s$; Launching Angle $\alpha_0=6.20°$; Impact Angle $\alpha_T=-10°$; Horizontal Range $x=10{,}000m$; Maximum Height of the Trajectory $y_m=345m$; Time of Flight: $t=17sec$. The Ballistics Coefficient is $c=0.2388$.

Solution

The Influence of Range Wind
In example 2 of section 6.1 for the same problem, we found that

$$\frac{\partial x}{\partial v_0}=15.75405 \quad , \quad \frac{\partial x}{\partial \alpha_0}=56043.52 \quad .$$

Substituting the above values in (6.5.15), we find that the horizontal correction due to the range wind is

$$\Delta X=w(t-\frac{\partial X_N}{\partial v_0}\cdot\cos\alpha_0+\frac{\partial X_N \cdot\sin\alpha_0}{\partial \alpha_0}\frac{}{v_0})$$

$$=10\cdot(17-15.75405\cos(6.20°)+56043.52\cdot\frac{\sin(6.20°)}{885})=81.77m$$

.

The actual horizontal range in presence of a range wind $w=10$ m/s is

$$X=X_N+\Delta X=(10000+81.77)=10,081.77m \quad .$$

Note: Using QBasic PC program RANGMET.BAS, we find that the actual range in presence of wind is $X=10,086m$.

Correction of the Launching Angle
To hit the target located at $X_T=10,000m$, we need to adjust the launching angle, reducing the range in presence of air by $\Delta X=-81.77$. Substituting, $\Delta X=-81.77$ in (6.1.5),

$$\Delta X=X_T-X=\frac{\partial X}{\partial \alpha_0}d\alpha_0$$

and

$$\frac{\partial x}{\partial \alpha_0}=56043.52$$

,

we can write

$$-81.77=(56043.52)d\alpha_0.$$

Hence,

$$d\alpha_0=-0.001459 \; radian=-0.0836°.$$

The launching angle needed to hit the target at the range $x=10,000m$ in presence of tailwind is

$$\alpha_0'=\alpha_0+d\alpha_0=6.20°-0.0836°=6.1164°.$$

Note: Using QBasic PC program ANGMET.BAS, we find that the value of the launching angle in presence of the tailwind is $\alpha_0'=6.113281°$.

Instruction: Use of ANGMET.BAS: standard atmosphere, tailwind present.

Input: Range = 10,000 m, Initial Velocity $v_0=885m/s$, Air Temperature = 15 Degree, Air Pressure = 750 mm, Pressure of Vapor = 6.35 mm, Projectile Mass = 27.30 kg, Change in Projectile Mass = 0, Propellant Temperature = 15 Degree, Tailwind = 10 m/s.

Output: Launching Angle $\alpha_0'=6.113281°$, Time of Flight = 16.57 [s], Terminal Speed 428 [m/s], Terminal Angle = -9.902186, Trajectory Vertex (5610, 340).

6.6 Projectile Motion in Presence of Crosswind. Crosswind Corrections

The crosswind component $\vec{w}_z$ of the wind exerts a perpendicular force on the projectile, and as a result, the projectile deviates from the launching plane (xy-plane) in the direction of the z-axis when the crosswind has the same direction as z-axis and opposite if the crosswind blows in the opposite direction of the z-axis (figure 2 and figure 3). The projectile will miss the target located on the x-axis.

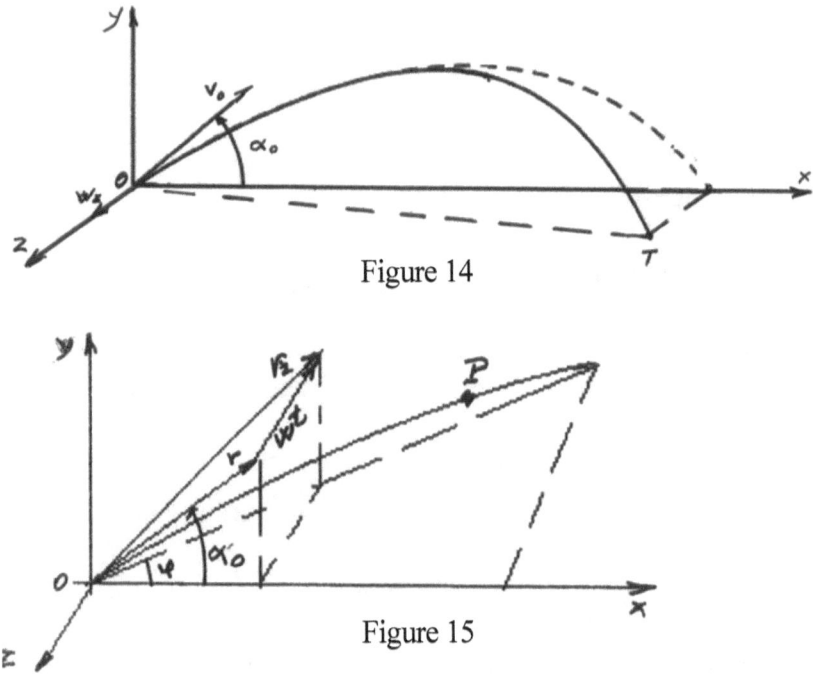

Figure 14

Figure 15

The deviation of the projectile from the target, along the z-axis, depends on the speed of the crosswind component of the wind velocity.

To hit the target in the presence of crosswind, we need to adjust the direction of fire deviating the launching plane in the opposite direction of the crosswind with an angle φ (figure 3).

Consider a crosswind that blows in the direction opposite the z-axis with velocity $\vec{w}_z$ and an observer that moves with the velocity of wind (figure 15).

For this observer, the relative velocity of the projectile in direction of the z-axis is zero. That means that for the moving observer, there is no drag force in the direction of wind.

The projectile trajectory in the system of coordinates associated with the moving observer can be studied using the same methods we have used in chapters 3 and 4, taking into consideration initial conditions of the projectile relative to the moving system of reference.

In other words, we need to solve the system of equations

$$\begin{cases} \dfrac{dv_x}{dt}=-ch(y)K_D(v)\dfrac{v_x}{(v_x^2+v_y^2)^{1/2}} \\ \dfrac{dp}{dt}=-\dfrac{g}{v_x} \\ \dfrac{dx}{dt}=v_x \\ \dfrac{dy}{dt}=v_y \end{cases}$$

or

$$\begin{cases} \dfrac{dv_x}{dx}=-ch_\tau(y)\dfrac{K(v)}{v} \\ \dfrac{dp}{dx}=-\dfrac{g}{v_x^2} \\ \dfrac{dt}{dx}=\dfrac{1}{v_x} \\ \dfrac{dy}{dx}=p \end{cases}$$

with modified initial conditions

$$v'_0=(v_0^2+w_z^2)^{1/2}, \qquad \tan\alpha'_0 = \frac{\tan\alpha_0}{[1+ w^2 / (v_0^2 \cos^2\alpha_0)]^{1/2}},$$

where

$$K_D(v)=\begin{cases} 1.212{\cdot}10^{-4}v^2 & \text{for} \quad v\le 256 m/s \\ (v-240)/3 & \text{for} \quad v>256 m/s \end{cases}.$$

Indeed. At a time t during the flight of the projectile, we can write for the vector positions of the projectile in both systems of coordinates

$$\vec{r}_2(t) = \vec{r}(t) + \vec{w}_z \cdot t .$$ (6.6.1)

Differentiating with respect to time, we get

$$\vec{v}_2(t) = \vec{v}(t) + \vec{w}_z .$$ (6.6.2)

Since from (6.6.1) and (6.6.2), we get respectively

$$x_2(t) = x(t), \quad y_2(t) = y(t), \quad z_2(t) = z(t) - w_z \cdot t ,$$ (6.6.3)

$$v_{2x}(t) = v_x(t), \quad v_{2y}(t) = v_y(t), \quad v_{2z}(t) = v_z(t) - w_z .$$ (6.6.4)

The first two equations of (6.6.3) and (6.6.4) show that the relative coordinates and relative velocity components of projectile along x and y directions are the same as the components in the absolute coordinate system.

Change of the Initial Conditions
The components of relative initial velocity are

$$v_{2x}(0) = v_x(0) = v_0 \cos\alpha_0, \quad v_{2y}(0) = v_y(0) = v_0 \sin\alpha_0 ,$$
$$v_{2z}(0) = -w_z .$$ (6.6.5)

The relative initial speed is

$$v_{20} = ((v_0 \cos\alpha_0)^2 + (v_0 \sin\alpha_0)^2 + (-w_z)^2)^{1/2} = (v_0^2 + w_z^2)^{1/2} .$$ (6.6.6)

The relative launching angle can be found using the relation

$$\tan\alpha_{20} = \frac{v_0 \sin\alpha_0}{(v_0^2 \cos^2\alpha_0 + w^2)^{1/2}} .$$ (6.6.7)

The first two equations of (6.6.3) and (6.6.4) show that the relative coordinates and relative velocity components of the projectile along the x- and y-axes are the same as the respective coordinates and velocity components of the projectile in the fixed system of coordinates.

Usually the crosswind speed is much lower than the projectile speed. So we can assume $w/v_0 \ll 1$.

The equation (6.6.6) and (6.6.7) can be written, respectively, as

$$v_{20} = v_0 (1+(w/v_0)^2)^{1/2} \qquad (6.6.8)$$

and

$$\tan \alpha_{20} = \frac{\tan \alpha_0}{[1 + w^2 / (v_0^2 \cos^2 \alpha_0)]^{1/2}}. \qquad (6.6.9)$$

Neglecting the term $(w/v_0)^2$ in (6.6.8), we find that the relative speed is approximately equal to the initial speed of projectile

$$v_{20} = v_0 (1+(w/v_0)^2)^{1/2} \approx v_0. \qquad (6.6.10)$$

Ignoring the term $w^2/(v_0^2 \cos^2 \alpha_0)$ in (6.6.9), we obtain

$$\tan \alpha_{20} = \frac{v_0 \sin \alpha_0}{v_0 \cos \alpha_0 (1+w^2/v_0^2 \cos^2 \alpha_0)^{1/2}} \approx \tan \alpha_0. \qquad (6.6.11)$$

Equation (6.6.11) shows that the relative launching angle is approximately equal to the initial launching angle,

$$\alpha_{20} \approx \alpha_0. \qquad (6.6.12)$$

The equations (6.6.10) and (6.6.12) show that the relative initial speed of the projectile and the relative launching angle are approximately equal to the respective values of the projectile in absence of wind, and as result, the relative horizontal range of projectile, flying in presence of wind, is approximately equal to the horizontal range of projectile flying in an atmosphere without wind.

We can say that the impact point of the projectile is shifted in the z-direction at the point with coordinates $(x, 0, z)$. The launching plane in the coordinate system related with the moving observer forms with xy-plane an angle φ determined by

$$\tan\varphi = -\frac{w_z}{v_0 \cos\alpha_0}.$$
(6.6.13)

Note: The angle φ is negative since the rotation of the firing plane is clockwise.

In figure 3, we see that the relative deviation of the projectile in the direction of the z-axis is

$$z_2(t) = x \cdot \tan\varphi = -\frac{w_z}{v_0 \cos\alpha_0}x.$$
(6.6.14)

Substituting in the last equation of (6.6.3),

$$z_2(t) = z(t) - w_z \cdot t,$$
(6.6.15)

we find that the z-coordinate of the projectile with respect to the fixed system of coordinates is

$$z(t) = w_z t - \frac{w_z}{v_0 \cos\alpha_0}x.$$
(6.6.16)

To hit the target, we need to rotate the launching plane in the opposite direction of the crosswind with an angle φ, determined by the absolute value of (6.6.13).

Example 1. Use the information in example 1, section 6.4, to adjust the initial data of the projectile if there is a crosswind of $w=10$ m/s.

Initial Speed: $v_0 = 855m/s$; Launching Angle $\alpha_0 = 6.20°$; Horizontal Range $x=10,000m$; Time of Flight $t=16.57sec$.

Note: The above data can be found using the PC program RangeC.Bas or can be found in the ballistics tables of the corresponding cannon.

Solution

Change in Direction of Fire

Because of the crosswind, the impact point deviates along the z-axis with a quantity

$$z(t)=w_z t-\frac{w_z}{v_0\cos\alpha_0}x=(10)(16.76)-\frac{10}{885\cdot\cos(6.20)}10,000=53.94m$$

To hit the target, we need to rotate the direction of fire in the opposite direction of the crosswind by an angular quantity φ determined using (6.6.13),

$$\tan\varphi =-\frac{w_z}{v_0\cos\alpha_0}.$$

Substituting, we find that we have to rotate the direction of shooting plane with an angle of

$$\varphi=\tan^{-1}(-\frac{w}{v_0\cos\alpha_0})=-\tan^{-1}\frac{10}{885\cos(14.483)}=-0.65°$$

in the opposite direction of the crosswind.

Example 2. In the table below are given the elements of the trajectory of a projectile of the 122 mm cannon, launched with initial speed of v_0=885m/s in standard conditions. Launching angle and the ballistics coefficient are respectively α_0=14.48333° and c=0.2548 (see example 2 of section 6.2).

Table 1. (v_0=885m/s , α_0=14.48333° , $c(14.48333°)$=0.2548)

Angle	Range	Height	Impact angle	Time of flight	Impact speed
α_0=14.48333	15,400 m	1455 m	- 28.50°	33.66 s	319.35 m/s

Find the trajectory elements when the same projectile is fired in the following atmospheric conditions: Temperature T=25° C, Pressure 762 mm, Projectile Speed v_0=879m/s .

The rangewind and the crosswind components are respectively $w_R=10m/s$ and $w_z=6m/s$.

Note: The range wind, $w_x=10m/s$, and the crosswind, $w_z=6m/s$, correspond to a wind with a speed of 11.66 m/s that blows in the direction that forms an angle of $30.96°$ with the direction of the x-axis.

Solution
   The changes in atmospheric conditions with respect to the standard ones alter the horizontal range and the direction of fire.

I.   Linear Change of Horizontal Range

The coefficient β
   We find

$$\bar{y}=(2/3)y_{max}=(2/3)\cdot(1455)=970m$$,

$$h(\bar{y})=(\frac{289.08-0.006328\bar{y}}{289.08})^{4.4}=(\frac{289.08-0.006328\cdot(970)}{289.08})^{4.4}=0.90988246$$,

$$\beta=\frac{h(\bar{y})}{\sqrt{\cos\alpha_0}}=\frac{0.90988246}{\sqrt{\cos(14.483°)}}=0.92469428$$.

(a) The change in initial speed

$$\frac{\partial x}{\partial v_0}=\frac{v_0}{\beta cK_D(v_0)}(1+\frac{\tan\alpha_0}{\tan|\alpha_T|}-\frac{gx}{v_0^2\cos^2\alpha_0\tan|\alpha_T|})$$

$$=\frac{885}{(0.92469428)(0.2458)(885-240)/3}(1+\frac{\tan(14.483°)}{\tan(28.50°)}-\frac{9.80665\cdot(15,400)}{(885)^2\cos^2(14.483°)\tan(28.50°)})=.$$

$$=19.8654$$

   The change in the horizontal range, which corresponds to the change in the initial speed $dv_0=879-885=-6$, is

$$\Delta x_1=\frac{\partial x}{\partial v_0}dv_0=(19.8645)(-6)=-119.20m$$.

(b) The change in the temperature of air is $dT_0=25°-15°=10°$.

Substituting in (6.1.7), we find that

$$\frac{\partial x}{\partial T_0}=(1-\frac{v_0}{2x}\frac{\partial x}{\partial v_0})\frac{x}{T_0}=(1-\frac{(885)}{2(15,400)}\cdot(19.8654))\frac{15,400}{298.15}=22.1699$$

The change in the horizontal range corresponds to the change of the air temperature $dT_0=10°$ is

$$\Delta x_2=\frac{\partial x}{\partial T_0}dT_0=(22.1699)(10)=221.699m$$

(c) The change in the atmospheric pressure p_0.

Substituting in (6.1.8), we find that

$$\frac{\partial x}{\partial p_0}=-(1-\frac{\tan\alpha_0}{\tan|\alpha_T|})\frac{x}{p_0}=-(1-\frac{\tan(14.483°)}{\tan(28°)})\frac{15400}{750}=-10.55843$$

The change in horizontal distance that corresponds only to the change $dp_0=10mm$ Hg is

$$\Delta x_3=\frac{\partial x}{\partial p_0}dp_0=(-10.55843)(10)=-105.584m$$

(d) The horizontal correction due to the given range wind is

$$\Delta x_w=w(t-\frac{\partial x_N}{\partial v_0}\cdot\cos\alpha_0+\frac{\partial x_N}{\partial\alpha_0}\cdot\frac{\sin\alpha_0}{v_0})$$

$$=10\cdot(33.66-19.864\cos(14.483°)+27030.6872\cdot\frac{\sin(14.483°)}{885})=220.66m$$

since

$$\frac{\partial x}{\partial v_0}=19.864 \quad \frac{\partial x}{\partial\alpha_0}=\frac{x\cos(2\alpha_0)}{\cos^2\alpha_0\tan|\alpha_T|}=\frac{15,400\cdot\cos(2\cdot14.483)}{\cos^2(14.483)\tan(28)}=27030.6872$$

The Total Change in Horizontal Range

Employing (6.1.2), we find the total change in distance as a result of the respective changes we estimated above is

$$\Delta x=-119.20+221.17-105.58+220.66=330.3306m .$$

Correction of the Launching Angle

It is obvious that the projectile in nonstandard conditions (temperature $T=25°C$, pressure 762 mmHg, projectile speed $v_0=879m/s$, and range wind $w_x=10m/s$) will strike at a point on the ground that is $\Delta x=330.3306m$ farther than the range table value of $x=15,400m$.

To hit the target located in the distance $x=15,400m$, we need to reduce the value of the launching angle, $\alpha_0=14.483°$, by a correction value $d\alpha_0$ that corresponds to the total change $\Delta x=330.3306m$ and is determined using the relation

$$\Delta x=\frac{\partial x}{\partial \alpha_0}d\alpha_0 .$$

Substituting

$$\frac{\partial x}{\partial \alpha_0}=27030.6872 \text{ and } \Delta x=330.3306m,$$

we find that correction in the launching angle is $d\alpha_0=0.7°$.

Hence, the new value of the launching angle needed to hit the target in distance 15.400 m is

$$\alpha_0'=\alpha_0-d\alpha_0=14.483-0.7=13.75352°.$$

The Change in the Direction of Fire

Because of the crosswind, the impact point deviates along the z-axis with a quantity

$$z(t)=w_zt-\frac{w_z}{v_0\cos\alpha_0}x=(10)(33.66)-\frac{10}{885\cdot\cos(14.483)}15,400=156.88 \text{ m.}$$

To hit the target, we need to rotate the direction of fire in the opposite direction of wind by an angular angle φ determined using (6.6.13),

$$\tan\varphi = -\frac{w}{v_0 \cos\alpha_0}.$$

Substituting, we find that we need to rotate the direction of shooting plane with an angle of

$$\varphi = \tan^{-1}\left(-\frac{w}{v_0 \cos\alpha_0}\right) = -\tan^{-1}\frac{10}{885\cos(14.483)} = -0.01167°$$

opposite the direction of the crosswind.

Note: Using QBasic PC program ANGMET.BAS, we find that the value of the launching angle in presence of the tailwind is $\alpha_0' = 13.82483°$.

6.7 Notes on the Spinning Projectile

The equations of projectile flight are obtained ignoring the spin of projectile and the relative effects. The axis of symmetry of the projectile was supposed to have the direction of the tangent to the trajectory.

As a matter of fact, for the stability in flight, the projectiles of artillery and small arms (except mortar mines, aircraft bombs, rockets) traveling inside the barrels obtain a spin characterized by a relatively great angular velocity "ω". The spinning of the projectile is associated with three effects:

- Deviation of projectile from the firing plane
- Lifting force
- Magnus pseudo-force effect

Deviation of Projectile from the Firing Plane

At the instant the spinning projectile leaves the barrel of the arm, its axis of symmetry forms an angle "δ" with the tangent of the projectile trajectory. As a result, the drag force $\vec{D}_S$ will makes an angle, usually greater than "δ", with the axis of projectile. The drag force has a component $\vec{D}$ along the tangent of the trajectory (direction of projectile velocity "v") and another one $\vec{D}_N$ perpendicular to the tangent applied at the center of the drag.

The construction of shells is such that the center of mass is located near the base of the projectile, while the center of the drag force is on the opposite side next to the projectile nose.

The component $\vec{D}_N$ of the drag tends to lift up the center of mass of the projectile, and at the same time, the torque of the drag force $\vec{D}_N$ tends to overturn the projectile that rotates with angular velocity $\vec{\omega}$ directed along the projectile axis of symmetry.

Because of the rotation the projectile axis will make a precession motion (like a gyroscope) with angular velocity $\vec{\omega}_1$ around the trajectory tangent opposing the overturning torque and so tending to keep unchanged the angle "δ". As a result, the axis of the projectile will "follow" the trajectory tangent aiming to keep a narrow yaw angle "δ" and so will fall on the objective with its nose.

Since the direction of trajectory tangent changes continuously aiming to go down, the "drop" of the tangent is equivalent to a "rotation" of the projectile with angular velocity $\vec{\omega}_2$ perpendicular to the plane of the figure and directed toward us.

As a result of the two abovementioned rotation motions, the resultant precession motion of the projectile will be performed around an axis that is to the right of the tangent of the trajectory when the projectile rotation observed from behind is clockwise (or to the left when projectile rotation is counterclockwise).

Thus, during the flight, the projectile will follow the direction of the tangent while the front part of the projectile (the part above the center of mass) and the center of drag force will be on average most of the time on the right of the direction of the tangent. The outcome is an unbalanced component of the drag force perpendicular to the tangent of the trajectory and directed to the right of the firing plane, i.e., in the direction of the z-axis.

Under the influence of the abovementioned force, the projectile deviates to the right of xy-plane in the direction of the z-axis when rotation of projectile is clockwise (or to the left, in the opposite direction of the z-axis when projectile rotation is counterclockwise).

The "deviation distance" along the z-axis can be estimated using the empirical formula

$$z=k \cdot t^2,$$ (6.7.1)

where "k" is a constant determined experimentally, while "t" is the time of flight to the target.

The deviation $z=k \cdot t^2$ can be considered as a correction to the trajectory of projectile flight.

When the deviation of the projectile is to the right of the firing plane, the launching plane has to be deviated to the left.

Magnus Force

The axis of symmetry of the projectile forms an angle "δ" with the tangent of the trajectory. As a result for a system of reference traveling with the projectile, there is an upward transversal flow of air surrounding the projectile. During the revolution about the axis of symmetry, the spinning projectile slows down the speed of flow on one side and accelerates the flow on the opposite side. As a result, the regime of flow changes and becomes asymmetric and deflected toward the opposite. The outcome of this flow is the Magnus force directed to the opposite side.

Magnus force does not appear if the axis of symmetry of the projectile coincides with the projectile velocity.

Note: We should point out the presence of the viscosity force that air exerts on the rotating surface of the projectile. The viscosity force tends to reduce the angular velocity of the spinning projectile and, as a result, the stability of the projectile flight.

7

Compilation of Range Tables

Introduction

O ne of the main tasks of exterior ballistics is the compilation of range tables for each firearm and projectile.

Range tables for a firearm allows the personnel to set launching (or projection) angle to fire and hit a given target with an acceptable accuracy. They are very useful to solve problems encountered in practice of shooting.

The range tables are compiled for the standard atmosphere and for projectiles that have standard characteristics. They usually contain as well the "corrections" for small changes of the characteristics of the atmosphere and projectile from the standard ones.

The trajectory of a given projectile flight is determined uniquely by the initial speed v_0, launching angle α_0, and the ballistics coefficient of projectile c (or equivalently, the form coefficient i).

While the projectile speed and launching angle can be measured (and set up) with great accuracy, the ballistics coefficient needs to be determined experimentally.

For a given initial speed and a given launching angle, the ballistics coefficient c can be seen as "a parameter" that matches the theoretical results (usually the horizontal range) obtained from the solution of differential equations of flight with the results (the horizontal range) obtained in practice during the firearm tests.

In other words, the ballistics coefficient c, or equivalently the coefficient of form i, of a given projectile, usually plays an "adjustable" role to meet the theoretical horizontal range with results of experiments with acceptable approximation.

7.1 The Ballistics Coefficient and the Projectile Velocity

Range tables of a given projectile can be obtained with satisfactory accuracy using the differential equations that describe the projectile flight employing a ballistics coefficient c of projectile obtained from experiments.

First Approach
Ballistics coefficient as a function of launching angle

In chapters 3 and chapter 4, we use an average ballistics coefficient for any launching angle to solve the ballistics problems by integrating the differential equations of the point-of-mass projectile. The theoretical projectile trajectory we obtain using an average ballistics coefficient matches satisfactorily with the experimental trajectory for certain launching angles, but in general, the approximation is not good if the same ballistics coefficient c is used for other launching angles.

The ballistics coefficient c is a function of the projectile speed, i.e., $c=c(v)$. For a given launching angle α_0, the projectile speed v changes from the initial value of the speed v_0 to the value of speed v_T at the impact point located at the horizontal range x_T. Thus, the ballistics coefficient can be considered a function of launching angle α_0, $c=c(\alpha_0)$ because the set of speeds of the projectile along trajectory is different for different launching angles.

Thus, exterior ballistics assumes that for a given launching angle α_0 (or for a short interval of launching angles), the ballistics coefficient c can be assumed constant, and based on that value we can calculate the corresponding elements of the trajectory.

Based on the fact that the ballistics coefficient c can be seen as a function of launching angle α_0, we follow the procedure described below to construct the range table of a given projectile:

- We consider the point of mass differential equations (chapter 2) for a nonrotating projectile, whose axis of symmetry is along the projectile velocity.
- Using the experimental data (usually the experimental horizontal range) obtained for different values of launching angle α_0 and the corresponding theoretical results we determine the ballistics coefficient $c=c(\alpha_0)$ of a given projectile as a function of the launching angle α_0.
- Employing $c=c(\alpha_0)$ in the differential equations of the projectile flight, we are able to find, with satisfactory accuracy, the elements of the trajectory and to prepare the corresponding range tables.
- We make corrections to match the other calculated elements of trajectory with the experimental data and construct the range table.

Note: The use of an average ballistics coefficient for all launching angles is not practical since it requires time-consuming procedures to match the theoretical results with tests results.

Second Approach

Ballistics coefficient as a function of projectile speed

One of the procedures we use to solve the ballistics problems of a given projectile and to construct the range tables can be summarized as follows:

- We consider the differential equations (shown in chapter 2) for a nonrotating, point-mass projectile, whose axis of symmetry is along the projectile velocity.
- The ballistics coefficient c and, related with it, the form factor i are determined experimentally as functions of projectile speed, i.e., $c=c(v)$ and $i=i(v)$, respectively, for example using wind tunnels or firearm tests.
- Employing the obtained function $c=c(v)$ in the differential equations that describe the projectile flight, we find with satisfactory accuracy the elements of the trajectory that corresponds to some different launching angles and prepare the corresponding range table.
- Using the data obtained from practical tests and appropriate methods, we make corrections to match the other calculated elements of trajectory with experimental data and construct the range tables.

Since the form factor (or ballistics coefficient) is a function of the projectile speed, we have to use the methods of numerical integration of differential equations to solve the equations of motion of the given projectile.

In this chapter, we illustrate the methods of compilation of the range tables using the Siacci method and the numeric integration of differential equations of projectile motion.

7.2 Determining the Ballistics Coefficients. Siacci's Method

Examples 1–2 illustrate the way we construct the range tables using the experimental data to find the ballistics coefficient c as a function of the launching angle α_0.

Since in practice it is difficult to have standard atmospheric conditions, the experimental data obtained in nonstandard conditions need to be brought into standard values.

Using the correction theory, we are able to bring the experimental data into the standard conditions (see example 1 of section 6.2).

Example 1. Some projectiles of mass $m=6.5kg$ were fired from $d=76.2mm$ cannon, with an initial speed $v_0=660m/s$ under an angle of $\alpha_0=10°$. The experimental horizontal range, brought into standard conditions, is $x=6427m$.

Use the experimental range, $x=6427m$, to determine all elements of trajectory at the point of impact. Find the ballistics coefficient and the form coefficient.

Solution

We need to find the **form coefficient** i or the **ballistics coefficient** c in order to match the experimental data with the theoretical ones.

To find the ballistics coefficient $c=c(10°)$, appropriate for the launching angle $\alpha_0=10°$, and the corresponding form coefficient $i=i(10°)$, we use the Siacci method.

Approximate value of the pseudo-speed at the experimental point of impact

Substituting $y=0$ in (5.2.24),

$$y=p_0x+\frac{x}{240B\cos^2\alpha_0}\cdot[\frac{A(u)-A(v_0)}{D(u)-D(v_0)}-J(v_0)] \quad , \qquad (7.2.1)$$

we can write

$$p_0+\frac{1}{240B\cos^2\alpha_0}\cdot[\frac{A(u)-A(v_0)}{D(u)-D(v_0)}-J(v_0)]=0 \quad . \qquad (7.2.2)$$

Hence,

$$p_0=-\frac{1}{240B\cos^2\alpha_0}\cdot[\frac{A(u)-A(v_0)}{D(u)-D(v_0)}-J(v_0)] \quad . \qquad (7.2.3)$$

We eliminate the unknown B from the above equation. For that, we consider the equation (5.2.12),

$$x=-\frac{1}{Bg}[D(u)-D(v_0)] \tag{7.2.4}$$

where

$$B=\beta \cdot b , \quad b=c/(3g) , \quad \beta=h(\overline{y})/\sqrt{\cos\alpha_0} , \quad \overline{y}=(2/3)y_{max}$$

and

$$h(\overline{y})=(\frac{289.08-0.006328\overline{y}}{289.08})^{4.4} .$$

Dividing (7.2.3) and (7.2.4), we write

$$p_0=\frac{gx}{240\cos^2\alpha_0}\cdot\frac{1}{D(u)-D(v_0)}\cdot[\frac{A(u)-A(v_0)}{D(u)-D(v_0)}-J(v_0)] \tag{7.2.5}$$

Hence

$$A(u)-A(v_0)=\frac{120\sin(2\alpha_0)}{gx}\cdot[D(u)-D(v_0)]+J(v_0)\}[D(u)-D(v_0)] \tag{7.2.6}$$

where

$$D(v_0)=[v_0+240\cdot\ln(v_0-240)]=660+240\cdot\ln9660-240)=2109.661131$$

and

$$J(v_0)=\ln(\frac{v_0-240}{v_0})=\ln\frac{660-240}{660}=-0.45199 .$$

Substituting, we have

$$A(u)-A(v_0)=[(0.000651184)\cdot(D(u)-2109.6611)-0.45199\}[D(u)-2109.6611)] \tag{7.2.7}$$

where

$$A(u)-A(v_0)=\int_{v_0}^{u}\frac{u}{u-240}\ln\frac{u-240}{u}du$$

$$(7.2.8)$$

and

$$D(u)=[u+240{\cdot}\ln(u-240)].$$

$$(7.2.9)$$

Solving the above equation for u by trial and error using a TI-83 Plus graphing calculator, we find the value of the pseudo-speed at the point of impact

$$u_T=261.7457.$$

The value of B

Substituting the above value of the pseudo-speed in (7.2.4), we can write

$$B(9.80665)(6427)=-[D(261.7457)-D(660)]=1108.85558$$

$$(7.2.10)$$

since

$$D(261.7457)=[261.7457+240{\cdot}\ln(261.7457-240)]=1000.80555.$$

From (7.2.10), we find that

$$B=0.01759325.$$

Angle of Impact

For the point of impact, we can write

$$p_T=p_0+\frac{1}{240B\cos^2\alpha_0}[\ln(\frac{u_T-240}{u_T})-\ln(\frac{v_0-240}{v_0})]=$$

$$=\tan(10°)+\frac{1}{4.095059}[\ln(\frac{261.7457-240}{261.7457})-\ln(\frac{660-240}{660})]=-0.32085076.$$

Hence, because $p_T=\tan\alpha_T$, for the angle of impact, we obtain

$$\alpha_T=\tan^{-1}(-0.32085076)=-17.78888°$$.

The Speed of Impact

Using the formula for the projectile pseudo-speed (see [5.2.3]),

$$u=v\frac{\cos\alpha}{\cos\alpha_0}$$,

we find that the speed of projectile at the impact point is

$$v_T=u_T\frac{\cos\alpha_0}{\cos\alpha_T}=(261.7457)\frac{\cos(10°)}{\cos(-17.78888)}=270.71m/s$$.

Time of Flight to the Target

Substituting $u_T=270.71m/s$, in equation (5.2.31), we find that the projectile time of flight to the target is

$$t=-\frac{1}{Bg\cos\alpha_0}[\ln(u-240)-\ln(v_0-240)]=$$

$$=-\frac{1}{(0.01759325)(9.80665)\cdot\cos(10°)}[\ln(261.7457-240)-\ln(660-240)]$$

$$=17.43 \text{ sec.}$$

Trajectory Vertex

At the point where the vertex is located, the projectile velocity is parallel to the x-axis. Thus,

$$p=\tan\alpha=0$$.

Using (7.2.33),

$$p_0+\frac{1}{240B\cos^2\alpha_0}[J(u)-J(v_0)]=0$$

,

we find

$$J(u)=J(v_0)-240Bp_0\cos^2\alpha_0.$$

(7.2.4)

Substituting in the above equation

$$J(u)=\ln(\frac{u-240}{u}),$$

$$J(v_0)=\ln(\frac{v_0-240}{v_0})=\ln(\frac{660-240}{660})=-0.4519851237,$$

$$p_0=\tan\alpha_0=\tan(10°)=0.1763269807$$

and

$$240Bp_0\cos^2\alpha_0=240(0.0175932475)(0.17632698)\cos^2(10°)=0.7220694,$$

we have

$$\ln\frac{u-240}{u}=J(v_0)-240Bp_0\cos^2\alpha_0=(-0.4519851237-0.7220694=-1.174054527.$$

Hence,

$$\frac{u-240}{u}=e^{-1.7405452}=0.3091111.$$

Solving the last equation for u, we obtain the pseudo-speed

$$u_m=347.3785716m/s$$

at the point where the vertex of the trajectory is located.
The speed at the vertex of trajectory is

$$v_m=u_m\frac{\cos\alpha_0}{\cos\alpha_m}=(347.378571)\frac{\cos(10°)}{\cos(0)}=342.10m/s.$$

Coordinates of the Vertex
   Substituting in (5.2.30), we find that the x-coordinate of the trajectory vertex is

$$x_m=-\frac{1}{Bg}[(u_m+240\cdot\ln(u_m-240))-(v_0+240\cdot\ln(v_0-240)]=$$

$$=-\frac{1}{0.1725308206}[1469.705124-2109.66113]=3709.227$$
.

We find as well

$$D(u_m)-D(v_0)=-Bgx_m=-(0.01759325)(9.80665)(3709.227)=-639.9559781$$
.

Then, substituting

$$x_m=3709.227 \quad,\quad J(v_0)=\ln(\frac{v_0-240}{v_0})=-0.4519851237 \quad,$$

$$A(u_m)-A(v_0)=\int_{v_0}^{u}\frac{u}{u-240}\ln\frac{u-240}{u}du$$

$$=\int_{660}^{347.38}\frac{u}{u-240}\ln\frac{u-240}{u}=471.57981 \quad,$$

$$p_0=\tan\alpha_0=\tan(10°)=0.1763269807$$

and

$$B=0.01759325$$

into the trajectory equation (5.2.24), we find that the maximum height of the projectile trajectory is

$$y_{max}=p_0x+\frac{x}{240B\cos^2\alpha_0}\cdot[\frac{A(u)-A(v_0)}{D(u)-D(v_0)}-J(v_0)]=628.288m$$
.

Correction Factor, Ballistics Coefficient, or Form Coefficient
  Correction factor is

$$\beta=h(\bar{y})/\sqrt{\cos\alpha_0}=0.9602784886/\sqrt{\cos(10°)}=0.96765706$$

since

$$\bar{y}=(2/3)y_{max}=(2/3)(628.288)=418.86m \; ,$$

and

$$h(\bar{y})=(\frac{289.08-0.006328\bar{y}}{289.08})^{4.4}=0.9602784886 \; .$$

To find ballistics coefficient c, we substitute in

$$B=\beta b=\beta c/(3g) \; .$$

We can write

$$(0.01759325)=(0.96765706)\cdot c/(3g) \; .$$

Hence, we find that

$$c=0.53489245 \; .$$

Employing the formula

$$c=\frac{id^2}{m}1000 \; ,$$

we find the coefficient of the form

$$i=0.598783585 \; .$$

Using the PC Programs to Verify the Siacci Method

Using the QBasic PC program COEFF.BAS (See appendix E), we find that the value of the ballistics coefficient for the same projectile is $c=0.52644$.
Employing PC program COEFF.BAS to find ballistics coefficient:

First, we input an approximate range = 6,430 and a corresponding (guessed) ballistics coefficient to get fast results. (We can start as well with range = 6,400).

1. INPUT DATA

Cancel Data File: Print y
Input Guessed Coefficient = 0.535
Input Projectile Speed = 660
Input Launching Angle = 10.00 =10 (read 10 Degree. 00 Minute)
Input Corresponding Range = 6,430
Input Longest Range = 6,430
Input Number of Steps = 10

RESULT: Ballistics Coefficient is (ko) = **0.5260879**
The ballistics coefficient is reserved as well in file C:/koef.dat

2. REPEAT the above steps, considering ballistics coefficient found above.

INPUT DATA
Cancel Data File: Print y
Input Guessed Coefficient = **0.5260879**
Input Projectile Speed = 660
Input Launching Angle = 10.00 =10
Input Corresponding Range = 6,430
Input Longest Range = 6,430
Input Number of Steps = 1

RESULT: Ballistics Coefficient is (ko) =**0.5264406**

3. Using PC QBasic program RANGEC.BAS to verify the results obtained above using the Siacci method

INPUT DATA:
Initial Coordinates $x = 0$, $y = 0$
Launching Angle: $\alpha_0 = 10°$, Speed $= 660$ m/s, Ballistics Coefficient = 0.5264384,
Time $t = 0$, Integration Step $h = 1$

RESULTS:
Range = 6,428, Impact Speed = 269, Impact Angle = - 17.41, Time = 17.41
Coordinates of the trajectory vertex ($x = 3699$, $y = 394$)

In table 1, given (for comparison) are the solution of the problem using Siacci's functions and the solution of the same problem obtained using the above ballistics coefficient ($c = 0.5260912$) and the QBasic PC program RANGEC.BAS (see appendix C).

Table 1

	Ballistics Coefficient	Range	Launching Angle	Time of Flight	Impact Speed	Impact Angle
Siacci	$c = 0.53489$	6427	10°	17.43	270.71	- 17.90°
RANGEC	$c = 0.52644$	6428	10°	17.41	269.30	- 17.70°

Example 2. The horizontal range $x_T = 15,502.55m$ obtained in example 5 of section 5.2 using the Siacci method and an average ballistics coefficient is different from the experimental value (range table value) $x_E = 15,400m$.

Use the experimental value $x_E = 15,400m$ to find an appropriate ballistics coefficient c.

Solution
(a) The pseudo-speed at the point of impact
    Consider (5.2.26) and (5.2.12),

$$\sin(2\alpha_0)=-\frac{1}{120B}[\frac{A(u)-A(v_0)}{D(u)-D(v_0)}-J(v_0)]$$

(1)

and

$$x=-\frac{1}{Bg}[D(u)-D(v_0)]$$.

(2)

Dividing (1) and (2), we can write

$$\frac{\sin(2\alpha_0)}{x}=\frac{g}{120\cdot[D(u)-D(v_0)]}[\frac{A(u)-A(v_0)}{D(u)-D(v_0)}-J(v_0)]$$.

(3)

Hence, we find the same equation we have used in example 3

$$A(u)-A(v_0)=\frac{120\sin(2\alpha_0)}{gx}[D(u)-D(v_0)]+J(v_0)\cdot[D(u)-D(v_0)]$$,

(4)

where

$$A(u)-A(v_0)=\int_{v_0}^{u}\frac{u}{u-240}\ln\frac{u-240}{u}du$$

(5)

and

$$D(u)=[u+240\cdot\ln(u-240)]$$.

(6)

Substituting in (4),

$$x_E=15,400m, \quad v_0=885m/s, \quad \alpha_0=14.48333°,$$

(7)

$$J(v_0)=J(885)=-0.3163373282, \quad D(v_0)=2437.620076,$$

(8)

we obtain an equation where the unknown is the pseudo-speed u_T at the point of impact of projectile on the horizontal ground

$$A(u_T)-A(v_0)=0.000384817\cdot(u_T+240\cdot\ln(u_T-240)-2437.62)^2-0.3163373282\cdot[u_T+240\ln(u_T-240)]$$

Using a trial-and-error procedure and a graphing calculator TI-83 Plus, we find that the value of the pseudo-speed is $u_T=289.87m/s$.

Value of Parameter B=βb

Substituting $u_T=289.87m/s$, $x_E=15,400m$ in (2),

$$x=-\frac{1}{Bg}[D(u)-D(v_0)]$$

,

and solving the obtained equation, we find that

$$B=0.008008674787$$.

Time of Flight

$$t=-\frac{1}{Bg\cos\alpha_0}[\ln(u-240)-\ln(v_0-240)]$$

$$=-\frac{1}{(0.00800867)g\cos(14.48333°)}[\ln(289.87-240)-\ln(855-240)]=33.663$$.

Impact Angle

Substituting in (5.2.6), we find

$$p=p_0+\frac{1}{240B\cos^2\alpha_0}[J(u)-J(v_0)]=-0.5429080271$$

,

where

$$J(u)=\ln(\frac{u-240}{u}), \quad J(v_0)=\ln(\frac{v_0-240}{v_0})$$.

Since $p=\tan\alpha$, we find that the impact angle is

$$\alpha_T=\tan^{-1}(-0.5429080271)=-28.49789°$$.

Impact Speed

$$v_T=u_T\frac{\cos\alpha_0}{\cos\alpha}=(289.87)\frac{\cos(14.48333°)}{\cos(28.49789°)}=319.35m/s$$

Coordinates of Trajectory Vertex

At the point where the vertex is located, the projectile velocity is parallel to the x-axis. Thus, $p=\tan\alpha=0$.

Using (5.2.33),

$$p_0+\frac{1}{240B\cos^2\alpha_0}[J(u)-J(v_0)]=0$$

we find

$$J(u)=J(v_0)-240Bp_0\cos^2\alpha_0.$$

Substituting in the above equation

$$B=0.008008674787,$$

$$J(u)=\ln(\frac{u-240}{u}),$$

$$J(v_0)=\ln(\frac{v_0-240}{v_0})=\ln(\frac{885-240}{885})=-0.3163373282,$$

$$p_0=\tan\alpha_0=\tan(14.48333°),$$

and

$$240Bp_0\cos^2\alpha_0=240(0.008008674787)\tan(14.48333°)\cos^2(14.48333°)=0.4654328206,$$

we obtain

$$\ln\frac{u-240}{u}=J(v_0)-240Bp_0\cos^2\alpha_0=(-0.3163373282-0.4654328206=-0.7817701488.$$

Hence,

$$\frac{u-240}{u}=e^{-0.7817701488}=0.4575952822.$$

Solving the last equation for u, we obtain the pseudo-speed at the point where the vertex of the trajectory is located:

$$u_m = 442.470 m/s .$$

Substituting the above value of the pseudo-speed in (5.2.30), we find that the x-coordinate of the trajectory vertex is

$$x_m = -\frac{1}{Bg}[(u+240 \cdot \ln(u-240)) - (v_0 + 240 \cdot \ln(v_0 - 240)] = 9175.135298 .$$

We find as well

$$D(u_m) - D(v_0) = -Bgx_m = -720.59926 .$$

Then, substituting

$$x_m = 9175.135298 ,$$

$$J(v_0) = J(885) = \ln(\frac{885 - 240}{885}) = -0.3163373282 ,$$

$$A(u_m) - A(v_0) = \int_{v_0}^{u_m} \frac{u}{u-240} \ln\frac{u-240}{u} du$$

$$= \int_{885}^{442.47} \frac{u}{u-240} \ln\frac{u-240}{u} = 357.45048 ,$$

$$D(u_m) - D(v_0) = -720.59926 ,$$

$$p_0 = \tan\alpha_0 = \tan(14.48333°) ,$$

and

$$B = 0.008008674787$$

into the trajectory equation (5.2.24), we find that the maximum height of the projectile trajectory is

$$y_{max}=p_0 x+\frac{x}{240 B \cos^2 \alpha_0} \cdot [\frac{A(u)-A(v_0)}{D(u)-D(v_0)}-J(v_0)]=1454.92m$$

Correction Coefficient β

Substituting the maximum height $y_{max}=1454.92m$ in $\bar{y}=(2/3)y_{max}$, we find that

$$\bar{y}=(2/3)y_{max}=969.946m$$

and

$$h(\bar{y})=(\frac{289.08-0.006328\bar{y}}{289.08})^{4.4}=(\frac{289.08-0.006328(969.946)}{289.08})^{4.4}=0.9098872911$$

Thus,

$$\beta=h(\bar{y})/\sqrt{\cos\alpha_0}=(0.9088872911)/\sqrt{\cos(14.4833333°)}=0.9246991905$$

Substituting $\beta=0.9246991905$, $B=0.0080086748$ in $B=\beta b$, we find

$$b=B/\beta=(0.0080086748)/0.9246991905=0.0086608433$$

Ballistics Coefficient

Substituting the above value in $b=c/(3g)$, we find the ballistics coefficient

$$c=3bg=3(0.0086608433)g=0.2548$$

Form Coefficient

Substituting the caliber $d=122mm=0.122m$, the projectile mass $m=27.3$ and $c=0.2548$ in

$$c=\frac{id^2}{m}1000 ,$$

we find that the form coefficient is

$$i=0.46735.$$

Note: The ballistics coefficients as well as the form coefficient are functions of the launching angle. Thus, we can write that

$$c(14.48333°)=0.2548.$$

Hereafter, in Table 1, are presented the elements of trajectory of a 122 mm cannon obtained above using the Siacci method as well as the corresponding range table value of the respective cannon.

Table 1. ($v_0=885m/s$, $\alpha_0=14.48333°$, $c(14.48333°)=0.2548$)

	Range	Height	Impact angle	Time of flight	Impact speed
Siacci	15,400 m	1455.00 m	- 28.50°	33.66 sec.	319.35 m/s
Range Table	15,400 m	1450.00 m	- 28.00°	34.0 sec.	314.00 m/s

7.3 Construction of the Range Table for the 122 mm Cannon

The following example illustrates the first method we use to find the ballistics coefficient as a function of the launching angle using the experimental data of shootings and then using that ballistics coefficient to construct the range tables.

We construct a partial table, just to illustrate the method.

Example 1. The following table gives the horizontal ranges obtained experimentally firing some projectiles with 122 mm field artillery cannon ($v_0=885m/s$). The horizontal ranges are brought into standard conditions.

Table 1. Experimental values of 122 mm cannon

Angle	1.017°	5.050°	10.083°	14.49°	20.25°	25.17°	30.42°
Range[m]	2,400	8,800	13,000	15,400	18,000	19,800	21,400

(a) Use the experimental range values to find an appropriate ballistics coefficient c for each range.
(b) Express the ballistics coefficient as a function of launching angle, i.e., $c=c(\alpha_0)$

(c) Make corrections to eliminate the discrepancies observed for other elements of trajectory: trajectory height and time of flight.

(d) Find the elements of the projectile trajectory for any other angle.

Solution

For demonstration purposes, we will consider only a few values of the launching angle α_0. For better results, we need to consider values of the launching angles that are relatively close to each other.

(a) Employing the same method we used in example 1, or example 2, of section 7.2, we find the elements of the trajectory and the corresponding ballistics coefficients.

The obtained results are presented in the following table.

Table2. Ballistics Range (Cannon 122 mm)

Angle α_0	Impact angle	Range x_T	Impact speed	Time t_T	Height y_{max}	Distance to y_{max}	c, Ballistics Coefficient	i, Form coefficient
1.017°	−1.20°	2,400	691.50	3.07	22.14	1250	0.34950	0.641047
5.050°	−7.85°	8,800	457.83	13.99	433.91	5463.08	0.24248	0.444751
10.083°	−18.95°	13,000	344.82	25.24	800.57	7588.31	0.2467	0.452503
14.483°	−28.50°	15,400	319.35	33.66	1454.9	9175.13	0.2548	0.467350
20.25°	−38.95°	18,000	324.13	43.80	2519.25	10,837.81	0.25954	0.476039
25.17°	−46.16°	19,800	340.58	51.91	3589.46	11,963.43	0.26489	0.485850
30.42°	−52.56°	21,400	363.80	60.32	8090.32	12946.17	0.34142	0.626223

(b) In table 3 are given the values of the launching angle α_0 in radian and the values of corresponding ballistics coefficients $c=c(\alpha_0)$, taken out from table 2.

Table 3

Range	2,400	8,800	13,000	15,400	18,000	19,800	21,400
α_0	0.017744	0.088139	0.179587	0.252782	0.353429	0.439241	0.530871
$c=c(\alpha_0)$	0.34950	0.24248	0.2467	0.2548	0.25954	0.26489	0.34142

The following chart (regression plot) in figure 1, constructed using the data of table 3, shows that $c=c(\alpha_0)$ can be approximated by a linear regression line

$$c(\alpha_0)=0.0653026\alpha_0+0.236578$$

$$(7.3.1)$$

for angles corresponding to ranges in the interval 2,400–19,800 m.

We need to have more experimental data to find a regression line (or curve) that will give more accurate data.

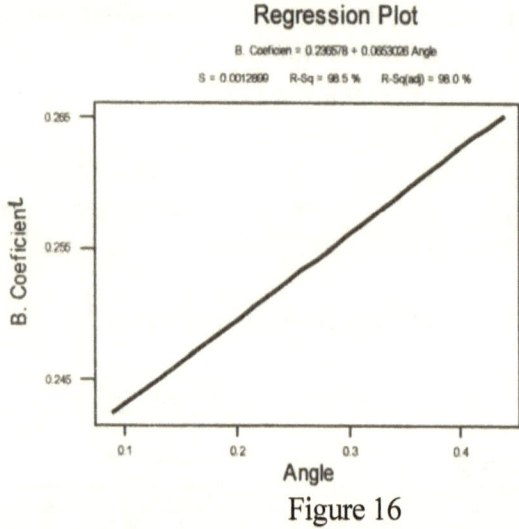

Figure 16

Comment: The function (7.3.1) does not give good values for the ballistics coefficient. We need to consider some more firing test to obtain a function $c=c(\alpha_0)$ that will give better values for the "adjustable" ballistics coefficients.

(c) Corrections due to discrepancies in trajectory height and time of flight

Table 4 shows that there are no differences in time of flight, but there are discrepancies observed between the experimental and theoretical data of projectile altitude and time of flight.

Table 4

Angle	Range	Time		Height		Difference
		Test	Theory	Test	Theory	$\Delta(height)$
5.050	8,800	14	13.99	240	433.91	- 193.91
10.083	13,000	25	25.24	800	800.57	- 0.57
14.483	15,400	34	33.66	1450	1454.9	-5
20.25	18,000	44	43.80	2480	2519.25	-39
25.17	19,800	52	51.91	3500	3589.46	-89.46

We can use the interpolation to find the intermediate values of $\Delta(height)$ and so to correct the altitude obtained theoretically.

(d) Find the horizontal range for the launching angle $\alpha_0=11.11\overline{6}°=0.194022radian$ using the Siacci method.

Ballistics Coefficient
Substituting in (7.3.1) the value of the launching angle $\alpha_0=.194022radian$, we find

$$c(0.194)=0.0653026(0.194)+0.236578=0.2492482.$$

Since the value of the launching angle $\alpha_0=11.11\overline{6}°$ is close to the table value 10.083° (see table 4), using interpolation, we find that the maximum height of the trajectory is 950 m.

Thus,

$$\bar{y}=(2/3)y_{max}=(2/3)(950)=633m,$$

$$h(\bar{y})=(\frac{289.08-0.006328\bar{y}}{289.08})^{4.4}=(\frac{289.08-0.006328(633)}{289.08})^{4.4}=0.9404173,$$

$$\beta=h(\bar{y})/\sqrt{\cos\alpha_0}=(0.9404173)/\sqrt{\cos(11.11\overline{6}°)}=0.931553,$$

and

$$B=\beta \cdot b=\beta \cdot c/(3g)=0.0078922 .$$

Using the Siacci method for relatively high speeds and the above value of B, we can find all the trajectory elements that correspond to launching angle $\alpha_0=11.11\overline{6}°$ and display them in the range table.

Hereafter, we will find only the horizontal range.

Horizontal Range

Approximate Value of the Pseudo-speed at the Point of Impact

Substituting $y=0$ in (7.2.24),

$$y=p_0 x+\frac{x}{240Bcos^2\alpha_0}\cdot[\frac{A(u)-A(v_0)}{D(u)-D(v_0)}-J(v_0)] \tag{1}$$

we can write

$$p_0+\frac{1}{240Bcos^2\alpha_0}\cdot[\frac{A(u)-A(v_0)}{D(u)-D(v_0)}-J(v_0)]=0 . \tag{2}$$

Hence,

$$p_0=-\frac{1}{240Bcos^2\alpha_0}\cdot[\frac{A(u)-A(v_0)}{D(u)-D(v_0)}-J(v_0)] , \tag{3}$$

where

$$D(v_0)=[v_0+240\cdot\ln(v_0-240)]=885+240\cdot\ln(885-240)=2437.62 ,$$

$$J(v_0)=\ln(\frac{v_0-240}{v_0})=\ln\frac{660-240}{660}=-0.316337 ,$$

$$A(u)-A(v_0)=\int_{v_0}^{u}\frac{u}{u-240}\ln\frac{u-240}{u}du , \tag{4}$$

and

$$D(u)=[u+240 \cdot \ln(u-240)] \qquad (5)$$

Substituting in (3), we have

$$A(u)-A(v_0)=-0.674686[u+240\ln(u-240)-2437.62] \qquad (6)$$

Solving the above equation for u by trial and error using a TI-83 Plus graphing calculator, we find the value of the pseudo-speed at the point of impact

$$u_T=320.503 \,.$$

The range that corresponds to $\alpha_0=11.11\overline{6}°$ is

$$x=-\frac{1}{Bg}[(u+240 \cdot \ln(u-240))-(v_0+240 \cdot \ln(v_0-240)]$$

$$=-\frac{1}{(0.0078922)g}[320.503+240\ln(320.503-240)-(885+240\ln(885-240)=13{,}746.54m \,.$$

7.4 The Range Tables of 0.30 M2 Ball Projectiles

Knowing the ballistics coefficient, $c=c(v)$, or the form factor, $i=i(v)$, of a projectile as a function of its speed allows us to solve the main problems of exterior ballistics and to construct the range tables regardless of the launching angle.

This method has advantages over the other method since the ballistics coefficient changes as the projectile speed changes and so reflects more accurately the value of the drag force.

The following example illustrates the method we use to construct the tables using the ballistics coefficient of the projectile that is a function of the projectile speed along the trajectory:

(a) Solution of the Differential Equations of Projectile Flight Using PC Programs

Example 1. In example 5 of section 2.6 for a 0.30 M2 ball bullet caliber $d=0.308in=0.00782m$, we found the ballistics coefficient as a function of projectile speed

$$c(v)=(0.913405-0.0009944v+0.00000062v^2)\cdot(6.2913992)_.\qquad (1)$$

Use the function $c=c(v)$ and the PC program **RANGEM30.BAS** to find the elements of the trajectory for the points with x-coordinates: 100, 200, 300, 400, 500, and 600 [meter], if the launching angle of the projectile is $\alpha_0=16.8'=0.28°$.

Mass of the bullet is 0.00972 kg. The speed of sound is $a=341.45$ m/s. The density of air at sea level is $\rho_0=1.205kg/m^3$.

Solution

Since the ballistics coefficient is a function of the projectile speed, we need to integrate the differential equations (3.1.10) using numerical methods.

Employing the system of differential equations (3.1.10) and the ballistics coefficient function (1), is prepared the QBasic PC program **RANGEM30.BAS** (see appendix E) based on Runge-Kutta numerical solution methods.

The PC program **RANGEM30.BAS** gives the elements of trajectory for a given launching angle (for example: $\alpha_0=16.8'=0.28°$) and a given initial speed (853.43 m/s).

The program can be easily modified to solve the problems related with other projectiles.

Using the **RANGEM30.BAS**, we find the following data (presented in table 2) for the projectile trajectory when launching angle is $\alpha_0=16.8'=0.28°$ and initial speed is 853.44 m/s.

Table 2

Elements of Trajectory of 0.30 M2 Ball Projectile				
Initial Speed v_0=853.43m/s; Launching Angle α_0=16.8'=0.28				
Estimated Coordinates of the Vertex x_m=279m, y_m=0.73m				
x-coordinate (meter)	y-coordinate (meter)	Time t (sec.)	Speed v (m/s)	Angle α (min.)
100	0.42	0.124	776.48	11.66
200	0.67	0.259	703.23	0.55
300	0.73	0.409	632.61	2.10
400	0.53	0.577	564.78	- 11.55
500	0.03	0.765	500.41	- 23.49

Comment. See table 3 of example 5 in section 3.3 to compare the results obtained above for the same projectile.

Instructions. Employing RangeM30.BAS program.

INPUT DATA
Initial x-coordinate = 0, Initial y-coordinate, Initial Time = 0
Input: Launching Angle [Deg.]: 0.328, Initial Speed = 853.43,
Input: x-coordinate of a Point [m] = 300, Integration Step = 1

RESULTS
x-coordinate of a given point on trajectory = 300 m, y-coordinate = 0.73 m,
Time 0.409 sec., Speed = 632.61 m/s, Angle = - 0.035°

Range = 506 m, y-coordinate = -0.005 m, Time of Flight − 0.775 s, Terminal Speed = 497.3 m/s, Terminal Angle = -0.40278°

Coordinates of trajectory vertex (x = 279 m, y = 0.73 m)

(b) Constructing the Ballistics Table of 0.30 M2 Ball Projectile

Example 2. Find the launching angle if given the range and projectile speed of the projectile.

Consider again the 0.30 M2 ball bullet, caliber 0.308".

Use the function (see example 5 of section 2.6)

$$c(v)=(0.913405-0.0009944v+0.00000062v^2)\cdot(6.2913992)$$

to find the launching angle and other elements of the trajectory for the following ranges: 300 ft. = 91.44 m, 600 ft. = 182.88 m, 900 ft. = 274.32 m, 1,200 ft. = 365.76 m, 1,500 ft. = 457.20 m, 1,800 ft. = 548.64 m.

Solution

Using the system of differential equations (3.1.10) and the ballistics coefficient (2), prepared is the QBasic PC program **ANGLEM30.BAS** (see the attached CD).

The results obtained using **ANGLEM30.BAS** are displayed in table 3.

Table 3

Range x (feet)	Coordinates of Vertex	Angle α_0 (min.)	Time t (sec.)	Speed v (m/s)	Angle α (min.)
\multicolumn{6}{c}{**Elements of Trajectory of 0.30 M2 Ball Projectile** Initial Speed v_0=853.43m/s}
300	(47, 0.015)	2.24	0.112	783	-2.37
600	(100, 0.067)	4.76	0.234	716	-5.35
900	(150, 0.168)	7.61	0.370	653	-9.11
1200	(198, 0.33)	10.86	0.515	591	- 13.87
1500	(250, 0.58)	14.59	0.678	534	-19.94
1800	(306, 0.95)	18.88	0.858	482	-27.66

Note: Employing PC program **ANGLEM30.BAS**

INPUT DATA (in metric system)
Range = 365.76 (in meters) – corresponds to 1,200 ft.

RESULTS
Launching Angle = 0.1809921, Time of Flight = 0.515,
Terminal Speed = 591, Terminal Angle = - 0.2312023
Trajectory Vertex is located at the point with coordinates (198 m, 0.33 m)

(c) Uphill Shooting with 0.30 M2 Ball Bullet

Example 3. Given x and y coordinates of the target, find the launching angle needed to hit the target.

Consider again the 0.30 M2 ball bullet, caliber 0.308".
Use the function

$$c(v)=(0.913405-0.0009944v+0.00000062v^2)\cdot(6.2966551198)$$

to find the launching angle α_0 and the projection angle A needed to hit a target located

(a) on a hill at a slant distance of $D=400m$ if the elevation angle is $E=20°$,
(b) at the horizontal distance of $R=400m$ from the gun,
(c) Use formula

$$\sin(2A+E)=\sin2\alpha_0\cdot\cos^2 E+\sin E$$

to find the projection angle for case (a).

Solution

(a) We use the QBasic PC program **ANTAIR.BAS** (see appendix E) to estimate the launching angle for uphill shooting.

USING **ANTAIR.BAS**:

INPUT DATA:
    x-coordinate of target is

$$x_T=D\cdot\cos(E)=400\cdot\cos(20)=376$$

y-coordinate of target is

$$y_T=D\cdot\sin(E)=400\cdot\sin(20)=136.81m$$

RESULTS: Launching Angle $\alpha_0=20.19629°$, Time of Flight $t=0.57s$, $v_T=570m/s$, Impact Angle $\alpha_T=26.17°$.

projetion angle is

$$A=\alpha_0-E=20.19629°-20°=0.19629°$$
.

(b) USING **ANTAIR.BAS** to find the launching angle corresponding to the given range $R=400m$

INPUT DATA:
  x-coordinate of target is

$$x_T=400m$$

  y-coordinate of target is

$$y_T=0$$

RESULTS: Launching Angle is $\alpha_0=0.20315°$, Time of Flight is $t=0.574s$, Terminal speed is $v_T=585m/s$.

(c) Substituting the values of launching angle, $\alpha_0=0.20315°$, and $E=20°$ the formula

$$\sin(2A+E)=\sin 2\alpha_0\cdot\cos^2 E+\sin E$$
,

we have

$$\sin(2A+20°)=\sin 2(0.20315°)\cdot\cos^2(20°)+\sin(20°)=0.348282$$
.

Hence, we find that

$$2A+20°=20.35269°$$
.

Solving for A, we find that the projection angle is $A=0.19113°$.

Remark: The results obtained in (a) and in (c) for the projection angle A show that the estimation of the projection angle employing the above formula are accurate. That confirms once again the accuracy of that formula for shooting in presence of air as well.

8

Numerical Integration of Projectile Equations.
PC Programs

Introduction

The projectile flight is affected by the variations of the meteorological factors: temperature, humidity, density, pressure of atmospheric air, and the wind. The effect of meteorological factors is reflected to the projectile flight mostly through the drag force that depends also on the density function and the speed of sound.

In chapter 6, and in chapter 7 as well, we have solved the problems of exterior ballistics for the projectile flying in a nonstandard atmosphere by "adjusting" the results obtained for the projectile motion in standard atmosphere, using the "correction" method, a traditional way that still is used in absence of computer technology.

The high-speed computers nowadays allow us to solve the differential equations of projectile flight in real time and with great accuracy.

We only need to reflect or include the nonstandard meteorological factors in the differential equations of projectile flight that describe the projectile trajectory in standard conditions, as well as the deviation of the projectile characteristics from the standard ones.

The nonstandard characteristics of the projectile (that are the result of small changes in projectile mass, the launching speed, the temperature of thrusting charge) also influence the projectile flight and need to be considered in the equations of the projectile flight to obtain more accurate solutions of differential equations of projectile.

In this chapter, we will see the motion of a projectile in a nonstandard atmosphere, as well as the motion of a nonstandard projectile.

We use some PC programs compiled using Runge-Kutta methods to solve the differential equations of the projectile motion.

8.1 The Humidity and the Virtual Temperature

The presence of humidity in atmosphere reduces the density of atmospheric air with respect to dry air. Diffusion of water vapor in dry air (at the same pressure and temperature as the dry air) decreases the molecules of air per unit volume. A number of air molecules in unit volume are replaced by the same number of water molecules. Because water molecules are lighter than the molecules of air (molecular weight of water vapor is 0.01802 kg/mol, while that of air is on average 0.0289644 kg/mol ; see example 4); the density of humid air is reduced compared to dry air.

Because the drag force varies with density of air, the humidity is a factor that affects the resistance a projectile encounters in flight.

The humidity is reflected to the drag through the virtual temperature we are going to study hereafter.

A measure of humidity in atmospheric air in a given temperature is the relative humidity that is the ratio of the quantity of water vapor in a meter cub of air to the quantity of saturated water vapor at the same temperature.[49]

Let's calculate the density of water in atmospheric air.

---

[49] Saturated water vapor represents the maximum quantity of water vapor that a meter cubic air can contain at a given temperature. Relative humidity for the saturated water vapor is 100%. The saturated water vapor does not obey the gas laws. The pressure of saturated water vapor depends only on the temperature. (See Table 1, page 59, Klimi, G. Exterior Ballistics – A New Approach, Xlibris, 2010)

Atmospheric air is a mix of dry air and water vapors. We assume that the atmospheric air, composed of dry air and water vapors, is an ideal gas. The density of the atmospheric air ρ is

$$\rho = \rho_A + \rho_W, \tag{8.1.1}$$

where ρ_A is the density of dry air, while ρ_W is the density of water vapors.

Suppose that p and T are respectively the pressure and temperature of atmospheric air.

To find the density of atmospheric air, we use the Dalton's law of partial pressures

$$p = (p_A + e), \tag{8.1.2}$$

where p_A and e are respectively the partial pressures of dry air and water vapor.

Employing the equation of state for an ideal gas, $p = \rho RT / \mu$, for the partial pressure of dry air and water vapors, we can write respectively

$$p_A = p - e = \frac{\rho_A}{\mu_A} RT \tag{8.1.3}$$

and

$$e = \frac{\rho_W}{\mu_W} RT, \tag{8.1.4}$$

where μ_A and μ_W are respectively the molar masses of dry air and water vapors.

Hence, for the density of dry air and water vapors we obtain respectively

$$\rho_A = (p - e) \frac{\mu_A}{RT} \tag{8.1.5}$$

and

$$\rho_W = e \frac{\mu_W}{RT},$$

(8.1.6)

where

$\mu_W = 18 \cdot 10^{-3}$ kg/mol, $\mu_A = 28.9644 \cdot 10^{-3}$ kg/mol, $R = 8.31441 J \cdot mol^{-1} K^{-1}$.

At the same pressure p_S and temperature T_S for the density of water vapors and dry air, we can write respectively

$$\rho_W = p_S \frac{\mu_W}{RT_S}, \quad \rho_A = p_S \frac{\mu_A}{RT_S}.$$

Dividing the above equations and simplifying, we write

$$\frac{\rho_W}{\rho_A} = \frac{\mu_W}{\mu_A}.$$

(8.1.7)

The molar mass of water and dry air are respectively

$$\mu_W = 18 \cdot 10^{-3} \text{ kg/mol}, \quad \mu_A = 28.9644 \cdot 10^{-3} \text{ kg/mol}.$$

Substituting in (8.1.7), we find that at the same temperature and pressure the density of water vapor is

$$\rho_W = 0.6215 \rho_A.$$

(8.1.8)

Substituting (8.1.8) on the left side of (8.1.6) and then adding (8.1.5) and (8.1.6), we find the density of atmospheric air

$$\rho = \rho_A + \rho_W = (p - e)\frac{\mu_A}{RT} + 0.621 \cdot e \frac{\mu_A}{RT} = (p - 0.3785 e)\frac{\mu_A}{RT}.$$

Hence,

$$\rho = (1 - 0.3785 \frac{e}{p})\frac{\mu_A p}{RT}.$$

(8.1.9)

The quantity

$$\tau=\frac{T}{1-0.3785(e/p)}$$
(8.1.10)

is called the "virtual temperature." From (8.1.10), it is obvious that the virtual temperature is greater than the temperature T of air.

Substituting (8.1.10) in (8.1.9), we express the density of atmospheric air through the virtual temperature

$$\rho=\frac{\mu_A}{R\tau}p,$$
(8.1.11)

where the molar mass of dry atmospheric air is $\mu_A=28.9644\cdot10^{-3}$ kg/mol^{-1}, while $R=8.31441 J\cdot mol^{-1}K^{-1}$ is the molar universal constant of the ideal gas.

The relationship (8.1.11) shows that the virtual temperature is equal to the temperature dry air with density ρ will have at pressure p.

Example 1. In exterior ballistics are accepted the following standard atmospheric characteristics of air on ground. At sea level: pressure $p=750$ mmHg, temperature $t=15$ °C, the partial pressure of water vapor that corresponds to a relative humidity of 50% is $e_{0N}=6.35$ mmHg. Find the virtual temperature τ, the density of water vapor, and the density ρ of atmospheric air.

Solution

The temperature in Kelvin is

$$T=273.15+t=273.15+15=288.15K.$$

The atmospheric pressure and the pressure of the water vapor in Pascal (Pa) are respectively

$$p=(750)\cdot(133.3)=99,975 Pa, \quad e=(6.35)\cdot(133.3)=846.455 Pa$$

Substituting in (8.1.10), we obtain the value of the virtual temperature for the standard atmosphere

$$\tau_{ON}=\frac{T}{1-0.3785\dfrac{e}{p}}=\frac{(288.15)}{1-0.3785\dfrac{(6.35)}{(750)}}=289.08K$$

Substituting the above value of the virtual temperature in (8.1.11), we find the density of atmospheric air (mix of dry air and water vapor) in standard conditions

$$\rho=\frac{\mu_A}{R\tau}p=\frac{28.9644\cdot10^{-3}}{(8.31441)(289.08)}(99,975)\approx1.205\cdot kg/m^3$$

Employing (8.1.7), for the density of water vapor in atmospheric air, we have

$$\rho_W=e\frac{\mu_W}{RT}=(846.455)\frac{0.018}{(8.31441)(288.15)}=0.0063595kg/m^3$$

Example 2. Use the data of example 1 to find the partial pressure of water vapor if the air humidity is 100% (saturated water vapor).

Solution

The density of saturated water vapor is twice the density we found in example 1, i.e.,

$$\rho_{W100\%}=2\rho_{W50\%}=2(0.00636)=0.012719kg/m^3$$

Employing (8.1.7), we find that the partial pressure is

$$e_{100\%}=\frac{\rho_{w100\%}RT}{\mu_w}=\frac{(0.01272)(8.31441)(288.15)}{(0.018)}=1693N/m^2$$

Note: The values obtained above for the density $\rho_{W100\%}=0.012719\,kg/m^3$ and for the partial pressure $e_{100\%}=1693\,N/m^2$ of saturated water vapor at $t=15\ °C$ are quite identical, respectively, to the values $\rho_{W100\%}=0.01283\,kg/m^3$ and $e_{100\%}=1705\,N/m^2$ given by Carl R. (Rod) Nave[50].

Example 3. The saturated vapor pressure at temperature $T=30°C$ is $e=4200\,N/m^2$.[51]

(a) Find the density and the partial pressure of water vapor if the relative humidity at $t=30°$ is 60%, while the atmospheric pressure is 760 mmHg.
(b) Find the virtual temperature and air density ρ of atmospheric air.

Solution

The temperature $t=30°C$ corresponds to a temperature $T=273.15+30=303.15\ K$.

The density of water that is contained in the saturated water vapor at $T=303.15$ is

$$\rho_W = e\frac{\mu_W}{RT} = (4200)\frac{0.018}{(8.31441)(303.15)} = 0.029994\,kg/m^3$$

The density of water vapor in air that has 60% humidity is

$$\rho_{w60\%} = 0.60\rho_w = 0.60(0.029994) = 0.0179964\,kg/m^3$$

The partial pressure of water vapor that corresponds to $\rho_{w60\%} = 0.0179964\,kg/m^3$ is

50   *http://hyperphysics.phy-astr.gsu.edu/hbase/kinetic/watvap.html#c1* [Web site]
51   Hurley, J. P., Garrod, C. *Principi Di Fisica*, P. 317. Zanichelli 1986, (Italian Edition)

$$e_{60\%} = \frac{\rho_{W60\%}}{\mu_W} RT = \frac{0.0179964}{0.018}(8.31441)(303.15) = 2520.01 N/m^2$$

The virtual temperature is

$$\tau_{60\%} = \frac{T}{1-0.379\frac{e_{60\%}}{p}} = \frac{(303.15)}{1-0.379\frac{(2520.01)}{(760.133)}} = 306.04 K$$

The density of air is

$$\rho = \frac{\mu_A}{R\tau}p = \frac{28.9644 \cdot 10^{-3}}{(8.31441)(306.04)}(101308) = 1.153 kg/m^3$$

Example 4. Dry atmospheric air contains approximately 23.1% oxygen (O_2), 75.6% nitrogen (N_2), and 1.3% argon (Ar.). The percentage of the remaining gases of atmospheric dry air is not significant. Determine the average molar mass of dry atmospheric air.

Solution
Employing Dalton's law of partial pressures, we have

$$p = p_{O_2} + p_{N_2} + p_{Ar},$$

where p is the pressure of atmospheric dry air of mass m located in a certain volume V and temperature T. Assuming that all three gases and the atmospheric air are ideal gasses and using the equation of state of an ideal gas, we can write

$$\frac{m}{\mu}RT = \frac{m_{O_2}}{\mu_{O_2}}RT + \frac{m_{N_2}}{\mu_{N_2}}RT + \frac{m_{Ar}}{\mu_{Ar}}RT.$$

Hence,

$$\frac{m}{\mu} = \frac{m_{O_2}}{\mu_{O_2}} + \frac{m_{N_2}}{\mu_{N_2}} + \frac{m_{Ar}}{\mu_{Ar}}.$$

Dividing both sides with the mass m and solving for the molar mass of atmospheric dry air μ we have

$$\frac{1}{\mu} = \frac{m_{O_2}/m}{\mu_{O_2}} + \frac{m_{N_2}/m}{\mu_{N_2}} + \frac{m_{Ar}/m}{\mu_{Ar}} \ .$$

Substituting, we have

$$\frac{1}{\mu} = \frac{0.231}{32 \cdot 10^{-3}} + \frac{0.756}{28.01 \cdot 10^{-3}} + \frac{0.013}{39.948 \cdot 10^{-3}} = 0.034544 \cdot 10^3$$

Hence for the average molar mass of dry atmospheric air we find:

$$\mu = 28.9565 \cdot 10^{-3} \, kg \, / \, mol \ .$$

8.2 Variation of Air Density and Pressure with Altitude

The (virtual) temperature τ as well as the density ρ and pressure p of atmospheric air varies with the altitude y above sea level.

The temperature variation with altitude in general depends on the weather. It is different for different seasons of the year and varies with location and time.

Though the atmosphere is unstable, exterior ballistics assumes an average stable atmosphere, where the air characteristics changes according to some laws.

The change of the virtual temperature τ with the altitude till around 10,000 m above sea level can be approximated by the formula[52]

$$\tau = \tau_0 - 0.006328 \cdot y \ , \tag{8.2.1}$$

where τ_0 is the temperature on the ground. The coefficient -0.006328 in Kelvin per meter is the average temperature gradient (temperature lapse rate) and is written as

---

[52]   Shapiro, J. M. *Vneshnaja Balistika*, p. 53. Oborongiz 50'.

$$d\tau/dy=-0.006328 \ ^{\circ}K/m .$$

Two other average values of the temperature gradient that are valid for altitudes till around 12,000–13,000 m are respectively[53, 54]

$$d\tau/dy=-0.00637 \ ^{\circ}K/m , \qquad\qquad d\tau/dy=-0.0065 \ ^{\circ}K/m .$$

We use the temperature gradient given by Shapiro, presented in (8.2.1), though the values are all averages, and approximately equal. The above formula is valid as well for the temperature of dry air,

$$T=T_0-0.006328 \cdot y ,$$

i.e., when the humidity is zero.

In exterior ballistics, at sea level, as we have seen, we assume the following average meteorological conditions that are accepted as standard atmospheric characteristics of air:

Pressure $p_{0N}=750$ mmHg, temperature $t_{0N}=15 \ ^{\circ}C$, the partial pressure of water vapor $e_{0N}=6.35$ mmHg (relative humidity 50%). There is no wind at all on the ground or in any other altitude.

In a standard atmosphere, the (virtual) temperature of air changes with altitude y over the ground according to the linear law

$$\tau=\tau_{0N}-0.006328 \cdot y , \qquad\qquad (8.2.2)$$

where $\tau_{0N}=289.08K$.

Since the (virtual) temperature varies with altitude according to (8.2.1), the density of atmospheric air ρ and the pressure p, according to (8.1.11),

---

[53]   Roller, D. E., Blum, R. *Fisica.* Vol. 1, p. 330. Zanichelli, 1984.

[54]   Granger, R. A. *Fluid Mechanic",* p. 109. Dover Publications, Inc., 1995.

$$\rho = \frac{\mu_A}{R\tau} p,$$

(8.2.3)

change as well.

Considering (8.2.2) and (8.2.3), we will show the law that the density and pressure of air change with altitude assuming the following:

- The atmospheric air is an ideal gas at rest.
- The acceleration of gravity g is constant, $g=9.80665$ m/s.
- The change of virtual temperature with altitude is given in (8.2.2).

Those assumptions are acceptable considering that for relatively low altitudes (less than 10,000 m), the acceleration of gravity can be considered constant and that for relatively low pressure the atmospheric air is approximately an ideal gas.

Consider a layer of thickness dy at a certain altitude y. For the pressure change dp through the layer dy, we can write

$$dp = -\rho \cdot g \cdot dy.$$

(8.2.4)

Solving (8.2.1) for y and substituting the results in the above equation, we have

$$dp = \frac{\rho g}{0.006328} \cdot d\tau.$$

Using (8.2.3), we can write

$$p = \frac{R}{\mu_A} \cdot \rho \cdot \tau.$$

(8.2.5)

Dividing the last two equations, we get the differential equation

$$\frac{dp}{p}=(\frac{g\cdot\mu_A}{0.006328\cdot R})\cdot\frac{d\tau}{\tau}.$$

(8.2.6)

Differentiating both sides of (8.2.5) and then dividing the result by p, we obtain

$$\frac{dp}{p}=\frac{d\rho}{\rho}+\frac{d\tau}{\tau}.$$

(8.2.7)

Substituting (8.2.7) on the left side of (8.2.6) and rearranging, we get the differential equation in the form

$$\frac{d\rho}{\rho}=(\frac{g\cdot\mu_A}{0.006328\cdot R}-1)\cdot\frac{d\tau}{\tau}.$$

(8.2.8)

If we denote ρ_0, τ_0, and p_0 as the density, the virtual temperature, and the pressure on the ground $y=0$, respectively, after integrating the last equation for the initial conditions $\rho=\rho_0$, $\tau=\tau_0$, $p=p_0$ when $y=0$, we obtain ρ as a function of the virtual temperature τ.

$$\rho=\rho_0(\frac{\tau}{\tau_0})^{g\mu_a/0.006328R-1}$$

(8.2.9)

Substituting $g=9.80665m/s^2$, $\mu_A=28.9644\cdot10^{-3}$ kg/mol^{-1}, and $R=8.31441Jmol^{-1}K^{-1}$, we find

$$\rho=\rho_0(\frac{\tau}{\tau_0})^{4.4},$$

(8.2.10)

where τ is given in (8.2.1).

Substituting the virtual temperature (8.2.1) in (8.2.10), we obtain the law of variation of the density of the atmospheric air with altitude

$$\rho(y)=\rho_0(\frac{\tau_0-0.006328\cdot y}{\tau_0})^{4.4}.$$

Hence, for the ratio $h(y)=\rho/\rho_0$, we find the density function

$$h(y)=(\frac{\tau_0-0.006328\cdot y}{\tau_0})^{4.4}$$

$$\text{(8.2.11)}$$

Employing (8.2.2) for the ratio ρ/ρ_0, we find

$$\frac{\rho}{\rho_0}=(\frac{\tau_0}{\tau})\frac{p}{p_0}.$$

$$\text{(8.2.12)}$$

Substituting (8.2.10) in (8.2.12) and rearranging, we get the law of variation of atmospheric pressure with altitude

$$p=p_0(\frac{\tau_0-0.006328\cdot y}{\tau_0})^{5.4}.$$

$$\text{(8.2.13)}$$

For the normal standard atmospheric data accepted in exterior ballistics,

$$\rho_{0N}=1.205kg/m^3, \quad \tau_{0N}=289.08K, \quad p_{0N}=750 \text{ mmHg,}$$

$$\text{(8.2.14)}$$

we can write (8.2.11) and (8.2.13) respectively as follows:

$$\rho=1.205(1-2.189\cdot10^{-5}y)^{4.4},$$

$$\text{(8.2.15)}$$

$$p=750(1-2.189\cdot10^{-5}y)^{5.4}.$$

$$\text{(8.2.16)}$$

Substituting $\tau_{0N}=289.08K$ in (8.2.11), we obtain the formula for the density function we have used in preceding chapters to study the projectile trajectory in normal standard conditions

$$h(y)=(\frac{289.08-0.006328\cdot y}{289.08})^{4.4}.$$

$$\text{(8.2.17)}$$

From (8.2.13), we can write

$$\frac{p}{p_0}=(\frac{\tau_0-0.006328\cdot y}{\tau_0})^{5.4}.$$

$$(8.2.18)$$

Denoting $H(y)=p/p_0$, we write the function of atmospheric pressure as follows:

$$H(y)=(\frac{\tau_0-0.006328\cdot y}{\tau_0})^{5.4}.$$

$$(8.2.19)$$

Considering (8.2.13), we have

$$H(y)=(\frac{\tau}{\tau_0})h(y).$$

$$(8.2.20)$$

The formulas (8.2.11) and (8.2.19) describe the change of density function and pressure function with altitude, while (8.2.20) shows the relationship between the pressure, the density, and the virtual temperature at a given altitude.

Example 1. Find the virtual temperature, the density, and the pressure of atmospheric air at the altitude $y=1,000$ m. Consider the standard atmosphere.

Solution

The virtual temperature is

$$\tau=\tau_{0N}-0.006328\cdot y=289.08-0.006328(1000)=282.752 \ Kelvin.$$

Substituting $y=1,000$ m in (8.2.15) and (8.2.16), we find

$$\rho_{1,000}=1.205(1-2.189\cdot10^{-5}\cdot(1,000))^{4.4}=1.093kg/m^3$$

and

$$p_{1,000}=750(1-2.189\cdot10^{-5}(1,000))^{5.4}=665.51\,mmHg.$$

Example 2. The temperature and pressure measured by a meteorological station located 650 m above sea level are respectively $10°\,C$ and 698 mmHg, while the relative humidity is 70%. Find the atmospheric pressure, the virtual temperature, the temperature, and the relative humidity of air at sea level.

Saturated water vapor pressure at $10°\,C$ is $1{,}227.693$ N/m².[55]

Solution

The temperature measured at the station location in Kelvin is

$$T=273.15+10°=283.15K.$$

The pressure in Pascal is

$$p=698\cdot(133.3)=93043.4\,N/m^2.$$

(a) The density, the partial pressure of water vapor, and virtual temperature at the station location

The pressure of saturated water vapor at $T=283.15K$ is $e=1227.693N/m^2$. Thus, using (8.1.6) for the density of saturated water vapor, we have

$$\rho_W=e\frac{\mu_W}{RT}=(1227.693)\frac{0.018}{(8.31441)(288.15)}=0.00922385kg/m^3.$$

The density of vapor water in the air with humidity 70% at the station location is

$$\rho_{W70\%}=0.70\rho_w=0.70(0.00922385)=0.0064567kg/m^3.$$

The partial pressure of water vapor at the station location is

---

[55] Carl R. (Rod) Nave,
      http://hyperphysics.phy-astr.gsu.edu/hbase/kinetic/watvap.html#c1. [Web site]

$$e_{70\%}=\frac{p_{W70\%}}{\mu_W}RT=\frac{0.0064567}{0.018}(8.31441)(283.15)=844.47\,N/m^2$$

Using (8.1.10), we find that the virtual temperature at the location of meteorological station is

$$\tau=\frac{T}{1-0.3785(e/p)}=\frac{283.15}{1-0.3785(844.47/93043.4)}=284.13K$$

Employing (8.1.11), we find that the density of the atmospheric air at the meteorological station is

$$\rho=\frac{\mu_A}{R\tau}p=\frac{28.9644\cdot10^{-3}}{(8.31441)(284.13)}(93043.4)=1.1408kg/m^3.$$

(b) The temperature, the density, and the pressure of air at sea level

Employing (8.2.1), we find that the virtual temperature at sea level is

$$\tau_0=\tau+0.006328\cdot y=(284.143)+0.006328(650)=288.26K=15.11\ ^\circ C.$$

Employing (8.2.11) and (8.2.18), we find that the density of atmospheric air and the pressure at sea level are respectively

$$\rho_0=\rho(\frac{\tau_0}{\tau})^{4.4}=1.1408\cdot(\frac{288.26}{284.143})^{4.4}=1.2153kg/m^3,$$

and

$$p_0=p(\frac{\tau_0}{\tau})^{5.4}=93043.4(\frac{288.26}{284.13})^{5.4}=100,584\,N/m^2=754.57mm\ Hg.$$

(c) The temperature of dry air and the relative humidity at sea level

Employing (8.2.1), for the temperature of dry air at sea level, we find

$$T_0 = T + 0.006328 \cdot y = 283.15 + 0.006328(650) = 287.2632\,K = 14.11\ ^\circ C$$.

Substituting in (8.1.10),

$$\tau_0 = \frac{T_0}{1 - 0.3785(e_0/p_0)},$$

we find the partial pressure of water vapor at sea level

$$e_0 = \frac{1}{0.3785}(1 - \frac{T_0}{\tau_0})p_0 = \frac{1}{0.3785}(1 - \frac{287.26}{288.15}) \cdot (754.57) = 6.158 mm\ Hg$$.

The pressure and the density of saturated water vapor at $T_0 = 14.11\ ^\circ C$ are respectively[56]

$$e_{100\%} = 12.078 mm\ Hg. \quad \text{and} \quad \rho_{100\%} = 0.01215 kg/m^3$$.

For the density of water vapor at $T_0 = 14.11\ ^\circ C$ and pressure $e_0 = 7 mm\ Hg$, using the ideal gas law, we find that the relative humidity is

$$\frac{\rho_{W0}}{\rho_{100\%}} = \frac{6.156}{12.078} = 50.97\%$$.

Hence, for the density of the water vapor at sea level, we have

$$\rho_{W0} = 50.97\%(0.01215) = 0.0062 kg/m^3$$.

8.3 The Isothermal Barometric Formula

The barometric formula we are going to present in this section is an approximate formula obtained assuming that the gravity is constant and the atmosphere is isothermal, i.e., that the atmospheric air

---

56 *http://hyperphysics.phy-astr.gsu.edu/hbase/kinetic/watvap.html#c1* [Web site]

temperature T does not change with altitude (temperature gradient is zero, i.e., $dT/dy=0$).

The isothermal barometric formula can be used when the maximum altitude of the projectile trajectory is relatively low so that the temperature of air can be considered constant for that layer of air the projectile flies into.

The atmospheric pressure is a function of time and location. For a given time and location, the pressure changes with altitude.

The change of pressure with altitude approximately can be estimated using the isothermal barometric formula

$$p = p_0 \cdot e^{-\mu g y/(RT)},$$

(8.3.1)

where p_0 is the pressure at a given altitude, p is the pressure at an altitude y above the ground, μ_A is the molar mass of air ($\mu_A = 28.9644 \cdot 10^{-3}$ kg/mol^{-1}), R is the universal constant of ideal gas $R = 8.31441 Jmol^{-1}K^{-1}$, g is gravitational acceleration $g = 9.80665 m/s^2$, and T is the temperature of air in Kelvin.

Indeed, let's consider equation (8.2.5),

$$dp = -\rho \cdot g \cdot dy.$$

Considering the atmospheric air an ideal gas, we can write

$$p = \frac{R}{\mu_A} \rho T.$$

Dividing the above equations side by side, we have

$$\frac{dp}{p} = \left(\frac{g \cdot \mu_A}{R \cdot T}\right) \cdot dy.$$

(8.3.2)

Integrating, for the initial conditions $y=0$ (sea level), $p=p_0$, we obtain the barometric formula (8.3.1), which we will write in the following form:

$$p = p_0 \cdot e^{-(\mu g/(RT)) \cdot y}.$$

(8.3.3)

Since

$$\frac{g \cdot \mu_A}{R} = \frac{(9.80665) \cdot (28.9644 \cdot 10^{-3})}{8.31441} = 0.0341628 \quad , \tag{8.3.4}$$

the barometric formula (8.3.3) can be written as

$$p = p_0 \cdot e^{-0.0341628 \cdot y/T} . \tag{8.3.5}$$

Now we give to the barometric formula (8.3.5) a form suitable for use in exterior ballistics. In exterior ballistics, the range tables are compiled for the standard atmosphere.

The standard temperature of atmospheric air on the ground is $T=288.15$ K ($15°$ C). Since the temperature T is supposed to be constant with altitude y, we consider it equal to $T=288.15$. Substituting $T=288.15$ in (8.3.5), we obtain another form of the barometric formula for normal standard conditions

$$p = p_0 \cdot e^{-y/8434.6} . \tag{8.3.6}$$

As we have seen, any deviation in atmospheric factors at the location of artillery gun from accepted standard values is reflected in projectile flight through the so-called "corrections" (see chapter 7). The correction of the projectile flight is performed based on the altitude of the artillery gun. Thus, to make corrections, the artillery battery personnel need to know the pressure of atmospheric air at the location of artillery battery.

The data for the weather, including the atmospheric pressure, are obtained from a meteorological station that in general is located somewhere else and reports the corresponding value of pressure (or the pressure reduced to sea level).

Based on pressure p_s measured at the station location that is at altitude y_s, we find the pressure p_b at the altitude y_b where the artillery battery (gun) is located.

Using the barometric formula (8.3.5), we can write for p_b and p_S respectively

$$p_b = p_0 \cdot e^{-0.03416 \cdot y_b / T}$$

(8.3.7)

and

$$p_S = p_0 \cdot e^{-0.03416 \cdot y_S / T}.$$

(8.3.8)

Dividing (8.3.8) into (8.3.7), we can write

$$p_b / p_S = e^{-0.03416(y_b - y_S)/T}.$$

(8.3.9)

Hence,

$$p_b = p_S \cdot e^{-0.03416(y_b - y_S)/T}.$$

(8.3.10)

We can obtain an approximate formula in case

$$(0.03416) \frac{|y_b - y_s|}{T} \ll 1$$

or the difference between the altitude of artillery battery and station in absolute value is relatively very small:

$$|y_b - y_S| \ll (29.274/T).$$

Expanding the right side in Maclaurin series, i.e., at $|y_b - y_s| = 0$, and considering the first two terms of the expansion, we have

$$\frac{p_b}{p_s} = 1 - 0.03416 \frac{y_b - y_s}{T}.$$

(8.3.11)

Hence, for the pressure at the artillery battery location, we have

$$p_b = p_s (1 - 0.03416 \frac{y_b - y_s}{T}).$$

(8.3.12)

Example 1. The change in altitude between the meteorological station and the battery of artillery is $(y_b - y_s) = -300m$. The pressure measured at the station location is p_s=720 mmHg. Find the pressure at the location of the battery if the temperature is 15° C.

Solution

The temperature at the battery location and at the meteorological station is assumed constant. It is T=273.15+15°=288.15 K.

Employing (8.3.12), we find that the pressure at the battery location is

$$p_b=p_s(1-0.03416\frac{y_b-y_s}{T})=720(1+0.03416\frac{300}{288.15})=745.56mm \; Hg.$$

Example 2. The temperature and pressure measured by a meteorological station located 650 m above sea level are respectively 10° C and 698 mmHg (the humidity is 70%). Use the barometric formula to find the atmospheric pressure at the sea level.

Solution

The temperature in Kelvin is

$$T=273.15+10°=283.15 \; K.$$

Assuming the temperature constant and equal to that measured by the meteorological station, using (8.3.5), we find that the pressure of air at sea level is

$$p_0=p_s \cdot e^{-0.03416 \cdot (y_b-y_S)/T}=698 \cdot e^{-0.03416(0-650)/283.15}=754.94mm \; Hg.$$

Note: The value of atmospheric pressure we found is approximately equal to the value of the atmospheric pressure we found in example 2 of section 8.2, using the barometric formula.

8.4 The Siacci Function of Resistance and the Speed of Sound

In preceding chapters, we obtained the projectile trajectory elements solving the differential equation of flight, (2.2.12),

$$\frac{d\vec{v}}{dt}=\vec{g}-c \cdot h(y)f_D(v)\frac{\vec{v}}{v} ,$$

$$(8.4.1)$$

assuming a standard atmosphere:

- Air density $\rho_{0N}=1.205kg/m^3$
- Temperature of air $T_{0N}=288.15K=15\ °C$
- Atmospheric pressure $p_{0N}=750$ mm Hg.
- Air virtual temperature $\tau_{0N}=289.08$, which corresponds to a 50% humidity and a water vapor pressure of $e_{0N}=6.35mm\ Hg$.
- Speed of sound in air on ground level is $a_0=a_{0N}=340.83$,

where $f_D(v)=4.732{\cdot}10^{-4}{\cdot}v^2 C_D(v/a)$ is the function of resistance.

For the Siacci function $f_D(v)=K_D(v)$,

where

$$K_D(v)=\begin{cases} 1.212{\cdot}10^{-4}v^2 & for \quad v{\le}256m/s \\ (v-240)/3 & for \quad v{>}256m/s \end{cases}$$

(8.4.2)

and

$$h(y)=\frac{\rho}{\rho_{0N}}=(\frac{289.08-0.006328y}{289.08})^{4.4}$$

(8.4.3)

For the standard atmosphere, the function of resistance, $f_D(v)$, can be written

$$f_D(v/a_{0N})=4.732{\cdot}10^{-4}{\cdot}v^2 C_D(v/a_{0N})$$

(8.4.4)

Because the standard atmospheric conditions are not always met in practice of firearm shooting, we need to reflect the influence of meteorological factors in the drag acceleration of a projectile when their values are different from those accepted as standard.

When the value of the speed of sound on the ground at sea level is different from the standard speed of sound, i.e., when $a_0{\ne}a_{0N}=340.83$, then the function of resistance

$$f_D(v/a_0)=4.732{\cdot}10^{-4}{\cdot}v^2 C_D(v/a_0)$$

(8.4.5)

is also a function of the speed of sound a_0.

For that reason, the drag coefficient $C_D(v/a_0)$—and as a result, the function of resistance,[57] $f_D(v/a_0)$—need to be determined experimentally for different values of the projectile speed and the speed of sound as well.

That will complicate the solution of any ballistics problem since the function of resistance (8.4.5), till now considered a function of projectile speed only, will be a function of two independent variables: the projectile speed v and the speed of sound a_0.

To simplify the study of the projectile flight when the speed of sound, a_0, on the ground, at sea level, is not equal to the standard value ($a_0 \neq a_{0N} = 340.83$), exterior ballistics uses the same experimental drag coefficient function, determined for the standard conditions, when the speed of sound is $a_{0N} = 340.83$, but modifies the function of resistance (8.4.4).

Indeed, we can write

$$C_D(v/a_0) = C_D(v \cdot \frac{a_{0N}/a_0}{a_{0N}})$$

(8.4.6)

We denote

$$u = \frac{a_{0N}}{a_0} v$$

(8.4.7)

Solving (8.4.7) for v, we find

$$v = a_0 u / a_{0N}$$

(8.4.8)

Substituting (8.4.8) in (8.4.5), we have

---

[57] After the First World War, it was realized that the function of resistance is a function of speed of sound as well. But when we study the projectile flight in standard atmosphere, the fact that the function of resistance is actually a function of Mach number is irrelevant.

$$f_D(v/a_0) = 4.732 \cdot 10^{-4} \cdot v^2 C_D \left(\frac{v \cdot a_{0N}/a_0}{a_{0N}} \right). \tag{8.4.9}$$

We will give (8.4.9) an appropriate form for use in the differential equations of projectile motion (8.4.1).

The Speed of Sound

The speed of sound in air is a function of the (virtual) temperature τ and can be estimated using the following formula:

$$a = \sqrt{\frac{\gamma R}{\mu_A} \tau}, \tag{8.4.10}$$

where $\gamma = 1.4$, $R = 8.31441 J \cdot mol^{-1} K^{-1}$, $\mu_A = 28.9644 \cdot 10^{-3} kg/mol$.

Substituting the above values in (8.4.10) for the speed of sound, we have

$$a = \sqrt{\frac{\gamma R}{\mu_A} \tau} = \sqrt{\frac{(1.4)(8.31441)}{(28.9644 \cdot 10^{-3})} \tau} = 20.05\sqrt{\tau} \tag{8.4.11}$$

Thus,

$$a = 20.05\sqrt{\tau}. \tag{8.4.12}$$

We denote a_0 and τ_0 respectively as the speed of sound and the virtual temperature on the ground, at sea level, when the atmosphere is not standard. Using (8.4.11) or (8.4.12), we write

$$a_0 = \sqrt{\frac{\gamma R}{\mu_A} \tau_0}, \quad a_0 = 20.05\sqrt{\tau_0} \tag{8.4.13}$$

and

$$a_{0N} = \sqrt{\frac{\gamma R}{\mu_A} \tau_{0N}}, \quad a_{0N} = 20.05\sqrt{\tau_{0N}}, \tag{8.4.14}$$

when the atmosphere on the ground is assumed to be standard. Substituting (8.4.13) and (8.4.14) in (8.4.8), we find

$$\frac{v}{u} = \frac{a_0}{a_{0N}} = \sqrt{\frac{\tau_0}{\tau_{0N}}}.$$ (8.4.15)

Considering (8.4.15) for the function of resistance (8.4.10), we write

$$f_D(v/a_0) = 4.732 \cdot 10^{-4} \cdot v^2 C_D \left(\frac{v \cdot a_{0N}/a_0}{a_{0N}} \right) = 4.732 \cdot 10^{-4} v^2 \cdot C_D \left(\frac{v\sqrt{\tau_{0N}/\tau_0}}{a_{0N}} \right).$$ (8.4.16)

Hence, and considering $f_D(v/a_{0N})$ given in (8.4.4), it follows that

$$f_D(v/a_0) = f_D \left(\frac{v\sqrt{\tau_{0N}/\tau_0}}{a_{0N}} \right).$$ (8.4.17)

Thus, (8.4.17) shows that we can use the same function of resistance determined for the standard atmosphere by formally replacing v with $v\sqrt{\tau_{0N}/\tau_0}$.

The differential equation of the projectile flight (8.4.1) can be written

$$\frac{d\vec{v}}{dt} = \vec{g} - c \cdot h_\tau(y) f_D \left(\frac{v\sqrt{\tau_{0N}/\tau_0}}{a_{0N}} \right) \cdot \frac{\vec{v}}{v},$$ (8.4.18)

where

$$h_\tau(y) = \frac{\rho}{\rho_0} = \left(\frac{\tau_0 - 0.006328y}{\tau_0} \right)^{4.4}, \quad \tau_{0N} = 289.08°K, \text{ and}$$

$$\tau_0 = \frac{T_0}{1 - 0.3785(e_0/p_0)}.$$ (8.4.19)

As a matter of fact γ (considered as constant and equal to 1.4) decreases with temperature and relative humidity. For example, at 15°c and 78% relative humidity, $\gamma = 1.3988$, while at the same temperature and 0% relative humidity, $\gamma = 1.4003$. This slight change affects the sonic velocity negligibly. We can say that neglecting the temperature and relative humidity affects γ inconsequently.

Example 1. Find the speed of sound in a standard atmosphere.

Solution
   The speed of sound on the ground, at sea level, for the standard atmosphere is

$$a_{0N}=\sqrt{\frac{\gamma R}{\mu_A}\tau_{0N}}=\sqrt{\frac{(1.4)(8.31441)}{(28.9644\cdot 10^{-3})}289.08}=340.84 m/s$$
.

Example 2.   Find v_{0N} if the projectile speed and the virtual temperature on the ground are $v_0=735 m/s$ and $\tau_0=303.15K$.

Solution
   Using (8.4.13), we find that

$$v_{0N}=v\sqrt{\frac{\tau_{0N}}{\tau_0}}=735\cdot\sqrt{(289.08)/(303.15)}=717.74 m/s$$
.

Example 3. Find the speed of sound at the altitude $y=1200m$ in standard atmospheric conditions.

Solution
   The virtual temperature when $y=1200m$ is

$$\tau=\tau_0-0.006328\cdot y=289.08-0.006328(1200)=281.49\ {}^\circ K$$
.

   The speed of sound is

$$a=\sqrt{\gamma R\tau/\mu_A}=\sqrt{(1.4)(8.31441)(281.49)/(28.9644\cdot 10^{-3})}=336.34 m/s$$
.

Example 4. Find the speed of sound, the value of the density function $h(y)$, and v_N at the altitude $y=1200m$ if the virtual temperature on the ground is $\tau_0=303.15K$ and the speed of projectile is $v=780m/s$. Find as well the value of the speed of sound on the ground.

Solution

The virtual temperature at altitude $y=1200m$ is

$$\tau=\tau_0-0.006328 \cdot y=(303.15)-0.006328(1200)=295.56K$$.

Substituting in (8.4.11), we find the speed of sound at $y=1200m$ is

$$a=20.05\sqrt{\tau}=20.05\sqrt{(295.56)}=344.70$$.

The value of the density function is

$$h_\tau(y)=(\frac{\tau_0-0.006328 \cdot y}{\tau_0})^{4.4}=0.89438$$

and

$$v_{ON} = v\sqrt{\frac{\tau_{ON}}{\tau}} = 780\sqrt{\frac{289.08}{295.56}} = 771.40m/s.$$

8.5 The Differential Equations of Projectile Motion in Nonstandard Atmosphere

To study the projectile flight in nonstandard atmosphere, we need to write the differential equations of flight that describe the projectile motion in nonstandard atmosphere.

The differential equation of projectile flight in vector form (8.4.18) for the Siacci function can be written as

$$\frac{d\vec{v}}{dt} = \vec{g}-c \cdot h_\tau(y)K_D(v \cdot \sqrt{\tau_{ON}/\tau_0}\ )\frac{\vec{v}}{v}, \qquad (8.5.1)$$

where

$$c=\frac{id^2}{m}1000$$
,

$$K_D(u)=K_D(v\sqrt{\tau_{0N}/\tau_0})=\begin{cases} 1.212\cdot10^{-4}v^2\tau_{0N}/\tau_0 & for \quad v(\tau_{0N}/\tau_0)^{1/2}\le256m/s \\ [v(\tau_{0N}/\tau_0)^{1/2}-240]/3 & for \quad v(\tau_{0N}/\tau_N)^{1/2}>256m/s \end{cases}, \qquad (8.5.2)$$

and

$$h_\tau(y)=\frac{\rho}{\rho_0}=(\frac{\tau_0-0.006328y}{\tau_0})^{4.4}, \quad \tau_{0N}=289.08K , \tau_0=\frac{T_0}{1-0.3785(e_0/p_0)} \quad (8.5.3)$$

Note that e_0 and p_0 are respectively the partial pressure of vapor present in air and the atmospheric pressure at sea level.

Using the same procedure we have used in chapters 2, from the differential equation (8.5.1), we obtain the following systems of differential equations that describe the projectile flight in a nonstandard atmosphere:

(a) For the speed of projectile $v>256m/s$,

$$\begin{cases} \dfrac{dv_x}{dt}=-ch_\tau(y)\cdot[v(\tau_{0N}/\tau_0)^{1/2}-240]\dfrac{v_x}{3v} \\[2mm] \dfrac{dp}{dt}=-\dfrac{g}{v_x} \\[2mm] \dfrac{dx}{dt}=v_x \\[2mm] \dfrac{dy}{dt}=v_y \end{cases} \qquad (8.5.4)$$

or

$$\left\{ \begin{array}{l} \dfrac{dv_x}{dx} = -ch_\tau(y) \cdot \dfrac{v(\tau_{0N}/\tau_0)^{1/2} - 240}{3v} \\[3mm] \dfrac{dp}{dx} = -\dfrac{g}{v_x^2} \\[3mm] \dfrac{dt}{dx} = \dfrac{1}{v_x} \\[3mm] \dfrac{dy}{dx} = p \end{array} \right. \qquad (8.5.5)$$

(b) For the speed of projectile $v \le 256 m/s$,

$$\left\{ \begin{array}{l} \dfrac{dv_x}{dt} = -1.212 \cdot 10^{-4} c \cdot h_\tau(y) v^2 \dfrac{\tau_{0N}}{\tau_0} \cdot \dfrac{v_x}{v} \\[3mm] \dfrac{dp}{dt} = -\dfrac{g}{v_x} \\[3mm] \dfrac{dx}{dt} = v_x \\[3mm] \dfrac{dy}{dt} = v_y \end{array} \right. \qquad (8.5.6)$$

or

$$\left\{ \begin{array}{l} \dfrac{dv_x}{dx} = -1.212 \cdot 10^{-4} ch_\tau(y) \cdot (v^2 \dfrac{\tau_{0N}}{\tau_0}) \dfrac{1}{v} \\[3mm] \dfrac{dp}{dx} = -\dfrac{g}{v_x^2} \\[3mm] \dfrac{dt}{dx} = \dfrac{1}{v_x} \\[3mm] \dfrac{dy}{dx} = p \end{array} \right. \qquad (8.5.7)$$

To solve the ballistics problems of projectiles flying in nonstandard atmosphere, we will use the differential equations of projectile flight (8.5.5) and (8.5.7) and employ the numerical solution method.

Systems (8.5.5) and (8.5.7), together with the initial conditions, consider the deviations of characteristics of the atmosphere from the characteristics of the standard atmosphere, but they do not take into account the deviation of firearm and projectile characteristics from handbook characteristics, for example,

- the small variations in projectile speed, projectile mass, launching angle, and propellant temperature (which is 15°C in standard atmosphere);
- the wind (the rangewind and the crosswind).

To consider all those factors in the projectile flight, we need to modify the differential equations (8.5.5) and (8.5.7) and define the initial conditions of the projectile flight.

1. Changes in the Launching Speed and Launching Angle

A small change in projectile speed dv_0 and a small change in the launching angle $d\alpha_0$ alter respectively the value of the launching speed from the standard speed v_0 to v_0+dv_0 and the value of launching angle from α_0 to $\alpha_0+d\alpha_0$.

For example, the change in the launching speed of a projectile for a firearm might be different from the standard speed of the firearm due to the consumption of the particular firearm.

2. Change in Projectile Mass

A small change dm in the projectile mass (different lots of projectile might have different masses) is reflected as a change in the ballistics coefficient c, as well as a change in the initial speed of the projectile.

The differential of the ballistics coefficient,

$$c=\frac{id^2}{m}1000,$$

(8.5.8)

is

$$dc=-1000\frac{id^2}{m^2}\cdot dm$$

$$(8.5.9)$$

Dividing (8.5.8) into (8.5.9) and then multiplying both sides by c, we find

$$dc=-\frac{dm}{m}\cdot c$$

$$(8.5.10)$$

Assuming we know the ballistics coefficient c of the projectile, we find that the modified coefficient that will formally replace c in (8.5.5) and (8.5.7) is

$$c'=c+dc=c-\frac{dm}{m}\cdot c=c(1-\frac{dm}{m})$$

$$(8.5.11)$$

In exterior ballistics, it is considered that a change dm in the projectile mass decreases the initial speed of the projectile with a quantity

$$dv_0=-0.4\frac{dm}{m},$$

$$(8.5.12)$$

where m is the standard (handbook) value of the projectile mass.

Thus, the actual initial speed corresponding to a change dm in the projectile mass only is

$$v'=v_0-0.4\frac{dm}{m}.$$

3. Change in the Propellant Temperature

A small change of the temperature of propellant dT_c (which normally is the same as the change of the temperature of air of the area where the projectiles are stored to be fired) from the standard value ($T_{c0}=15°Celsius$) increases the launching speed v_0 of the projectile according to the relation

$$dv_0=0.001{\cdot}v_0 dT_c.$$

$$(8.5.13)$$

Thus, the modified launching speed of the projectile can be estimated using the equation

$$v_0'=v_0(1+0.001{\cdot}dT_c).$$

$$(8.5.14)$$

4. Change in the Atmospheric Pressure

The equations of projectile flight contain the density function $h_\tau(y)$. The equation (8.2.20),

$$H(y)=(\frac{\tau}{\tau_0})h_\tau(y),$$

$$(8.5.15)$$

which relates the pressure function $H(y)$ and the density function $h(y)$, shows that a small change only in the atmospheric pressure affects the density function $h(y)$. For the density function, we can write

$$h_\tau(y)=(\frac{\tau_0}{\tau})H(y).$$

$$(8.5.16)$$

A small change dH in $H(y)$ causes a small change dh in the density function value (8.5.16). Differentiating (8.5.16), we find that

$$dh=(\frac{\tau_0}{\tau})dH.$$

$$(8.5.17)$$

Dividing (8.5.17) and (8.5.16), we have

$$\frac{dh}{h} = \frac{dH}{H}.$$

Hence,

$$dh = \frac{h(y)}{H(y)} dH.$$

(8.5.18)

Considering $dh = h' - h_\tau$, from (8.5.18), we obtain

$$h' = h_\tau(y)(1 + \frac{p_0 - p_{0N}}{p_{0N}}) = h_\tau(y) \cdot \frac{p_0}{p_{0N}}$$

(8.5.19)

To introduce the density changes as a result of the changes in atmospheric pressure, we substitute in (8.5.5) and (8.5.7) instead of $h_\tau(y)$ the expression on the right side of (8.5.19).

5. The Influence of Wind

The influence of the velocity of wind $\vec{w}$ in the projectile flight can be studied using the vector differential equation of the projectile flight in a three-dimensional coordinate system (figure 13)

$$\frac{d\vec{V}}{dt} = \vec{g} - c \cdot h(y) K_D(V) \frac{\vec{V}}{V},$$

(8.5.20)

where

$$V = \sqrt{(\vec{v} - \vec{w})^2}$$

(8.5.21)

is the relative speed of the projectile with respect to air. The solution of the system of the differential equations that can be obtained from (8.5.20) requires numerical integration.

To simplify the solution of (8.5.`20) and to use the results already obtained for the projectile flight in absence of wind, we use the traditional method that considers the components of wind: the range wind $\vec{w}_x$ and the crosswind $\vec{w}_z$.

The Influence of the Range Wind

Consider a range wind that blows horizontally in the direction of the x-axis with a constant velocity $\vec{w}_x$ with respect to the coordinative system x, y, and z with the center located at the launching point of the projectile.

The air resistance to the projectile, as well as the Siacci function of resistance (8.5.2), are functions of the speed of the projectile $V(t)$ relative to the moving air, i.e.,

$$K_D(u)=K_D(V) .$$

(8.5.22)

The projectile relative velocity $\vec{V}(t)$, at a given point P on the trajectory, can be expressed as the vector sum of the projectile velocity $\vec{v}(t)$ and the velocity $(-\vec{w}_x)$,

$$\vec{V}(t)=\vec{v}(t)-\vec{w}_x .$$

(8.5.23)

The relative speed of the projectile V, which is present in (8.5.22), is

$$V=[(\vec{v}-\vec{w}_x)^2]^{1/2}=(v^2+w_x^2-2v \cdot w_x \cdot \cos\alpha)^{1/2} .$$

(8.5.24)

To reflect the influence of wind in the function of resistance, we replace, in the Siacci function of resistance (8.5.22), V with the expression that is on the right side of (8.5.24).

The Influence of the Crosswind

Consider the crosswind that blows in the direction of the z-axis with velocity $\vec{w}_z$ (figure 17). The resultant velocity of the projectile at a point P on the trajectory is

$$\vec{v}_w(t)=\vec{v}+\vec{w}_z=v_x \vec{i}+v_y \vec{j}+w_z \vec{k} ,$$

where $\vec{v}$ is the projectile velocity in absence of wind.

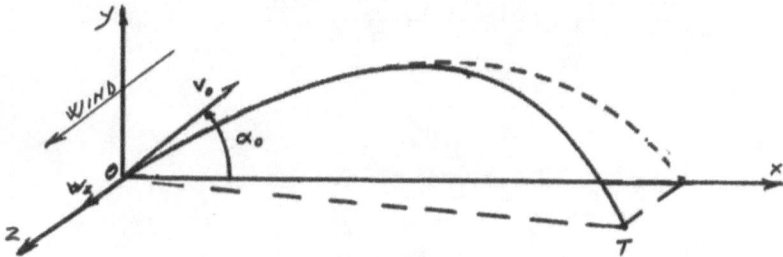

Figure 17

We will use the results obtained in section 6.7 to take into account the influence of the crosswind, which deviates the plane of the projectile flight to the right (or to the left) of the xy-plane with rotation angle φ determined by equation (6.7.6),

$$\tan\varphi = -\frac{w_z}{v_0\cos\alpha_0}.$$

$$(8.5.25)$$

The z-coordinate of the projectile during the flight is determined by equation (6.79),

$$z(t)=w_z t-\frac{w_z}{v_0\cos\alpha_0}x,$$

$$(8.5.26)$$

where x and t are respectively the x-coordinate of a point on the projectile trajectory and the time the projectile reaches that point.

Summary:

To solve the problems of exterior ballistics in nonstandard conditions, we use the differential equations of projectile flight (8.5.5) and (8.5.7),

$$\begin{cases} \dfrac{dv_x}{dx} = -ch_\tau(y) \cdot \dfrac{v(\tau_{0N}/\tau_0)^{1/2} - 240}{3v} \\[2mm] \dfrac{dp}{dx} = -\dfrac{g}{v_x^2} \\[2mm] \dfrac{dt}{dx} = \dfrac{1}{v_x} \\[2mm] \dfrac{dy}{dx} = p \end{cases} \qquad (8.5.27)$$

and

$$\begin{cases} \dfrac{dv_x}{dx} = -1.212 \cdot 10^{-4} ch_\tau(y) \cdot \dfrac{(v \cdot \tau/\tau_{0N})^2}{v} \\[2mm] \dfrac{dp}{dx} = -\dfrac{g}{v_x^2} \\[2mm] \dfrac{dt}{dx} = \dfrac{1}{v_x} \\[2mm] \dfrac{dy}{dx} = p \end{cases} \qquad (8.5.28)$$

where

$$v = [(\vec{v}' - \vec{w}_x)^2]^{1/2} = (v'^2 + w_x^2 - 2v' \cdot w_x \cdot \cos\alpha)^{1/2}, \qquad (8.5.29)$$

(v' is the projectile speed in the fixed coordinate system.)

The initial conditions in the absolute system of the Cartesian coordinate centered at the launching point of the projectile are as follows:

Initial Coordinates

- Coordinates of the launching point and initial time are respectively

$$x_0=0, \quad y(0)=y_0, \quad z(0)=0, \quad t(0)=0.$$

- Initial Speed

The actual launching speed of the projectile due to the factors shown in (8.5.1)–(8.5.3) is

$$v_0'=(v_0+dv_0)-0.4\frac{dm}{m}+0.001\cdot v_0 dT. \tag{8.5.30}$$

The components of the actual velocity are

$$v_{0x}'=v_0'\cos\alpha_0, \quad v_{0y}'=v_0'\sin\alpha_0, \quad v_{0z}'=0 \tag{8.5.31}$$

The range wind is w_x and is constant during the flight.

Modifications and Corrections

- Change in the Atmospheric Pressure

The density function $h_\tau(y)$ on the right side of (8.5.27) and (8.5.28) must be replaced by

$$h'=h_\tau(y)\cdot(1+\frac{dp}{p}). \tag{8.5.32}$$

- Modified Ballistics Coefficient

The ballistics coefficient c must be replaced by

$$c'=c(1-\frac{dm}{m}). \tag{8.5.33}$$

- Crosswind Corrections
  Angle of rotation of the launching plane in absence of wind is

$$\tan\varphi = -\frac{w_z}{v_0 \cos\alpha_0}.$$

(8.5.34)

The coordinate z of a point on the trajectory is

$$z = w_z t - \frac{w_z}{v_0 \cos\alpha_0} x,$$

(8.5.35)

where t is the time of flight to the target.

8.6 Programs. Compilation Remarks

In preceding chapters, we solved the differential equations of projectile flight based on the Siacci function of resistance $K_D(v)$ when the projectile is launched in standard atmosphere using approximate methods, as well as the remarkable Siacci method that uses the pseudo-speed u.

The trajectory elements of a projectile flying in nonstandard atmospheric conditions till now were estimated using the Siacci correction method. We have employed in chapter 7, as well as in other chapters, some QBasic PC programs, to verify the accuracy of results obtained using approximate methods.

The PC programs we present in this book can be classified into three categories[58]:

- **Some PC programs use an average ballistics coefficient, $c=\bar{c}$, which is constant for any trajectory.**
- **Another category of programs use a ballistics coefficient that is a function of projectile speed, $c=c(v)$.**

---

[58] For technical reason it was not possible to accompany the book with a CD that contains the PC program. Request an electronic copy or a CD with all PC programs to the author at: *gklimi@pace.edu*, or *iven24@aol.com*

- The other category of programs is based on the idea that considers the ballistic coefficient c to be an "adjusting" factor and a function of the launching angle $c=c(\alpha_0)$.

  The "adjusting" ballistics coefficient $c=c(\alpha_0)$ can be found from firearm testing or, as in our case, using the range tables of the given projectile and firearm.

The PC programs can as well be classified according to the main three types of ballistics problems they solve:

- PC programs that are used to find the range given the launching angle, the launching speed, and the ballistics coefficient.
- PC Programs that are used to find the launching angle needed to hit a given target when the projectile speed and the ballistics coefficient are known.
- A PC program that is used to find the ballistics coefficient for a given projectile launched with a known launching angle and speed.[59]

There are shown also some PC programs, used for uphill or downhill shooting, skydiving and parachute fall that are derived from the two first main types of PC programs.

The equations used to program the solution of ballistics problems are shown hereafter.

To determine the trajectory of a projectile flying in nonstandard conditions, we employ the numerical methods (specifically Runge-Kutta methods) facilitated by QBasic PC programs to integrate the systems of differential equations of projectile flight (8.5.5) and (8.5.7), considering as well the appropriate modifications shown in that section,

---

[59] We avoid the solution of differential equations related with rotation of projectile, and the gyroscopic effects of projectile.

$$
\begin{cases}
\dfrac{dv_x}{dx} = -c'h_\tau(y)\dfrac{p_0}{p_{0N}}\cdot\dfrac{v(\tau_{0N}/\tau_0)^{1/2}-240}{3v} \\[3mm]
\dfrac{dp}{dx} = -\dfrac{g}{v_x^2} \\[3mm]
\dfrac{dt}{dx} = \dfrac{1}{v_x} \\[3mm]
\dfrac{dy}{dx} = p
\end{cases}
\qquad , \qquad (8.6.1)
$$

or

$$
\begin{cases}
\dfrac{dv_x}{dx} = -1.212\cdot10^{-4}c'h_\tau(y)\dfrac{p_0}{p_{0N}}\cdot\dfrac{v^2\tau_{0N}/\tau_0}{v} \\[3mm]
\dfrac{dp}{dx} = -\dfrac{g}{v_x^2} \\[3mm]
\dfrac{dt}{dx} = \dfrac{1}{v_x} \\[3mm]
\dfrac{dy}{dx} = p
\end{cases}
\qquad , \qquad (8.6.2)
$$

where

$$
c'=c(1-\dfrac{dm}{m}) , \qquad\qquad (8.6.3)
$$

$$
h_\tau(y)=\dfrac{\rho}{\rho_0}=(\dfrac{\tau_0-0.006328y}{\tau_0})^{4.4} , \qquad\qquad (8.6.4)
$$

$$
\tau_{0N}=289.08K , \quad \tau_0 = \dfrac{T_0}{1-0.3785(e_0/p_0)} , \qquad (8.6.5)
$$

and

$$
v = [(\vec{v}'-\vec{w}_x)^2]^{1/2} = (v'^2 + w_x^2 - 2v'\cdot w_x\cdot\cos\alpha)^{1/2} . \qquad (8.6.6)
$$

The initial conditions are

$$x_0=0, \quad y(0)=y_0, \quad z(0)=0, \quad t(0)=0, \quad v'_{0x}=v'_0\cos\alpha_0,$$
$$v'_{0y}=v'_0\sin\alpha_0, \quad v'_{0z}=0, \tag{8.6.7}$$

where

$$v'_0=(v_0+dv_0)-0.4\frac{dm}{m}+0.001\cdot v_0 dT. \tag{8.6.8}$$

For the standard atmosphere, we use the differential equations of flight (8.6.1) and (8.6.2), considering

Air temperature $T_0=15°Celsius$, virtual temperature $\tau_0=\tau_{0N}=289.08K$, atmospheric pressure $p_0=p_{0N}=750mm$, pressure of air vapors $e_0=e_{0N}=6.35mm$, propellant temperature $T_{c0}=15°Celsius$ (8.6.9)

The modified equations of flight (8.6.1) and (8.6.2), together with the initial conditions, take into account the following:

- Coordinates of the firearm and the target in the two- or three-dimensional space (i.e., practically all possible relative locations of artillery gun and target)
- Characteristics of the firearm and the projectile (i.e., the projectile caliber, weight, load, initial velocity, etc., and their deviations from normal standard values) and the ballistics characteristics of the projectile as well
- All meteorological factors (wind velocity, barometric pressure, humidity, and their deviations from the normal standard values as well)

NOTE: Unfortunately, for technical reasons, the CD that contains the PC programs is not attached with the book.

The CD with all the twenty PC programs can be ordered by writing to the author at these e-mail addresses: gklimi@pace.edu or iven24@aol.com.

Only three PC programs are shown in the appendixes B–D of the book.

8.7 Using PC Programs to Solve Exterior Ballistics Problems

The QBasic programs presented in this section, and used somewhat in the preceding chapters, are prepared mostly during 1992–1994 with the assistance of my friends and colleagues Col. Genc Kokoshi, Prof. Guido LoVechio, Prof. Carlo Ferrario, and Dr. Marco Merli, to whom I dedicate this chapter.[60]

I. Estimating the Ballistics Coefficient

QBasic Program COEFF.Bas: Estimates automatically the ballistics coefficients of the projectile in the standard atmosphere using fire testing results or firing tables.

QBasic program COEFF.BAS is prepared in (1992–93) to find the ballistics coefficients of cannons. Coeff.Bas can be used to find the ballistics coefficient of any firearm projectile, given range.

Example 1 and the examples we have seen in preceding sections show the use of the PC program Coeff.Bas to find the ballistics coefficient for a given launching angle (given range).

Example 1. Use the QBasic PC program COEFF.BAS to calculate the ballistics coefficient for the following data:

Initial Speed of Projectile $v_0 = 885 m/s$, Launching Angle $\alpha_0 = 6.20$,

Corresponding Range $x = 10,000$ m.
Solution Instructions
Open the QBasic PC program COEFF.BAS.

---

[60] Prof. G. Kokoshi, is artillery Officer and PC Programmer at Military Academy of General Staff, Tirana.
Guido LoVechio, PhD, Carlo Ferrario, PhD, and Dr. Marco Merli, outstanding Italian Professors of Physics and good friends of mine, Universita Di Ferrara, Italy.

Input the above data, as well as a guessed ballistics coefficient—for example, the value $c=0.20$
Execute the program.
Output: Ballistics Coefficient is $ko=0.2388549$.

The ballistics coefficient is saved as well in the file: c:\ koef.dat.

II. Determining the Launching Angle Needed to Hit a Given Target

The following two programs in QBasic are prepared for the Russian 122 mm field artillery cannon.

They can be used to solve the problems of artillery fire control for the 122 mm artillery cannon, model 1960, projectile OF-472 (made in Russia, or China, actually in use by armies of the East European countries, Asia, Africa, etc.).

The PC programs, written in QBasic, can compute in a few seconds the firing data of the artillery cannon model 1960, and in general, with some modifications, it can be used for any model of a nonreactive firearm or projectile.

The ballistics coefficient we use in the PC program is a function of the launching angle, and it is found using the ballistics coefficients obtained by PC program Coeff.Bas.

Projectile Data: Launching Speed $v_0=885m/s$;
Diameter $d=121.92mm$; Mass $m=27.30kg$

1. **QBasic PC Program ANGLE122.BAS**:

Estimates the launching angle as well as other elements of the projectile trajectory of the 122 mm Russian field cannon for a given range in the standard atmosphere.

The program can be used as well to find the elements of the trajectory that corresponds to any value of the x-coordinate.

Note: The QBasic program ANGLEC.BAS (see appendix B), used in chapter 3 and 4, is the modified program ANGLE122.BAS that can be used for any projectile with known ballistics coefficient or with a ballistics coefficient that can be determined using methods presented in chapter 2 or by using the QBasic program Coeff.Bas.

The QBasic program ANTAIRC.BAS used for uphill shooting represent the modified program ANGLEC.BAS in standard atmosphere.

Example 2. Use the QBasic PC program ANGLE122.BAS to estimate the elements of the trajectory of 122 mm Russian field cannon for the following data: launching speed of the projectile is $v_0=885m/s$; range is $x=7,200m$.

Find as well the elements of the trajectory at a point on the trajectory that corresponds to $x=2,600m$.

Solution Instructions

Input: Range $x=7,200m$; Projectile Speed $v_0=885m/s$; x-coordinate of a Point $x=2,600m$.

Results: Launching Angle: $\alpha_0=3.76477$; Time of Flight $t=10.72$; Terminal Speed $v_T=515m/s$; Terminal Angle $\alpha_T=-5.40°$; Trajectory Vertex (3930, 141).

The y-coordinate of the point with abscissa $x=2,600m$ is $y=123m$; Corresponding time of flight is $t=3.22s$; Corresponding speed is $v=737m/s$; Corresponding angle is $\alpha=71.5179°$.

Note: The tableau of the 122 mm field cannon gives the following results, corresponding to the range $x=7,200m$:

Launching Angle: $\alpha_0=3.7\overline{6}$; Time of Flight $t=11$; Terminal Speed $v_T=512m/s$; Terminal Angle $\alpha_T=-5.2°$; Trajectory Altitude 140 m.

2. **QBasic PC program ANGEMET.BAS:**

Estimates the launching angle and other elements of the projectile trajectory of the 122 mm Russian field cannon for a given range, in nonstandard atmospheric conditions, considering as well the small deviations of the projectile parameters: initial speed, projectile mass, and temperature of propellant (black powder charge). The launching point and the target are at sea level.

The program can be used to find the elements of the trajectory that corresponds to a given x-coordinate. It also displays the ballistics coefficient that corresponds to the launching angle.

Example 3. Use QBasic program ANGEMET.BAS to find the launching angle needed to hit a target located in the horizontal distance $x=8200m$ for the following changes in atmospheric conditions (with respect to the standard ones), firearm, and projectile standard characteristics:

Change in Initial Speed $dv_0=-8m/s$; Change in the temperature of air $dT_0=10°\,C$:

Change in atmospheric pressure $dp_0=12mm\;Hg$; Change in propellant temperature is $dT_c=10°$; Change in relative projectile mass: $dm=-0.190kg$; Range wind (headwind) $w_x=-15m/s$.

Find as well the elements of the trajectory at the point with abscissa $x=6100m$.

Solution
We will solve the differential equations of flight (8.6.1), employing the PC program ANGEMET.BAS for the following initial conditions:

Initial Data
Range is $x=8200m$
The initial speed is $v_0+dv_0=885-8=877m/s$.
The temperature of air is $T_0+dT_0=15+10=25°$.
The atmospheric pressure is $p_0+dp_0=750+12=762mm$.
The pressure of air vapor is $e_0=6.35mm$ (standard value)

Projectile mass $m=27.30kg$

Change in projectile mass $dm=-0.19$

Propellant temperature $T_c+dT_c=15°+10°=25°$

Range wind (headwind) $w_x=-15m/s$

Solution Instructions

Input: Input the above data in the ANGEMET.BAS program, as well as $x=6100$.

Results: We find that the launching angle is $\alpha_0=4.6617°$.

We find as well the other elements of the trajectory at the target:

Time of Flight $t=12.97s$, Terminal Speed $v_T=458m/s$, Impact Angle $\alpha_T=-7.25°$,

The coordinates of the vertex are (4560, 207).

The y-coordinate of the point with abscissa $x=6,100m$ is $y=175m$;

Corresponding time of flight is $t=12.54s$; Corresponding speed is $v=483m/s$; Corresponding angle is $\alpha=1.943°$.

The ballistics coefficient that corresponds to $\alpha_0=4.6617°$ is $c=0.242$.

Example 4. Use QBasic program ANGMET.BAS to find the launching angle needed to hit the target located in the horizontal distance $x=6200m$ in standard atmosphere conditions when the characteristics of the projectile are the handbook ones.

Solution Instructions

Input:

Range $x=6,200m$; the initial speed is v_0 $885m/s$; the temperature of air is $T_0=15°$; the atmospheric pressure is $p_0=750mm$; the pressure of air vapor is $e_0=6.35mm$ (standard value); projectile mass $m=27.30kg$;

change in projectile mass $dm=0$; propellant temperature $T_c=15°$; range wind (headwind) $w_x=0m/s$.

Results:

Launching angle is $\alpha_0=3.0719°$; time of flight $t=8.88s$; terminal speed $v_T=553/s$; impact angle $\alpha_T=-4.202°$.

The coordinates of the vertex are (3340, 97).

Note: The data given in firing tables of the 122 mm cannon are $\alpha_0=3.0\overline{6}°$, $t=8.6s$, $v_T=564m/s$, $\alpha_T=-4°$, projectile maximum altitude = 96.

3. QBasic PC Program ANTA122.BAS

Estimates the launching angle as well as other elements of the projectile trajectory of the 122 mm Russian field cannon for a given location of target and cannon (cannon and target are located in different altitudes above sea level), in nonstandard atmosphere, for small changes in projectile characteristics.

Example 5. Use QBasic program ANTA122.BAS to find the launching angle needed to hit the target located on a hill at a point with coordinates (7800, 300) when the location of the cannon is at the point (0, 50). The coordinate system is at sea level at the point (0, 0).

The characteristics of the atmosphere and the projectile are as follows:

The initial speed is $v_0=880m/s$; the temperature of air is $T_0=8°$; the atmospheric pressure is $p_0=760mm$; the pressure of air vapor is $e=7mm$; projectile mass $m=27.30kg$; change in projectile mass $dm=-0.25$; propellant temperature $T_c=8°$; range wind (headwind) $w_x=-12m/s$.

Solution Instructions

Input:

x-coordinate of the target $x_T=7800m$; y-coordinate of target $y_T=300m$; the initial speed is $v_0=880m/s$; y-coordinate of the cannon $y_0=50m$; the temperature of air is $T_0=8°$; the atmospheric pressure is $p_0=760mm$; the pressure of air vapor is $e=7mm$ (standard value); projectile mass is $m=27.30kg$; change in projectile mass is $dm=-0.25$; propellant temperature $T_c=8°$; range wind (headwind) $w_x=-12m/s$.

Results:

Launching angle is $\alpha_0=6.202°$; time of flight $t=12.15s$; terminal speed $v_T=477m/s$; impact angle $\alpha_T=-4..729°$.

Example 6. Use QBasic program ANTA122.BAS to find the launching angle needed to hit the target located at the point (4560, 207) (which is the trajectory vertex of the projectile of example 3), when the location of the cannon is at the origin of the coordinates.

Initial Data: The initial speed is $v_0=877m/s$; the temperature of air is $T=25°$; the atmospheric pressure is $p=762mm$; the pressure of air vapor is $e_0=6.35mm$ (standard value); projectile mass $m=27.30kg$; change in projectile mass $dm=-0.19$; propellant temperature $T_c=25°$; range wind (headwind) $w_x=-15m/s$.

Solution Instructions

Input:

x-coordinate of target $x_T=7060$; y-coordinate of target $y_T=651$; projectile speed $v_0=877m/s$; y-coordinate of cannon $y_0=0$; the temperature of air $T=25°$; atmospheric pressure $p=762mm$; pressure of air vapor $e_0=6.35mm$; projectile mass $m=27.30kg$; change in projectile mass $dm=-0.19$; propellant temperature $T_c=25°$; range wind (headwind) $w_x=-15m/s$.

Results:

Launching angle $\alpha_0 = 4.6593°$; time of flight $t = 6.13s$; terminal speed $v_T = 625 m/s$; impact angle $\alpha_T = -0.007° \approx 0°$.

Note: In our example, the target is located at the vertex of the trajectory of example 3.

The result obtained in this example is compatible with the result obtained for the trajectory vertex of example 3, i.e., the trajectory vertex (4560, 207) that corresponds to the launching angle $\alpha_0 = 4.6617°$.

There is a negligible discrepancy ($d\alpha_0 = 4.6617° - 4.6593° = 0.0024° = 0.14'$) between the values of the launching angle, found in example 3 ($\alpha_0 = 4.6617°$) and the launching angle ($\alpha_0 = 4.6593°$) found in this example (example 6).

III. Determining the Range for a Given Launching Angle

4. *QBasic PC Program RANGE122.BAS*: For a given launching angle, it estimates the range as well as the other elements of the projectile trajectory in standard atmosphere. It can be used only for the 122 mm Russian field cannon.

Example 7. Find the launching angle needed to hit a target located in a horizontal distance of $x_T = 5300m$.

Solution

Input:

Initial x-coordinate $= 0$; initial y-coordinate $= 0$; initial time $= 0$; projectile speed $= 885$; launching angle $= 5.0$; step $= 10$

Output:

Range $= 8,740$; terminal speed $= 599.70$; terminal angle $= -7.74$; coordinates of vertex (4850, 236.60).

5. The QBasic program RANGEC.BAS (see appendix C) is the modified program RANGE122.BAS that can be used for any projectile with known ballistics coefficient.

Example 8. Find the range for a projectile launched with an initial speed of 853 m/s with an angle of 0.25° if the ballistics coefficient of the projectile is 3.385.

Solution

Input:
Initial x-coordinate = 0; initial y-coordinate =0; initial time =0; projectile speed = 853; launching angle = 0.25; step = 10; abscissa of a point = 100

Output:
Range = 470; terminal speed = 515; terminal angle = -0.366; coordinates of vertex (250, 0.59).

Abscissa of a point = 100; corresponding ordinate = 0.364; speed = 773.50; time = 0.123; angle = 0.165°.

6. *QBasic PC program RANGMET.BAS*: Estimates the projectile range for a given launching angle for nonstandard atmospheric conditions considering as well the small deviations in projectile parameters: initial speed, projectile mass, and temperature of propellant. It can be used for any projectile with known ballistics coefficient.
7. The QBasic program RANBALL.BAS represent a modification of the PC program RANGMET.BAS that can used to find the elements of a trajectory when it is known the launching angle of the projectile caliber 0.30 M2 ball.

Both programs can be used as well to estimate the shooting "corrections" (see chapter 6).

Note: PC Programs in English Units

In appendix D is given a PC programs that use the Ingalls-Siacci function in English units (see the second chapter).

8.8 Using PC Programs to Determine the Corrections for the 0.30 M2 Ball Projectile

The QBasic PC programs can be used to estimate the corrections that are the results of small deviations of the atmospheric conditions and projectile characteristics respectively from the standard atmosphere and standard characteristics of the firearm and projectile (handbook characteristics).

Any handbook of a firearm contains tableaus of shooting "corrections" caused by small changes of meteorological conditions and ballistics characteristics of the projectile (see chapter 6).

Hereafter we illustrate the use of QBasic programs RANGMET.BAS and RANBALL.BAS to estimate the "shooting corrections," we have studied in chapter (6) using the Siacci method.

In example 2 of section 7.4, we constructed the ballistics tableau of 0.30 M2 ball bullet, caliber 0.308" and mass 0.00972 kg, using the ballistics coefficient function

$$c(v)=(0.913405-0.0009944v+0.00000062v^2)\cdot(6.2913992) \tag{1}$$

and the QBasic PC program ANGLEM30.BAS:

Table 1

Elements of the Trajectory of 0.30 M2 Ball Projectile Initial Speed v_0=853.43m/s					
Range x	Coordinates of Vertex	Angle a_0 in minutes	Time t	Speed v	Angle a in minutes
91.44	(47, 0.015)	2.24	0.112	783	-2.37
182.88	(100, 0.067	4.76	0.234	716	-5.35
274.32	(150, 0.168)	7.61	0.370	653	-9.11
365.76	(198, 0.33)	10.86	0.515	591	- 13.87
457.20	(250, 0.58)	14.59	0.678	534	-19.94
548.64	(306, 0.95)	18.88	0.858	482	-27.66

Now we will use the QBasic PC program RANBALL.BAS to estimate the corrections in range and in the vertical direction for 0.30 M2 ball projectiles for ranges 274 m, 366 m, and 549 m (see table 1), which correspond to the following small deviations of the atmospheric air and projectile characteristics from standard values:

Change in Initial Speed dv_0=10m/s; change in the temperature of air dT_0=10° C; change in atmospheric pressure dp_0=10mm Hg; change in propellant temperature is dT_c=10°; Rangewind (tailwind) is w_x=10m/s.

Considering the above changes, we find the following initial values we use in RANBALL.BAS:

Initial Values
(a) The initial speed is v_0+dv_0=853.43+10=863.43m/s.
(b) The temperature of air is T_0+dT_0=15+10=25°.
(c) The atmospheric pressure is p_0+dp_0=750+10=760mm.
(d) The pressure of air vapor is e_0=6.35mm (standard value).
(e) Projectile mass m=0.00972kg
(f) Change in projectile mass dm=0
(g) Propellant temperature T_c+dT_c=15°+10°=25°
(h) Range wind (tailwind) w_x=10m/s

Note: We will consider latter a crosswind of 10 m/s.

Estimation of Corrections

To estimate corrections caused by the above small changes, we use RANBALL.BAS PC program considering one change at a time (simultaneously all the other small changes will be considered zero).

Example 1. (0.30 Ball M2 Projectile) Use RANBALL.BAS to find the corrections in range and vertical direction of fire corresponding only to a change of $dv_0=10m/s$ in initial speed for the following values of the launching angle:

I. $\alpha_0=7.61'=0.12683°$ (corresponds to a standard range of $x=300 yards=274.32m$)

II.   (corresponds to a standard range of $x=400 yards=365.76m$)

III. (corresponds to a standard range of $x=600 yards=549m$)

Solution Instructions

Execute the PC program.

Input

(a) The modified initial speed of projectile is $v_0=863.43m/s$ (considering a change of $dv_0=10m/s$); all other atmospheric and projectile characteristics remain unchanged, that is,

(b) The temperature of air is $T_0=15°$ (standard);

(c) The atmospheric pressure is $p_0=750mm$ (standard);

(d) The pressure of air vapor is $e_0=6.35mm$ (standard);

(e) Projectile Mass $m=0.00972kg$; (f) Change in projectile mass $dm=0$;

(g) Propellant temperature is $T_c=15°$ (standard); and

(h) Range wind (tailwind) is $w_x=0m/s$.

Input as well the value of range in standard atmosphere:

x-coordinate of a point on trajectory $x=274m$
Executing RANBALL.BAS, we find the following values:

Results

I.   Actual range $x=280m$ ($y=0$); trajectory vertex (152, 0.169); time of flight $t=0.372s$; y-coordinate that corresponds to the range in standard conditions $x=274m$ is $y=0.015m$.

Using the above results, we find that the change in range is $dx=280-274=6m$, while the change in vertical direction is $dy=0.015-0=0.015m$.

In the same way, we find the following results for II and III:

II.   Actual range $x=373m$ ($y=0$); trajectory vertex (203, 0.332); time of flight $t=0.52s$; y-coordinate that corresponds to the range in standard conditions $x=366m$ is $y=0.0275m$.

Using the above results, we find that the change in range is $dx=373-366=7m$, while the change in vertical direction is $dy=0.0275m$.

III   Actual range $x=561m$ ($y=0$); trajectory vertex (312, 0.936); time of flight $t=0.872s$; y-coordinate that corresponds to the range in standard conditions $x=549m$ is $y=0.0945m$.

Using the above results, we find that the change in range is $dx=561-549=12m$, while the change in vertical direction is $dy=0.0945m$.

Example 2. (0.30 M2 ball projectile) Use the RANBALL.BAS to find the corrections in range and vertical directions corresponding only to a change of $d\alpha_0=0.5'$ in launching angle for the following values of launching angle in standard conditions:

I. $\alpha_0=7.61'=0.12683°$ (corresponds to a standard range of $x=300yards=274.32m$)

II.   (corresponds to a standard range of $x=400yards=365.76m$)

III. (corresponds to a standard range of $x=600yards=549m$)

Solution
The only change is the change of launching angle:

I. $\alpha_0 + d\alpha_0 = 7.61' + 0.5' = 0.135\overline{16}°$, $v_0 = 853.43m/s$ (standard value of launching speed)

II. $\alpha_0 + d\alpha_0 = 10.86' + 0.5' = 0.189\overline{3}°$, $v_0 = 853.43m/s$ (standard value of launching speed)

III. $\alpha_0 + d\alpha_0 = 19' + 0.5' = 0.325°$, $v_0 = 853.43m/s$ (standard value of launching speed)

Employing RANBALL.BAS, we find the following:

I.   Actual range $x = 289m$ $(y=0)$, trajectory vertex $(157, 0.187)$; time of flight $t = 0.39s$; y-coordinate that corresponds to the range $x = 274m$ (in standard conditions) is $y = 0.041m$.

Using the above results, we find that the change in range is $dx = 15m$, while the change in vertical direction is $dy = 0.041m$.

II. Actual range $x = 379m$ $(y=0)$, trajectory vertex $(206, 0.354)$; time of flight $t = 0.538s$; y-coordinate that corresponds to the range $x = 366m$ (in standard conditions) is $y = 0.053m$.

Using the above results, we find that the change in range is $dx = 13m$, while the change in vertical direction is $dy = 0.053m$.

III. Actual range $x = 561m$ $(y=0)$, trajectory vertex $(312, 0.96)$; time of flight $t = 0.89s$; y-coordinate that corresponds to the range $x = 549m$ (in standard conditions) is $y = 0.10m$.

Using the above results, we find that the change in range is $dx = 561 - 549 = 12m$, while the change in vertical direction is $dy = 0.10m$.

Example 3. (0.30 M2 ball projectile) Use the RANBALL.BAS to find the corrections in range and vertical directions corresponding only to a change of range wind of $dw_x = 10m/s$ in launching angle for the following values of launching angle:

I. $\alpha_0 = 7.61' = 0.12683°$ (corresponds to a standard range of $x = 300 \, yards = 274.32 m$)

II. $\alpha_0 = 10.86' = 0.181°$ (corresponds to a standard range of $x = 400 \, yards = 365.76 m$)

III. $\alpha_0 = 19' = 0.316°$ (corresponds to a standard range of $x = 600 \, yards = 549 m$)

Solution

Employing RANBALL.BAS, we find the following:

I.   Actual range $x = 275 m$ ($y = 0$), trajectory vertex (149, 0.166); time of flight $t = 0.37 s$; y-coordinate that corresponds to the range $x = 274 m$ (in standard conditions) is $y = 0.003 m$.

Using the above results, we find that the change in range is $dx = 1m$, while the change in vertical direction is $dy = 0.003 m$.

II. Actual range $x = 368 m$ ($y = 0$), trajectory vertex (199, 0.326); time of flight $t = 0.517 s$; y-coordinate that corresponds to the range $x = 366 m$ (in standard conditions) is $y = 0.005 m$.

Using the above results, we find that the change in range is $dx = 2m$, while the change in vertical direction is $dy = 0.005 m$.

III. Actual range $x = 555 m$ ($y = 0$), trajectory vertex (307, 0.92); time of flight $t = 0.865 s$; y-coordinate that corresponds to the range $x = 549 m$ (in standard conditions) is $y = 0.05 m$.

Using the above results, we find that the change in range is $dx = 6m$, while the change in vertical direction is $dy = 0.05 m$.

Example 4. (0.30 M2 ball projectile) Use the RANBALL.BAS to find the corrections in range and vertical directions corresponding only to

(a) Change in atmosphere temperature $dT_0 = 10°$

(b) Change in temperature of propellant $dT_c=10°$

(c) Change in atmospheric pressure $dp=10mm$ Hg

for the following value of launching angle in standard conditions:

I. $\alpha_0=7.61'=0.12683°$ (corresponds to a standard range of $x=300yards=274.32m$)

II $\alpha_0=10.86'=0.181°$ (corresponds to a standard range of $x=400yards=365.76m$)

III. $\alpha_0=19'=0.316°$ (corresponds to a standard range of $x=600yards=549m$)

Solution Instructions

Employing RANBALL.BAS, we find

(a) Change in Atmosphere Temperature Only, $dT_0=10°$

I. $\alpha_0=7.61'=0.12683°$:

Actual range $x=275m$ ($y=0$); trajectory vertex (149, 0.166); time of flight $t=0.367s$; y-coordinate that corresponds to the range $x=274m$ (in standard conditions) is $y=0.0036m$. Using the above results, we find that the change in range is $dx=-1m$, while the change in vertical direction is $dy=0.0036m$.

II. $\alpha_0=10.86'=0.181°$:

Actual range $x=368m$ ($y=0$); trajectory vertex (200, 0.33); time of flight $t=0.516s$; y-coordinate that corresponds to the range $x=366m$ (in standard conditions) is $y=0.007m$.

Using the above results, we find that the change in range is $dx=2m$, while the change in vertical direction is $dy=0.007m$.

III. $\alpha_0 = 19' = 0.31\overline{6}°$:

Actual range $x=556m$ ($y=0$); trajectory vertex $(308, 0.92)$; time of flight $t=0.866s$; y-coordinate that corresponds to the range $x=549m$ (in standard conditions) is $y=0.05m$.

Using the above results, we find that the change in range is $dx=7m$, while the change in vertical direction is $dy=0.05m$.

(b) Change in Propellant Temperature Only, $dT_c = 10°$

(I) $\alpha_0 = 7.61' = 0.1268\overline{3}°$:

Actual range $x=279m$ ($y=0$); trajectory vertex $(151, 0.167)$; time of flight $t=0.371s$; y-coordinate that corresponds to the range $x=274m$ (in standard conditions) is $y=0.013m$.

Using the above results, we find that the change in range is $dx=5m$, while the change in vertical direction is $dy=0.013m$.

II. $\alpha_0 = 10.86' = 0.181°$:

Actual range $x=372m$ ($y=0$); trajectory vertex $(202, 0.33)$; time of flight $t=0.52s$;
y-coordinate that corresponds to the range $x=366m$ (in standard conditions) is $y=0.0023m$.

Using the above results, we find that the change in range is $dx=6m$, while the change in vertical direction is $dy=0.0023m$.

III. $\alpha_0 = 19' = 0.31\overline{6}°$:

Actual range x=560m (y=0); trajectory vertex (311, 0.934); time of flight t=0.866s; y-coordinate that corresponds to the range x=549m (in standard conditions) is y=0.084m.

Using the above results, we find that the change in range is dx=11m, while the change in vertical direction is dy=0.084m.

(c) Change in the Atmospheric Pressure Only, dp=10mm Hg

I. α_0=7.61'=0.12683̄° :

Actual range x=275m (y=0); trajectory vertex (149, 0.166); time of flight t=0.368s; y-coordinate that corresponds to the range x=274m (in standard conditions) is y=0.

Using the above results, we find that the change in range is dx=1m, while the change in vertical direction is dy=0m.

II. α_0=10.86'=0.181° :

Actual range x=367m (y=0); trajectory vertex (199, 0.33); time of flight t=0.52s; y-coordinate that corresponds to the range x=366m (in standard conditions) is y=0m.

Using the above results, we find that the change in range is dx=1m, while the change in vertical direction is dy=0m.

III. α_0=19'=0.316̄° :

Actual range x=554m (y=0); trajectory vertex (307, 0.92); time of flight t=0.865s; y-coordinate that corresponds to the range x=549m (in standard conditions) is y=0.037m.

Using the above results, we find that the change in range is dx=5m, while the change in vertical direction is dy=0.037m.

Example 5. (0.30 M2 ball projectile) Use the results obtained in the above exercises, (1)–(4) to find the correction in launching angle $\alpha_0 = 10.86' = 0.181°$ in order to hit a target located at the range $x = 400\,yards = 365.76m$ for the following changes:

Change in initial speed $dv_0 = -8m/s$; change in range wind $dw_x = -12m/s$; change in atmosphere temperature $dT_0 = -15°$; change in temperature of propellant $dT_c = -15°$.

Solution

Range Corrections
(a) In (II) of example 1, we found that to a change in the launching speed $dv_0 = 10m/s$, the change in range is $dx = 7m$, while the change in vertical direction is $dy = 0.0275m$.

Thus, to the actual change in launching speed of projectile, $dv_0 = -8m/s$, corresponds to a change in range of

$$dx_v = (-8)(7)/(10) = -5.6m.$$

In a similar way, we find the following:

(b) The change $dw_x = -12m/s$ corresponds to a change in range of

$$dx_w = (-12)(2)/(10) = -2.4m$$

(because in II, example 3, we found that to $dw_x = 10m/s$ corresponds $dx_w = 2m$)

(c) The change in temperature $dT_0 = -15°$ corresponds to a change in range of

$$dx_T = (-15)(2)/(10) = -3m$$

(because in (a), II, of example 4, we found that $dT_0 = 10°$ corresponds to $dx = 2m$)

(d) The change in temperature of the propellant $dT_c=-15°$ corresponds to a change in range of

$$dx_T=(-15)(6)/(10)=-9m$$

(because in (a), II of example 4, we found that to $dT_c=10°$ corresponds to $dx=6m$)

Total Range Correction
    Total range correction is

$$dx=(-5.6)+(-2.4)+(-3)+(-9)=-20m$$.

Actually the projectile launched with angle $\alpha_0=10.86'=0.181°$ will miss the target located at $x=366m$ and will hit the ground at the horizontal distance

$$x=366-20=346m$$.

Correction in Launching Angle
    To find the correction in launching angle that results from the change $dx=-20m$ of range, we need to increase the launching angle $\alpha_0=10.86'=0.181°$.
    Using the result obtained in II of example 2:

To the change in launching angle $d\alpha_0=0.5'$ corresponds a change in range of $dx=13m$.
    Thus, we find that the correction in launching angle is

$$d\alpha_0=(20)(0.5')/(13)=0.76923'$$.

Launching Angle
    The launching angle needed to hit the target located at the range $x=400yards=365.76m$ is

$$\alpha_0+d\alpha_0==10.86'+0.76923'=11.62923'$$.

Note: Use RANBALL.BAS to verify the results obtained in this example.

Input

    (a)   The initial speed is $v_0 + dv_0 = 853.43 - 8 = 845.43 m/s$.

    (b)   The launching angle is $\alpha_0 = 11.62923' = 0.1938205°$.

    (c)   The temperature of air is $T_0 + dT_0 = 15° - 15° = 0°$.

    (d)   The atmospheric pressure is $p_0 + dp_0 = 750 + 0 = 750 mm$.

    (e)   The pressure of air vapor is $e_0 = 6.35 mm$ (standard value).

    (f)   Projectile mass $m = 0.00972 kg$.

    (g)   Change in projectile mass $dm = 0$.

    (h)   Propellant temperature $T_c + dT_c = 15° - 15° = 0°$.

    (i)   Range wind (tailwind) $w_x = -12 m/s$.

Results

    Range $x = 366 m$; terminal speed $v_T = 558 m/s$; terminal angle $\alpha_T = -0.25426$; trajectory vertex (199, 0.355)

Comment: The results obtained using the QBasic program RANBALL.BAS are identical to the results obtained in this example using the corrections corresponding to small changes of atmosphere and projectile characteristics.

Example 6. (0.30 M2 ball projectile) Find the correction in direction of fire (firing plane) caused by a crosswind that blows perpendicular to the launching plane (xoy), in the direction of the z-axis with a speed $w_z = 10 m/s$, as well as the z-coordinate of the point of impact on the ground if (a) the launching speed is $v_0 = 853.43 m/s$, (b) the launching angle is $\alpha_0 = 10.86' = 0.181°$ that correspond to the range $x = 274.32 m$, and (c) time of flight is $t = 0.37 s$ (see table 1).

    All other atmosphere characteristics and ballistics characteristics of projectile are standard.

Solution

    Employing (6.6.13), we find

$$\tan\varphi = -\frac{w_z}{v_0\cos\alpha_0} = -\frac{10}{853.43\cdot\cos(0.181°)} = -0.1171748$$

Hence, the clockwise deviation of the shooting direction of the point of impact is

$$\varphi=\tan^{-1}(-0.1171748)=-0.67133°.$$

Using (6.6.15), we find the corresponding z-coordinate of the deviation of projectile from tha launching plane

$$z(t)=w_z t-\frac{w_z}{v_0\cos\alpha_0}x=(10)\cdot(0.37)-\frac{(10)}{(853.43)\cdot\cos(0.181°)}\cdot(274.32)=0.486m.$$

Thus, to hit the target located on the x-axis at the range $x=274.32m$, we need to rotate the rifle counterclockwise with an angle $\varphi=0.67133°$.

8.9 Determining the Corrections for the 122 mm Field Cannon Using PC Programs

The QBasic program **RANGMET.BAS** can be used in the same way as RANBALL.BAS to find the corrections of the projectile of the 122 mm Russian field cannon.

We will illustrate with examples the use of the software.

Example 1. (Projectile of 122 mm cannon) Use RANGMET.BAS to find the corrections in range caused by the following small changes:

(a) Change in launching speed is $dv_0=9m/s$
(b) Change in launching angle is $d\alpha_0=0.06°$
(c) Change in temperature of air is $dT_0=10°$
(d) Change in the atmospheric pressure is $dH_0=10mm$
(e) The pressure of air vapor is $e_0=6.35mm$ (standard value)
(f) Projectile mass $m=27.3kg$
(g) Change in projectile mass $dm=0.20$

(h) Propellant temperature $dT_c=10°$

(i) Range wind (tailwind) $w_x=10m/s$

Projectile   Data: Launching   Speed $v_0=885m/s$;   diameter $d=121.92mm$; mass $m=27.30kg$; range $x=11200m$; launching angle $\alpha_0=7.5\overline{6}°$.

Solution
Considering the above changes, we find the following initial values to be used in RANBALL.BAS:

Initial Values
To estimate corrections caused by the above small changes, we use RANGMET.BAS PC program considering one change at a time (simultaneously all the other changes will be considered zero).

(a) The initial speed is $v_0+dv_0=885+9=894m/s$.

(b) Launching angle is $\alpha_0+d\alpha_0=7.5\overline{6}+0.06°=7.62\overline{6}$.

(c) The temperature of air is $T_0+dT_0=15+10=25°$.

(d) The atmospheric pressure is $H_0+dH_0=750+10=760mm$.

(e) The pressure of air vapor is $e_0=6.35mm$ (standard value).

(f) Projectile mass $m=27.3kg$.

(g) Change in projectile mass $dm=0.200$.

(h) Propellant temperature $T_c+dT_c=15°+10°=25°$.

(i) Range wind (tailwind) $w_x=10m/s$.

Using RANGMET.BAS, we find the following range corrections:

(a) Range correction corresponding to the change in launching speed

Input Data
The initial speed is $v_0+dv_0=894m/s$; launching angle is $\alpha_0=7.5\overline{6}$; the temperature of air is $T_0=15°$ (standard value); the atmospheric pressure is $H_0=750mm$ (standard value); the pressure of air vapor is

$e_0=6.35mm$ (standard value); projectile mass $m=27.3kg$; the change in projectile mass $dm=0.200$; the propellant temperature $T_c=15°$; range wind (tailwind) $w_x=10m/s$.

Input as well the coordinate of a point on the trajectory $x=11200m$ in order to find the vertical deviation of the projectile at the range $x=11200m$.

Program Output
Actual range is $x=11360m$.
The y-coordinate that corresponds to the range $x=11200m$ is $y=35.78m$.

The change in range and in y-coordinate that correspond to the change $dv_0=9m/s$ in launching speed are respectively

$$dx=11356-11200=156m, \text{ and } dy=35.78m$$

In a similar way, we find the following results:
(b) Range correction corresponding only to the change of the launching angle is $d\alpha_0=0.06°$. Launching angle is $\alpha_0+d\alpha_0=7.5\overline{6}+0.06°=7.62\overline{6}$.

Actual range is $x=11252m$.
The y-coordinate that corresponds to the range $x=11200m$ in standard conditions is $y=12m$.

The change in range and in y-coordinate that correspond to the change $d\alpha_0=0.06°$ in launching angle are respectively

$$dx=11252-11200=52m, \text{ and } dy=12m.$$

(c) Range correction corresponding only to the change of the temperature of air is $dT_0=10°$. The temperature of air is $T_0+dT_0=15+10=25°$.

Actual range is $x=11329m$.

The y-coordinate that corresponds to the range $x=11200m$ in standard conditions is $y=29.15m$.

The change in range and in y-coordinate that correspond to the change $dT_0=10°$ in air temperature are respectively

$$dx=11329-11200=139m, \text{ and } dy=29.15m.$$

(d) Range correction corresponding only to the change of the atmospheric pressure is $dH_0=10mm$.

The atmospheric pressure is $H_0+dH_0=750+10=760mm$. (Because of the program construction, instead of $760mm$, the input is $H_0-dH_0=750-10=740mm$ in order to evaluate the vertical change.)

Actual range is $x=11269m$.

The y-coordinate that corresponds to the range $x=11200m$ in standard conditions is $y=15.80m$.

The change in range (x-coordinate) and in y-coordinate that correspond to the change $dH_0=10mm$ in air temperature are respectively

$$dx=11269-11200=69m \text{ , and } dy=16m.$$

(g) Range correction corresponding only to the change of the projectile mass is $dm=0.200kg$ (projectile mass $m=27.3kg$).

Because of the program construction, instead of $dm=0.200$, the input will be $dm=-0.200$ in order to evaluate the vertical change.

The actual range is $x=11211m$.

The y-coordinate that corresponds to the range $x=11200m$ in standard conditions is $y=2.5m$.

The change in range and in the y-coordinate that correspond to the change $dm=0.200kg$ in projectile mass are respectively

$dx=11211-11200=11m$, and $dy=16m$.

(h) Range correction corresponding only to the change of the propellant temperature is $dT_c=10°$.

The propellant temperature is $T_c+dT_c=15°+10°=25°$.
The actual range is $x=11354m$.
The y-coordinate that corresponds to the range $x=11200m$ in standard conditions is $y=35.20m$.

The change in range and in y-coordinate that correspond to the change $dT_c=10°$ in projectile propellant temperature are respectively

$dx=11354-11200=154m$, and $dy=35.20m$.

(i) The range correction corresponding only to the range wind (tailwind) is $w_x=10m/s$.

The tailwind speed is $w_x=10m/s$.

The actual range is $x=11317m$.
The y-coordinate that corresponds to the range $x=11200m$ in standard conditions is $y=26.42m$.
The change in range and in y-coordinate that correspond to tailwind speed ($w_x=10m/s$) are respectively

$dx=11317-11200=117m$, and $dy=26.42m$.

9

Skydiving and Parachute Falling

Introduction

The fall of personnel and objects with parachute are important activities in military and rescue operations, emergency situations, skydiving, etc.

A parachutist, in general, jumps from an airplane that flies horizontally with a relatively low speed, without deploying the parachute or deploying the parachute, after he/she jumps from airplane, at a safe distance.

We assume that the initial velocity of the parachutist is equal to the velocity of the airplane that can be considered horizontal.

In this chapter, we will apply the differential equations of projectile flight to describe the fall of parachutist jumping from a plane considering the parachutist as a particle of a given mass m flying with initial velocity $\vec{v}_0$ under the influence of gravity and drag.

The results we obtain for the parachutist can be employed as well for other bodies of a different shape—for example, unguided aviation bombs launched from low-speed flying airplanes.

9.1 Differential Equations of Skydiving and Parachute Fall

The fall of "bodies" with (or without) parachute "launched" from an airplane (figure 1) can be studied in a similar way we studied the ballistics of artillery projectiles.

The magnitude of drag force that the atmospheric air exerts on the body is

$$D = A\frac{\rho v^2}{2} C(\frac{\rho v d}{\eta}, \frac{v}{a})$$

(9.1.1)

where A is the cross-sectional area of the projectile, ρ is air density, v is the speed of projectile, η is the viscosity of air, and a is the speed of sound in air.

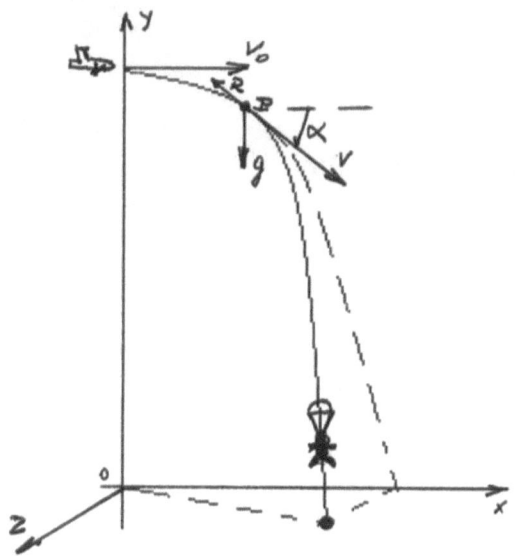

Figure 18

The density of air is a function of altitude of the projectile above sea level ($\rho = \rho(y)$); d is a characteristic dimension of the body— most of the time, the (average) diameter of a cross-sectional area. The quantity

$C(\frac{\rho v d}{\eta},\frac{v}{a})$ is the "drag function," $\frac{\rho v d}{\eta}$ is the Reynolds number

R_e, while $\frac{v}{a}$ is the Mach number M, i.e.,

$$R_e=\frac{\rho v d}{\eta}, \quad M=\frac{v}{a}.$$ (9.1.2)

The Reynolds number, as a characteristic of the kinematics viscosity of the flow, plays a significant role in the resistance of nonspinning projectiles (like spheres or other convex or concave bodies) that fly with relatively low speeds.

In many problems of ballistics of shaped bodies flying with relatively low velocities—for example, the game balls—the drag function is considered to be constant and is denoted as

$$c_D=C(\frac{\rho v d}{\eta},\frac{v}{a}).$$ (9.1.3)

The constant c_D is the "drag coefficient" or "coefficient of air resistance." Its value depends on the form and the orientation of the falling body with respect to the direction of flight.

Substituting (9.1.3) in (9.1.1), we see that the drag force is the Newton force of resistance

$$D=A\frac{\rho v^2}{2}c_D$$ (9.1.4)

or

$$D=c_D\frac{\pi d^2 \rho}{8}v^2,$$ (9.1.5)

where $A=\pi \cdot d^2/4$ is the cross-section area.

The acceleration of a point-mass projectile due to the drag force is

$$a_D = c_D \frac{\pi \cdot d^2 \rho}{8m} v^2 .$$

(9.1.6)

The values of drag coefficient for projectiles are determined experimentally using wind tunnels or other experiments with flying shaped bodies.

The differential equation of motion of a parachutist (projectile, aviation bomb, artillery projectile, etc) flying with relatively low speeds, less than 256 m/s, is

$$\frac{d\vec{v}}{dt} = \vec{g} - c_D \frac{\pi \cdot d^2 \rho}{8m} v^2 \frac{\vec{v}}{v} ,$$

(9.1.7)

where

$$A = \pi \cdot d^2 / 4$$

(9.1.8)

is the cross-section area of the parachute or the parachutist (projectile).

The drag acceleration is

$$a_D = c_D \frac{\pi \cdot d^2 \rho}{8m} v^2 , \text{ or } a_D = \frac{c_D A}{2} \rho v^2 .$$

(9.1.9)

- For the parachutist falling with a closed parachute (a skydiver), the coefficient of air resistance c_D varies during the flight and depends on parachutist orientation, jumping suit, or other equipments. The values of the drag coefficient [61] varies from 0.4 to 1.

---

[61] Halliday, Rescnik, Walker. *Fundamentals of Physics*, 6th edition, p. 104. John Wiley and Sons, 2001.

For approximate calculations, the value of c_D can be considered to be 0.8, while the averages cross-section area of the skydiver A can be taken equal to $0.5m^2$, i.e.,

$$c_D=0.8, \quad A=0.5m^2.$$ (9.1.10)

- The value of the cross-sectional area A of a skydiver changes as well and depends on the orientation of the body. It is reasonable to accept that, in a certain orientation of the body of skydiver, the value of the product $c_D A$ is a constant.
- For the parachutist falling with a fully deployed (circular) parachute the coefficient of resistance for a parachute that has a semispherical form can be considered equal to 1.33,

$$c_D=1.33.$$

We assume that the atmospheric air is in standard conditions:

- Air density on the ground is $\rho_0=1.205kg/m^3$.
- The temperature is $T_0=15^\circ$ C.
- The pressure $h_0=750mm$ Hg.
- The humidity is 50%.
- The weather is without wind.

Multiplying and dividing (9.1.7) by ρ_0, we can write

$$\frac{d\vec{v}}{dt}=\vec{g}-\pi\frac{c\cdot d^2}{8m}\frac{\rho}{\rho_0}\rho_0 v^2\frac{\vec{v}}{v}.$$ (9.1.11)

Substituting

$$h(y)=\frac{\rho}{\rho_0}, \text{ where } h(y)=(\frac{289.08-0.006328y}{289.08})^{4.4},$$ (9.1.12)

the vector differential equation (9.1.11) can be written as

$$\frac{d\vec{v}}{dt}=\vec{g}-c_D\frac{\pi \cdot d^2 \rho_0}{8m}h(y)v^2\frac{\vec{v}}{v}.$$

(9.1.13)

We denote

$$b=c_D\frac{\rho_0\pi \cdot d^2}{8}=c_D\frac{\rho_0 A}{2}.$$

(9.1.14)

Substituting (9.1.14) into (9.1.13), we can write equation (9.1.13) in the form

$$\frac{d\vec{v}}{dt}=\vec{g}-\frac{b\cdot h(y)}{m}v^2\frac{\vec{v}}{v},$$

(9.1.15)

Projecting (9.1.15) on the coordinate axis, we get the system of differential equations that describes the fall of bodies (personnel or cargo) with or without parachute from an airplane that flies horizontally with initial velocity $\vec{v}_0$,

$$\begin{cases} \dfrac{dv_x}{dt}=-\dfrac{b}{m}h(y)\cdot v^2\dfrac{v_x}{v} \\ \dfrac{dv_y}{dt}=-g-\dfrac{b}{m}h(y)v^2\dfrac{v_y}{v} \\ \dfrac{dx}{dt}=v_x \\ \dfrac{dy}{dt}=v_y \end{cases}$$

(9.1.16)

Introducing the variable $p=\tan\alpha$ defined by

$$v_y=pv_x,$$

(9.1.17)

system (9.1.16) can be written as

$$\begin{cases} \dfrac{dv_x}{dt} = -\dfrac{b}{m}h(y)v^2 \cdot \dfrac{v_x}{v} \\[2mm] \dfrac{dp}{dt} = -\dfrac{g}{v_x} \\[2mm] \dfrac{dx}{dt} = v_x \\[2mm] \dfrac{dy}{dt} = v_y \end{cases} \tag{9.1.18}$$

Expressing (9.1.18) through the independent variable x, we have

$$\begin{cases} \dfrac{dv_x}{dx} = -\dfrac{b}{m}h(y)\cdot v^2 \dfrac{1}{v} \\[2mm] \dfrac{dp}{dx} = -\dfrac{g}{v_x^2} \\[2mm] \dfrac{dt}{dx} = \dfrac{1}{v_x} \\[2mm] \dfrac{dy}{dx} = p \end{cases} \tag{9.1.19}$$

Another equivalent system of differential equations of the parachute fall similar to (4.5.7) is

$$\begin{cases} \dfrac{dv_x}{dp} = \dfrac{b}{mg}\cdot h(y)\cdot v_x^3(1+p^2)^{1/2} \\[2mm] \dfrac{dx}{dp} = -\dfrac{v_x^2}{g} \\[2mm] \dfrac{dy}{dp} = -\dfrac{p\cdot v_x^2}{g} \\[2mm] \dfrac{dt}{dp} = -\dfrac{v_x}{g} \end{cases} \tag{9.1.20}$$

where

$$b = c_D \frac{\rho_0 \pi \cdot d^2}{8} = c_D \frac{\rho_0 A}{2} \tag{9.1.21}$$

Example 1. Find the value of the parameter b, the drag acceleration, and the drag force the atmospheric air exerts on a parachutist falling with closed parachute when he is 1,000 m above the ground and has a speed of 30 m/s.

The mass of the parachutist (parachute and equipment included) is 100 kg.

Solution

For the parachutist (with closed parachute): $c=0.8$, $A=0.5m^2$.
Substituting in (9.1.21), we find

$$b_D=c_D\frac{P_0 A}{2}=(0.8)\frac{(1.205)(0.5)}{2}=0.241 \quad \text{kg/m.}$$

The numerical value a_D of the drag acceleration $\vec{a}_D$ can be estimated using (9.1.9), i.e.,

$$a_D=\frac{b_D\cdot h(y)}{m}\cdot v^2 .$$

Substituting $y=1,000$ in (9.1.12), we find

$$h(y)=(\frac{289.08-0.006328\cdot y}{289.08})^{4.4}=(\frac{289.08-0.006328\cdot(1,000)}{289.08})^{4.4}=0.907 .$$

The numerical value of drag acceleration is

$$a_D=\frac{b_D\cdot h(y)}{m}\cdot v^2=\frac{(0.241)\cdot(0.907)}{(100)}\cdot(30)^2=1.968m/s^2 .$$

For the drag force, we have

$$D=m\cdot a_D=(100)\cdot(1.967)=196.80N .$$

Example 2. Find the value of b and the numerical value of drag acceleration, as well as the drag force, that the atmospheric air exerts on a parachutist falling with open parachute when he is 1,000 m above the ground and has a speed of 10 m/s. The parachute has a semispherical form with diameter 7.5 m.

Solution

For the parachutist falling with an open parachute, the coefficient of resistance can be considered constant and equal to 1.33.

Employing (9.1.21), we find

$$b=c\frac{\rho_0 \pi \cdot d^2}{8}=(1.33)\frac{(1.205)\cdot\pi(7.5)^2}{8}=35.40$$.

The numerical value of drag acceleration is

$$a_D=\frac{b\cdot h(y)}{m}\cdot v^2=\frac{(35.40)\cdot(0.9072)}{(100)}\cdot(8)^2=20.55 m/s$$.

The numerical value of drag force is

$$D=m\cdot a_D=(100)\cdot(20.55)=2,055 N$$.

Example 3. A parachutist jumps from an airplane and opens the parachute when his falling speed is 70 m/s. After 3 sec. the speed of the parachutist drops to 10 m/s.

Find the dynamic force exerted on the parachutist as the result of deceleration.

Solution

We will use the impulse-momentum theorem to find an average force F_{ave}

$$F_{ave}\cdot\Delta t=mv_2-mv_1$$.

Substituting, we find that

$$F_{ave}=\frac{mv_2-mv_1}{\Delta t}=\frac{100(10-70)}{3}=-2000 N$$.

The average deceleration is

$$\bar{a}=F/m=-2000/100=-20m/s^2.$$

It is twice the acceleration of the gravity g.

9.2 Skydiver's Jump from the Airplane

In practice the jump of a parachutist from a horizontally flying plane is related with the following scenarios.

- A parachutist of mass m (including the mass of the parachute) jumps from an airplane, which flies horizontally with velocity $\vec{v}_0$ at an altitude y_0, with closed parachute that opens after a free fall in presence of air resistance.
- A parachutist jumps with an open parachute from a hovering helicopter or a parachute tower.

To study the fall of a parachutist without parachute, we can use the systems of differential equations (9.1.18), (9.1.19), or (9.1.20), supposing that the average coefficient of air resistance and the average cross-section area of a free-fall parachutist (skydiver) are respectively

$$c_D=0.8, \quad A=0.5m^2.$$

We will use the same equations to study the fall of the skydiver jumping from an airplane with an open parachute, taking into consideration that the coefficient of resistance for hemisphere parachutes is in general $c_D=1.33$.

For any particular parachute, or other shaped parachutes the coefficient of resistance can be determined experimentally.

Consider the system of differential equations (9.1.20) that describes the fall of a parachutist (skydiver) jumping from a horizontally flying airplane

$$\begin{cases} \dfrac{dv_x}{dp} = \dfrac{b}{mg} \cdot h(y) \cdot v_x^3 (1+p^2)^{1/2} \\[2mm] \dfrac{dx}{dp} = -\dfrac{v_x^2}{g} \\[2mm] \dfrac{dy}{dp} = -\dfrac{p \cdot v_x^2}{g} \\[2mm] \dfrac{dt}{dp} = -\dfrac{v_x}{g} \end{cases} \tag{9.2.1}$$

where

$$p = \tan\alpha$$

and

$$b = c_D \frac{\rho_0 \pi \cdot d^2}{8} = c_D \frac{\rho_0 A}{2} \tag{9.2.2}$$

For approximate solutions, we will consider the density function to be constant,

$$h(y) = h(\bar{y}) \tag{9.2.3}$$

where $\bar{y} = (1/2)y_m$, y_m is the maximum altitude of the trajectory that corresponds to the jumping altitude. The density function (9.2.3) can be estimated using

$$h(\bar{y}) = \left(\frac{289.08 - 0.006328\bar{y}}{289.08}\right)^{4.4} \tag{9.2.4}$$

Considering (9.2.3), we can write the system (9.2.1) as follows:

$$\begin{cases} \dfrac{dv_x}{dp} = B \cdot v_x^3 (1+p^2)^{1/2} \\[2mm] \dfrac{dx}{dp} = -\dfrac{v_x^2}{g} \\[2mm] \dfrac{dy}{dp} = -\dfrac{p \cdot v_x^2}{g} \\[2mm] \dfrac{dt}{dp} = -\dfrac{v_x}{g} \end{cases} \tag{9.2.5}$$

where

$$B = \frac{b}{m \cdot g} h(\bar{y}), \quad b = c_D \frac{\rho_0 \pi \cdot d^2}{8} = c_D \frac{\rho_0 A}{2} \tag{9.2.6}$$

or

$$B = c_D \frac{\rho_0 \pi \cdot d^2}{8mg} h(\bar{y}). \tag{9.2.7}$$

The initial conditions when the parachutist jumps from airplane flying horizontally are

$$y(0) = y_0, \quad t_0 = 0, \quad v(0) = v_0 \quad \text{when} \quad p_0 = \tan \alpha_0 = 0. \tag{9.2.8}$$

The above system has an identical form as system (4.5.7). That allows us to use the same solutions we have obtained in section 4.5.

Thus, adapting the solutions of (4.5.7), we find the solution of the differential equations (9.2.5) of parachute fall:

Components of Velocity

$$v_x = v_0 \cdot \frac{1}{f(p)}, \quad v_y = v_0 \cdot \frac{p}{f(p)} \tag{9.2.9}$$

Speed

$$v^2=v_x^2+v_y^2=v_0^2 \cdot \frac{1+p^2}{f^2(p)}$$

(9.2.10)

Coordinates of the Skydiver (Parachutist)
    Employing the second and the third equation of (9.2.5), we find that

$$x=-\frac{v_0^2}{g(1+p_0^2)} \cdot \int_{p_0}^{p} f^{-2}(p)dp \quad y=y_0-\frac{v_0^2}{g(1+p_0^2)} \int_{0}^{p} pf^{-2}(p)dp$$

,   (9.2.11)

where

$$f^{-2}(p)=[1-Bv_0^2 J(p)]^{-1}$$

(9.2.12)

and

$$J(p)=p\sqrt{1+p^2} +\ln(p+\sqrt{1+p^2})$$.

(9.2.13)

Time of Flight
    Using the fourth equation of (9.2.5), we obtain

$$t=-\frac{v_0}{g(1+p_0^2)^{1/2}} \cdot \int_{p_0}^{p} f^{-1}(p)dp$$

,   (9.2.14)

where

$$f^{-1}(p)=[1-Bv_0^2 J(p)]^{-1/2}$$.

(9.2.15)

    The integrals in (9.2.11) and (9.2.14) can be evaluated using numerical methods, using graphing calculators, or other math PC programs.
    The speed of the skydiver (9.2.10) as a function of $p=\tan\alpha$ is

$$v^2 = v_0^2 \cdot \frac{1+p^2}{1-B \cdot v_0^2 \cdot J(p)}.$$

(9.2.16)

Example 1. A parachutist with mass $m=100$ kg (parachute and equipment included) jumps from a plane flying horizontally at an altitude of 1,220 m above sea level with a speed of 100 m/s in the absence of wind.

The parachutist deploys his parachute after 10 sec.

Estimate the position of the parachutist and his speed at the time he deploys the parachute.

Assume weather with no wind.

Solution

I. Approximate Solution

Estimate the Parameter b:

$$b = c \frac{\rho_0 A}{2} = (0.8) \frac{(1.205)(0.5)}{2} = 0.241$$

Estimate B:

As a first approach, we consider $h(y)$ being a constant and equal to the value it has at the altitude 1,220 m, i.e., $h(y)=h(1,220)$.

$$h(y) = (\frac{289.08 - 0.006328 \cdot y}{289.08})^{4.4} = (\frac{289.08 - 0.006328 \cdot (1,220)}{289.08})^{4.4} = 0.8877$$

For B, we find

$$B = \frac{b}{m \cdot g} h(y) = \frac{(0.241)}{(100)(9.80665)} \cdot (0.8877) = 2.18156 \cdot 10^{-4}.$$

Initial Conditions:

Since the airplane is flying horizontally with speed 100 m/s at the time ($t=0$) when the parachutist jumps, he leaves the plane flying horizontally with initial speed $v_0=100 m/s$. The launching angle is $\alpha_0=0$.

We ignore the low velocity the parachutist has as a result of jumping out the airplane.

Thus, the initial conditions are

$$x_0=0, \ y_0=1220, \ t_0=0, \ v_0=100, \ p_0=\tan\alpha_0=0.$$

We find $p=\tan\alpha$ at time $t=10$ sec.

Substituting $\alpha_0=0$, $p_0=0$, $v_0=100$, $B=2.18156 \cdot 10^{-4}$ in (9.2.14), we obtain

$$f^{-1}(p)=[1-2.18156 \cdot J(p)]^{-1/2}.$$

Substituting in (9.2.14), for the time of flight, we can write

$$t=-\frac{v_0}{g} \cdot \int_{p_0}^{p} f^{-1}(p)dp=-\frac{100}{g} \int_{0}^{p}[1-2.18156 \cdot J(p)]^{-1/2} dp.$$

To find the coordinates and the velocity of the parachutist at the moment of parachute deployment, we give to $p=\tan\alpha$, i.e., to the angle of flight α, different (guessing) values till we find the corresponding value of "deployment" time $t=10$ sec.

The following table contains the estimation of time of flight t obtained using the above procedure.

Table 1

Angle	-15	-30	-45	-60	-65.80
Time	2.21	4.07	6.06	8.64	10.00

Table 1 shows that when the angle of parachutist velocity is $\alpha=-65.80°$ ($p=\tan\alpha=\tan-65.80°=-2.1622$), the estimated time of flight is $t=10$ sec., i.e.,

$$p=-2.1622, \text{ for } t=10.$$

Coordinates of the Parachutist at t=10 s

Substituting in both relations of (9.2.11): $p_0=0$, $v_0=100$, $p(10)=-1.75$, and

$$f^{-2}(p)=[1-2.18156 \cdot J(p)]^{-1}$$

and then calculating the obtained integral, we find the coordinates of the parachutist 10 sec. after he left the airplane:

$$x=-\frac{100^2}{g} \cdot \int_0^{-2.1622} f^{-2}(p)dp=-\frac{100^2}{g}(-0.5009)=510.78m$$

and

$$y=1{,}220-\frac{100^2}{g} \cdot \int_0^{-2.1622} p \cdot f^{-2}(p)dp=1220-\frac{100^2}{g}(0.32332)=890.31m.$$

II. Improved Solution

To get a more accurate solution than in I, we evaluate $h(y)$, considering an average value

$$h(\bar{y})=\frac{h(y_0)+h(y_{10})}{2},$$

where

$$h(y_0)=h(1220)=0.8877$$

and

$$h(y_{10})=h(890.31)=(\frac{289.08-0.006328\bar{y}}{289.08})^{4.4}=(\frac{289.08-0.006328(890.31)}{289.08})=0.91704.$$

Substituting, we find

$$h(\bar{y})=\frac{h(y_0)+h(y_{10})}{2}=\frac{(0.88771)+(0.91701)}{2}=0.902375.$$

For B, we find

$$B=\frac{b}{mg}h(y)=\frac{(0.241)}{(100)(9.80665)}\cdot(0.902375)=2.217601\times10^{-4}$$

.

We find as well

$$f^{-1}(p)=[1-2.2176\cdot J(p)]^{-1/2}, \quad f^{-2}(p)=[1-2.2176\cdot J(p)]^{-1}.$$

Using the new value for B and quite the same steps as in I, we find the following values:

$$p(10)=-\tan(66)=-2.2460, \quad x=507m, \quad y=892m,$$

at the instant $t=10\,\text{sec}$.

We can still improve the accuracy of our estimations. The solution we obtain repeating again the same procedure does not change significantly.

The components of velocity along x- and y-axes and the projectile speed are respectively

$$v_x(10)=v_0\frac{1}{f(p_{10})}=\frac{100}{f(-2.2460)}=100(0.24485)=24.49m/s$$

,

$$v_y(10)=v_0\cdot\frac{p_{10}}{f(p_{10})}=100\frac{(-2.460)}{f(-2..460)}=-55m/s$$

.

Hence, at the instant the parachute is deployed, the speed of the parachutist is

$$v(10)=(v_x^2+v_y^2)^{1/2}=[(24.49)^2+(-55)^2]=60.20m/s$$

.

The parachutist has fallen down vertically (from the jumping point) approximately

$$|\Delta y|=|y_{10}-y_0|=|892-1220|=328m$$

.

Verifying Results

To verify the solution obtained using the above approximate analytic method, we use the QBasic PC program Parach.Bas:

Input Data
Launching altitude = 1,220
Altitude of the parachute deployment = 700
Launching speed = 100
Launching angle = 0
Initial x-coordinate = 0
Initial time =0
Time of interest 10
Parameter b = 0.241
Mass of parachutist = 100
Integration step = 0.01

Output
At time t =10 sec.
Parachutist location is (510 m, 891 m)
Corresponding speed = 60.12 m/s
Corresponding angle = -65.92°

Comment: Comparing the results obtained using the analytical method with results obtained by numerical integration, we see that practically there is no difference between them.
That shows that the analytical approach is accurate.

9.3 Skydiver's Trajectory (Second Approach)

Hereafter, we show a different approach to solve the differential equations of the parachutist (9.1.15),

$$\frac{d\vec{v}}{dt} = \vec{g} - \frac{b \cdot h(y)}{m} v^2 \frac{\vec{v}}{v}.$$

(9.3.1)

Considering

$$h(y)=h(\bar{y}),$$

we find that the tangential acceleration of parachutist in the descending part of trajectory is

$$\frac{dv}{dt}=-g\cdot\sin\alpha-\frac{b\cdot h(\bar{y})}{m}v^2.$$

(9.3.2)

We denote

$$B_1=\frac{b}{m}h(\bar{y})$$

(9.3.3)

and consider as well the fourth differential equation of system (9.1.20),

$$\frac{dt}{dp}=-\frac{v_x}{g}.$$

(9.3.4)

But

$$\sin\alpha=\tan\alpha\cdot(1+\tan^2\alpha)^{-1/2}=p(1+p^2)^{-1/2}$$

(9.3.5)

and

$$v_x=v\cdot\cos\alpha=v\cdot(1+\tan^2\alpha)^{-1/2}=v\cdot(1+p^2)^{-1/2}.$$

(9.3.6)

Expressing (9.3.1) and (9.3.4) through $p=\tan\alpha$, using the above relations, we can write respectively

$$\frac{dv}{dt}=-g\cdot p\cdot(1+p^2)^{-1/2}-B_1v^2$$

(9.3.7)

and

$$\frac{dt}{dp}=-\frac{v}{g}(1+p^2)^{-1/2}.$$ (9.3.8)

From (9.3.5) and (9.3.6), we obtain

$$\frac{dv}{dp}=\cdot p\cdot(1+p^2)^{-1}v+\frac{B_1}{g}(1+p^2)^{-1/2}v^3$$ (9.3.9)

or

$$\frac{dv}{dp}=\cdot p\cdot(1+p^2)^{-1}v+B(1+p^2)^{-1/2}v^3,$$ (9.3.10)

where

$$B=\frac{B_1}{g}=\frac{b}{mg}h(\bar{y}).$$

We need to solve the differential equation (9.3.10) to find the parachutist's speed $v=v(p)$ as a function of $p=\tan\alpha$. We can write (9.3.10) in the form

$$\frac{dv}{dp}-p(1+p^2)^{-1}v=B(1+p^2)^{-1/2}\cdot v^3.$$ (9.3.11)

The last equation is a Bernoulli differential equation that can be transformed into a linear differential equation substituting

$$u=v^{-2}.$$ (9.3.12)

Thus, substituting (9.3.12) into (9.3.11), we obtain the following linear differential equation

$$\frac{du}{dp}+2p(1+p^2)^{-1}u=-2B\cdot(1+p^2)^{-1/2}.$$ (9.3.13)

The integrating factor of (9.3.13) is

$$I(p)=e^{\int 2p(1+p^2)^{-1}dp}=e^{\ln(1+p^2)}=(1+p^2).$$

We write (9.3.13)

$$\frac{d[(1+p^2)\cdot u]}{dp}=(1+p^2)\cdot(-2B\cdot(1+p^2)^{-1/2})$$

or

$$\frac{d[(1+p^2)\cdot u]}{dp}=-2B\cdot(1+p^2)^{1/2} \tag{9.3.14}$$

Integrating (9.3.14), we find

$$(1+p^2)u=-B[p(1+p^2)+\ln(p+(1+p^2)^{1/2})]+C.$$

Substituting (9.3.12) in (9.3.14), we obtain the general solution of the Bernoulli differential equation (9.3.11),

$$(1+p^2)v^{-2}=-B[p(1+p^2)+\ln(p+(1+p^2)^{1/2})]+C. \tag{9.3.15}$$

Employing the initial condition,

$$v=v_0, \text{ when } p=p_0, \tag{9.3.16}$$

we find

$$C=(1+p_0^2)v^{-2}+B[p_0(1+p_0^2)+\ln(p_0+(1+p_0^2)^{1/2})]. \tag{9.3.17}$$

Substituting (9.3.17) in (9.3.15), we find the speed of parachutist as a function of $p=\tan\alpha$,

$$\frac{1+p^2}{v^2}-\frac{1+p_0^2}{v_0^2}=-B[(p\sqrt{1+p^2}+\ln(p+\sqrt{1+p^2})-(p_0\sqrt{1+p_0^2}+\ln(p_0+\sqrt{1+p_0^2})]$$

$$\tag{9.3.18}$$

Hence,

$$v^2 = \frac{v_0^2(1+p^2)}{(1+p_0^2)+Bv_0^2[J(p_0)-J(p)]} ,$$

(9.3.19)

where

$$J(p)=p\sqrt{1+p^2}+\ln(p+\sqrt{1+p^2}) ,$$
$$J(p_0)=p_0\sqrt{1+p_0^2}+\ln(p_0+\sqrt{1+p_0^2})$$

(9.3.20)

and

$$B=\frac{b}{m\cdot g}h(\bar{y}) , \quad b=c_D\frac{\rho_0\pi\cdot d^2}{8}=c_D\frac{\rho_0 A}{2} .$$

(9.3.21)

For the parachutist that jumps from a horizontally flying airplane with initial velocity parallel to the ground, the initial conditions are

$$v(0)=v_0 , \text{ when } p_0=0 .$$

Substituting the above initial conditions in (9.3.19), we find the square of the speed of parachutist as a function of p

$$v^2 = \frac{v_0^2(1+p^2)}{1-Bv_0^2 J(p)} .$$

(9.3.22)

Components of the Parachutist's Velocity

Substituting $v=v_x/\cos\alpha=v_x(1+p^2)^{1/2}$ in (9.3.22), we obtain

$$v_x^2 = \frac{v_0^2}{1-Bv_0^2 J(p)} .$$

(9.3.23)

Hence, for the x-component of parachutist velocity, we have

$$v_x = \frac{v_0}{[1-Bv_0^2 J(p)]^{1/2}} .$$

(9.3.24)

Similarly, substituting in (9.3.22) $v=v_y/\sin\alpha = v_y p(1+p^2)^{1/2}$, we find

$$v_y^2 = \frac{p^2 v_0^2}{1-Bv_0^2 J(p)} .$$

(9.3.25)

For the y-component of velocity, we have

$$v_y = \frac{p v_0}{[1-Bv_0^2 J(p)]^{1/2}} .$$

(9.3.26)

We denote again

$$f(p)=[1-Bv_0^2 J(p)]^{1/2}, \quad f^2(p)=1-B\cdot v_0^2 J(p),$$

(9.3.27)

where

$$J(p)=p\sqrt{1+p^2}+\ln(p+\sqrt{1+p^2}) .$$

(9.3.28)

Substituting $v_x = dx/dt$ in (9.3.24) and $v_y = dy/dt$ in (9.3.26) and then integrating, we obtain the coordinates of the parachutist trajectory

$$x = -\frac{v_0^2}{g} \cdot \int_0^p f^{-2}(p)dp, \quad y = y_0 - \frac{v_0^2}{g} \int_0^p p f^{-2}(p)dp .$$

(9.3.29)

Substituting (9.3.24) in the differential equation (9.3.4) and integrating, we find the time of flight as a function of parameter p

$$t = \frac{v_0^2}{g} \int_0^p f^{-1}(p)dp .$$

(9.3.30)

Example 1. Show that that the limit of the square of the parachutist speed v^2 (equation (9.3.22)) when α approaches to $-\pi/2$ ($p=\tan\alpha \to -\infty$) is

$$v_l^2 = 1/B$$
(9.3.31)

where

$$B = \frac{b}{m\cdot g} h(\bar{y})$$

Solution

Indeed, dividing the numerator and denominator by $1+p^2$, we have

$$v^2 = \frac{v_0^2}{1/(1+p^2) - Bv_0^2 J(p)/(1+p^2)}$$

But

$$1/(1+p^2) \xrightarrow[p\to-\infty]{} 0$$

and

$$J(p)/(1+p^2) = p\sqrt{1+p^2}/(1+p^2) + \ln(p+\sqrt{1+p^2})/(1+p^2) \xrightarrow[p\to-\infty]{} -1$$

since $J(p)<0$.

Thus,

$$v^2 \to \frac{v_0^2}{Bv_0^2} = \frac{1}{B}, \quad \text{when } p=\tan\alpha \to -\infty.$$

The limiting speed (the so-called terminal speed) is given by the equation

$$v_l^2 = \frac{mg}{b\cdot h(\bar{y})}$$
(9.3.32)

where

$$b=c_D\frac{\rho_0\pi\cdot d^2}{8}=c_D\frac{\rho_0 A}{2}.$$

(9.3.33)

Substituting (9.3.33) in (9.3.32), we obtain the limiting speed of the parachutist[62]

$$v_l=\sqrt{2mg/(\rho_0 c_D A\cdot h(\overline{y}))}.$$

(9.3.34)

9.4 Skydiver's Minimum Speed. The Optimal Time of Parachute Deployment

We have shown in section 2.11 that the minimum speed of a projectile is reached on the descending part of the projectile trajectory. This conclusion is valid for the parachutist as well.

We need to find the minimum speed for a given parachutist, as well as the time and the point on the descending trajectory where the minimum speed is reached.

The tangential acceleration (9.3.2) of the parachutist in the descending part of the trajectory can be written as

$$\frac{dv}{dt}=-g\cdot\sin\alpha-B\cdot v^2,$$

(9.4.1)

where

$$B=bh(\overline{y})/(mg).$$

(9.4.2)

---

[62]   In the practice of skydiving the limiting speed is called the fictive speed that the skydiver reaches considering the density function $h(y)=1$. In fact, the limiting speed decreases as the skydiver approaches to the ground, since the density of air increases.

The minimum speed of parachutist is reached at a point on the descending part of the trajectory, where the derivative $dv/dt=0$, i.e., at the point where

$$-g \cdot \sin\alpha - B \cdot v^2 = 0 .$$

(9.4.3)

Hence, for the minimum speed, we have

$$v_m = \pm(-\frac{g \cdot \sin\alpha_m}{B})^{1/2} .$$

(9.4.4)

Since the negative speed value has no physical sense, we accept only the positive value. Using the first derivative test, we find that the minimum value is

$$v_m = (-\frac{g \cdot \sin\alpha_m}{B})^{1/2} .$$

(9.4.5)

Substituting B, for the minimum speed, we have

$$v_m = (-\frac{mg \cdot \sin\alpha_m}{bh(y)})^{1/2} .$$

(9.4.6)

Since $\sin\alpha = p(1+p^2)^{-1/2}$, we find

$$v_m^2 = -\frac{g \cdot p_m}{B \cdot (1+p_m^2)^{1/2}} .$$

(9.4.7)

We note that α_m and p_m are negative numbers.

We need to find the angle α_m or $p_m = \tan\alpha_m$ at the point of the trajectory where the parachutist speed has a minimum value. For any value of $p = \tan\alpha$, the speed of parachutist, given in (9.3.22), is

$$v^2 = \frac{v_0^2(1+p^2)}{1-Bv_0^2 J(p)} .$$

(9.4.8)

Substituting p_m in (9.4.8), we find that the value of the minimum speed is

$$v_m^2 = \frac{v_0^2 (1+p_m^2)}{1-Bv_0^2 J(p_m)} .$$

(9.4.9)

From (9.4.7) and (9.4.9), we find the following equation:

$$\frac{1}{Bv_0^2} + \frac{(1+p_m^2)^{3/2}}{g \cdot p_m} = J(p_m) ,$$

(9.4.10)

where

$$J(p_m) = p_m \sqrt{1+p_m^2} + \ln(p_m + \sqrt{1+p_m^2}) .$$

(9.4.11)

Solving the transcendental equation (9.4.10)—for example, by using a graphing calculator—we find $p_m = \tan\alpha_m$ and the angle α_m where the parachutist's speed takes the minimum value.

Then, substituting the obtained value of $p_m = \tan\alpha_m$ in (9.4.9), we find the value of the minimum speed.

Using (9.3.29), (9.3.30), we can find the coordinates where the parachutist speed is minimum, as well as the time the parachutist reaches the minimum speed.

Note: In example 1 of section 9.3, we proved that the limit of v^2 and v, when α approaches $-\pi/2$ (i.e. $p = \tan\alpha \cdot \to -\infty$), is given respectively by the equations

$$v_l^2 = \frac{mg}{b \cdot h(\overline{y})}, \; v_l = \sqrt{2mg / (\rho_0 c_D A \cdot h(\overline{y}))},$$

(9.4.12)

where

$$b=c_D\frac{\rho_0\pi\cdot d^2}{8}=c_D\frac{\rho_0 A}{2}$$

(9.4.13)

and

$$h(\bar{y})=(\frac{289.08-0.006328\bar{y}}{289.08})^{4.4}.$$

(9.4.14)

From (9.4.12), it follows that the limiting speed does not depend on the initial speed of the skydiving; it depends on the coordinate y of the parachutist since the average value of $\bar{y}$ depends on the altitude of the skydiving.

Because the density function $h(y)$ decreases with altitude, the higher the parachutist's altitude, the greater the limiting speed.

Comment: This is an important conclusion for practical use that follows from the existence of the minimum speed (assuming that the altitude is relatively high above the ground to allow the parachutist to open the parachute before falling on the ground):

• The parachutist should deploy the parachute at the time the descending speed is minimum in order to avoid relatively high dynamic deceleration forces acting on the parachutist as a result of parachute deployment.

After the parachutist reaches the minimum speed, his speed increases again and approaches asymptotically to the limit (terminal) speed, determined by (9.4.12).

Note that the limiting speed (terminal speed) is a theoretical concept; it depends on the altitude of flight since, in the formula (9.4.12), is present the density function $h(\bar{y})$.

Example 1. Find the minimum speed and the time needed to reach it if a parachutist of mass $m=80$ kg is launched from an airplane flying horizontally with speed $v_0=80 m/s$ at an altitude of $h=3,600$ m.

Solution

$$b=c_D\frac{\rho_0 A}{2}=(0.8)\frac{(1.205)(0.5)}{2}=0.241$$

As a first approach, we consider constant $h(y)=h(\bar{y})$, where $\bar{y}=3600$

$$h(3600)=(\frac{289.08-0.006328\cdot(3600)}{289.08})^{4.4}=0.696845787$$

and

$$B=\frac{b}{m\cdot g}h(\bar{y})=\frac{(0.241)}{(80)\cdot(9.80665)}(0.696846)=2.1406327\times10^{-4}$$

Minimum Speed
To find the minimum speed of the projectile, we employ equation (9.4.10),

$$\frac{1}{Bv_0^2}+\frac{(1+p_M^2)^{3/2}}{p_M}=J(p_M)$$

where

$$J(p_m)=p_m\sqrt{1+p_m^2}+\ln(p_m+\sqrt{1+p_m^2})$$

Substituting $B=2.1406327\times10^{-4}$ and $v_0=80 m/s$, we have

$$0.72992286+\frac{(1+p_m^2)^{3/2}}{p_m}=J(p_m)$$

Solving the above equation using a graphing calculator TI-83 Plus, we find

$$p_m = \tan\alpha_m = -0.8741368, \text{ and } J(p_m) = -1.950547 .$$

From the first relation, we find the angle of flight,

$$\alpha_m = \tan^{-1}(-0.8741368) = -41.158°$$

at the moment the parachutist speed is minimum.
We can write

$$v_m^2 = \frac{v_0^2(1+p_m^2)}{1-Bv_0^2 J(p_m)} = \frac{(80)^2(1+(-1.950547)^2)}{1-(2.1406327\times10^{-4})\cdot(80)^2\cdot(-1.950547)} = 3074.50 .$$

Hence, for the minimum speed, we get

$$v_m = \sqrt{374.50} = 55.45 m/s .$$

Vertical Drop

The y-coordinate of the parachutist at the point where the speed is minimum can be estimated using the second equation of (9.3.29) and a graphing calculator TI-83 Plus.

Thus, we find that the y-coordinate of the parachutist at the moment the speed is minimum is

$$y = y_0 - \frac{v_0^2}{g}\int_0^p pf^{-2}(p)dp = 3600 - \frac{(80)^2}{9.80665}\int_0^{-0.874} p(1-Bv_0^2 J(p))^{-1}dp = 3600 - 98.77 = 3501.22m$$

since

$$f^{-2}(p) = [1-Bv_0^2 J(p)]^{-1}, \quad \text{and}$$

$$\int_0^{-0.874}(1-Bv_0^2 J(p))^{-1}dp = 0.15134938 .$$

The vertical drop of the parachutist from the launching plane is 98.77 m.

Improving Results

Since the altitude of the parachutist changes during his free fall, we calculate the average altitude

$$\bar{y}=\frac{y_m+y_0}{2}=\frac{3,500.22+3,600}{2}=3,550.61m$$

and

$$h(\bar{y})=h(3,55.61)=(\frac{289.08-0.006328\cdot\bar{y}}{289.08})^{4.4}=(\frac{289.08-0.006328\cdot(3550.61)}{289.08})^{4.4}=0.70024$$

We find that

$$B=\frac{b}{m\cdot g}h(\bar{y})=\frac{(0.241)}{(80)(9.80665)}\cdot(0.70045)=2.15106\times10^{-4}$$

Repeating the steps we followed above to get the minimum speed, we find the angle for which the speed is minimum:

$$p_m=\tan\alpha_m=-0.8763028, \text{ and } \alpha_m=\tan(-0.8763028)=-41.23°.$$

Minimum Speed

Employing (9.4.9), for the square of the minimum speed of the parachutist, we have

$$v_m^2=\frac{v_0^2(1+p_m^2)}{1-Bv_0^2J(p_m)}=\frac{(80)^2(1+(-0.8763028^2))}{1-(2.1517144\cdot10^{-4})(80)^2\cdot J(-0.8763028)}=3062.95.$$

Hence, we find the minimum speed of the parachutist

$$v_m=55.34m/s.$$

Time of Flight to the Point the Parachutist Gets the Minimum Speed
 Substituting the new value of B, $B=2.1517144 \times 10^{-4}$, in (9.3.27), we have

$$f^{-2}(p)=[1-1.3770972 \cdot J(p)]^{-1}, \quad f^{-1}(p)=[1-1.3770972 \cdot J(p)]^{-1/2}.$$

The time of flight is

$$t=-\frac{v_0}{g} \cdot \int_0^p f^{-1}(p)dp=-\frac{80}{g} \int_0^{-0.876} f^{-1}(p)dp=-\frac{80}{g}(-0.6090748)=4.97s.$$

Coordinates of the point where the speed has the minimum value are respectively

$$x=-\frac{(80)^2}{g} \int_0^{-0.876} f^{-2}(p)dp=\frac{(80)^2}{g}(0.4381314)=285.93m$$

and

$$y=3,600-\frac{(80)^2}{g} \int_0^{-0.876} p \cdot f^{-2}(p)dp=3600-\frac{80^2}{g}(0.15142)=3501.18.$$

Conclusion: The parachutist should deploy the parachute around 5 sec. after he/she is launched from the plane. At that time the coordinates of the parachutist are respectively

$$x=286m \text{ and } y=3501m.$$

Example 2. Find the minimum speed and the time needed to reach it if a parachutist of mass $m=103$ kg is launched from an airplane flying horizontally with speed $v_0=115$ m/s, at an altitude of $h=4,000$ m.

Solution
 We find

$$b=c\frac{\rho_0 A}{2}=(0.8)\frac{(1.205)(0.5)}{2}=0.241$$

and

$$h(y)=(\frac{289.08-0.006328 \cdot y}{289.08})^{4.4}=(\frac{289.08-0.006328 \cdot (4,000)}{289.08})^{4.4}=0.66839 \ .$$

For B, we get the following value:

$$B=\frac{b}{m \cdot g}h(y)=\frac{(0.241)}{(103)(9.80665)} \cdot (0.66839)=1.5947 \cdot 10^{-4}$$

Minimum Speed

To find the minimum speed of the projectile, we employ equation (9.4.10),

$$\frac{1}{Bv_0^2}+\frac{(1+p_m^2)^{3/2}}{g \cdot p_m}=J(p_m)$$

where

$$J(p_m)=p_m\sqrt{1+p_m^2}+\ln(p_m+\sqrt{1+p_m^2}) \ ,$$

Solving the above equation using a graphing calculator, we find

$$p_m=\tan\alpha_m=-1.042704 \ , \quad \alpha_m=\tan^{-1}(-1.042704)=46.20° \ .$$

The square of the minimum speed is

$$v_m^2=\frac{v_0^2(1+p_m^2)}{1-Bv_0^2 J(p_m)}=\frac{(115)^2(1+(-1.042704^2)}{1-(1.5942 \cdot 10^{-4})(115)^2 \cdot J(-1.042704)}=4526.60$$

since

$$J(-1.042704)=-2.147669 \ .$$

Hence,

$$v_m=(4526.60)^{1/2}=67.28m/s \ .$$

Vertical Drop

The y-coordinate of the parachutist at the point where the speed is minimum can be estimated using the second equation of (9.3.29),

$$y=y_0-\frac{v_0^2}{g}\int_0^P pf^{-2}(p)dp \quad ,$$

where

$$f^{-2}(p)=[1-2.1083295\cdot J(p)]^{-1} \quad .$$

Integrating using graphing calculator TI-83 Plus, we have

$$y_m=y_0+\frac{(115)^2}{g}\int_0^{-1.04} p\cdot f^{-2}(p)dp=4{,}000-195.806=3{,}804.20m \quad .$$

The vertical drop of the parachutist from the launching plane is 195.81 m.

Improving Results

Since the altitude of the parachutist changes during his free fall, we calculate the average altitude

$$\bar{y}=\frac{y_m+y_0}{2}=\frac{3{,}804.20+4{,}000}{2}=3{,}902m$$

and

$$h(\bar{y})=h(3{,}902)=(\frac{289.08-0.006328\cdot y}{289.08})^{4.4}=(\frac{289.08-0.006328\cdot(3902)}{289.06})^{4.4}=0.6751 \quad .$$

We find that

$$B=\frac{b}{m\cdot g}h(\bar{y})=\frac{(0.241)}{(103)(9.80665)}\cdot(0.675101)=1.61075\cdot10^{-4} \quad .$$

Repeating the steps we followed above to find the minimum speed, we find that

$$p_m = \tan\alpha_m = -1.046379,$$

$$\alpha_m = \tan(-1.046379) = 46.30°.$$

Minimum speed is $v_m = 67 m/s$.

Time of Flight to the Point of Minimum Speed

Substituting the new value of B, $B = 1.61075 \cdot 10^{-4}$, in (9.4.28), we have

$$f^{-2}(p) = [1 - 2.1083295 \cdot J(p)]^{-1}, \quad f^{-1}(p) = [1 - 2.1083295 \cdot J(p)]^{-1/2}.$$

The time of flight is

$$t = -\frac{v_0}{g} \cdot \int_0^p f^{-1}(p)dp = -\frac{115}{g} \int_0^{-1.046} f^{-1}(p)dp = -\frac{115}{g}(-0.6189211) = 7.25s.$$

Coordinates of the point where the speed is minimum are

$$x = -\frac{(115)^2}{g} \int_0^{-1.046379} f^{-2}(p)dp = \frac{(115)^2}{g}(0.3905536) = 527.69$$

and

$$y = 4,000 - \frac{(115)^2}{g} \int_0^{-1.046} p \cdot f^{-2}(p)dp = 3803.85.$$

Conclusion. The parachutist should deploy the parachute around 7 sec. after he/she is launched from the plane.

Example 3. Parachutist's Trajectory

Using the estimation method shown above, in table 1 are displayed the characteristics of the trajectory of the parachutists falling without parachute from an altitude of 4,000 m.

The data displayed in table 1 are obtained using (9.2.11), (9.2.12), and (9.2.16). From table 1, it can be seen that the parachutist's speed within 5 sec. decreases from 115 m/s to around 69.50 m/s.

The coordinates of the parachutist at that moment are (408, 3900). That means that the parachutist has fallen vertically around 100 m from the launching point and has deviated horizontally in the direction of airplane velocity about 408 m.

The parachutist's speed reaches the minimum speed, $v_m=67m/s$, around 7.2 sec. after jumping from airplane. At this moment, the parachutist has fallen down for around 190 m and deviated horizontally for around 520 m in the direction of airplane flight.

Then, for about 3 sec., i.e. 7.2–11.3 sec., the parachutists speed remains under 70 m/s.

At time 11.3 sec., the coordinates of the parachutists are (675, 3575). The parachutist has fallen around 400 m and deviated horizontally 425 m.

After around 36 sec., the parachutist trajectory is approximately vertical. We can consider the speed $v_l=77.50m/s$ the parachutist has at time $t=36sec.$ to be approximately equal to limiting speed $v_l^2=mg/(bh(y))$.

The skydiver that fails to deploy the parachute will fall on the ground with a speed of $v_l=77.50m/s$ at the point with abscissa 1,315 m after flying for around 62.2 sec.

Table 1. Trajectory of the Parachutist

Angle	Altitude	Distance	Speed	Time
α	y	x	v	t
0	4000	0	115.00	0
-2	3999	44	107.43	0.4
-4	3997	83	101.30	0.8
-6	3994	117	96.22	1.1
-8	3990	149	91.96	1.5
-10	3986	178	88.33	1.8
-12	3981	204	85.21	2.1
-14	3975	229	82.51	2.4
-16	3969	253	80.17	2.7
-18	3962	275	78.12	3.0
-20	3954	296	76.33	3.3
-22	3947	317	74.76	3.6
-24	3938	336	73.39	3.9
-26	3930	355	72.19	4.1
-28	3920	373	71.15	4.4
-30	3910	391	70.25	4.7
-32	3900	408	69.48	5.0
-34	3889	425	68.84	5.3
-36	3877	442	68.30	5.6
-38	3865	458	67.87	5.9
-40	3852	474	67.54	6.2
-42	3838	490	67.30	6.5
-44	3823	506	67.14	6.9
-46	3807	522	67.08	7.2
-48	3790	538	67.09	7.5
-50	3771	554	67.18	7.9
-52	3752	570	67.35	8.3
-54	3730	586	67.58	8.7
-56	3707	602	67.89	9.1
-58	3682	618	68.27	9.5
-60	3654	635	68.71	10.0
-62	3624	651	69.21	10.5
-64	3590	668	69.78	11.1
-66	3553	685	70.39	11.6
-68	3512	702	71.05	12.3
-70	3465	720	71.76	13.0
-72	3412	738	72.50	13.8
-74	3350	756	73.26	14.7
-76	3279	775	74.04	15.7
-78	3195	794	74.80	16.8
-80	3092	813	75.54	18.2
-82	2964	832	76.23	19.9
-84	2795	852	76.83	22.2
-86	2554	871	77.31	25.3
-88	2136	890	77.58	30.8
-89	1721	899	77.53	36.2
-89.93	162	1134	77.15	58.3
-89.95	0	1315	76.86	62.2

Graphs. The three graphs presented in figure 19, figure 20, and figure 21 show the following:

- Graph 1: Variation of parachutists speed with time
- Graph 2: Variation of parachutist's speed with altitude
- Graph 3: Coordinates of the parachutist during the flight

Graph 1 and graph 2 show the variation of the parachutist's speed. The speed drops to the minimum value and then increases again and tends to reach the limiting speed (about 77 m/s).

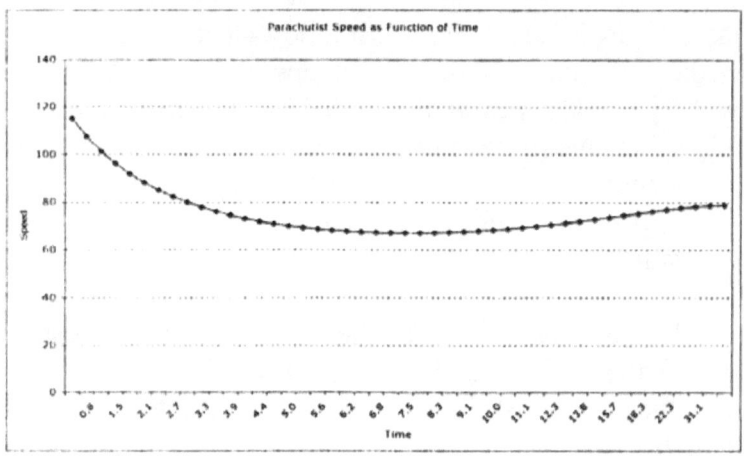

Figure 19

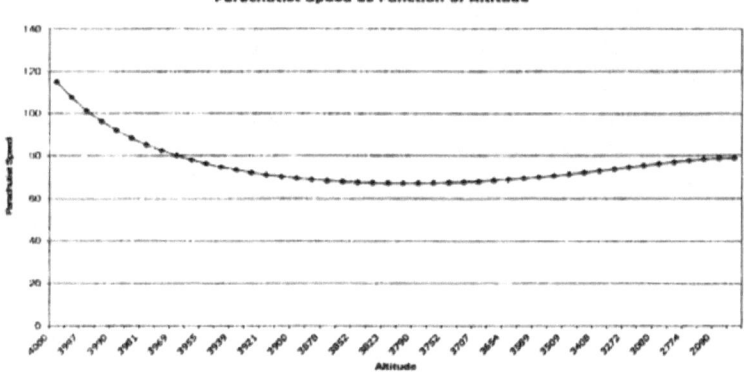

Figure 20

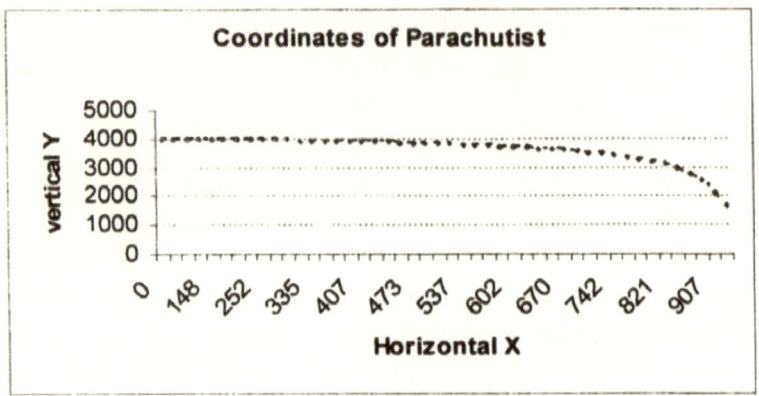

Figure 21

From table 1, graph 1, and graph 2, it follows that the parachutist should open the parachute 6–10 sec. after jumping from the plane (that is flying with speed 115 m/s at the altitude 4,000 m) in order to experience less dynamic (deceleration) force.

The graphs show that the speed of the parachutist in the time interval 6–10 sec. can be considered constant and approximately close to the minimum speed.

Example 4. (Use of table 1 in example 3) For the parachutist of example 2, use the data of table 1 (see example 2) to find the coordinates of the parachutists if he deploys the parachute 8 sec. after jumping from the plane.

Solution

Using Table 1, and interpolating, we find that after 8 sec., the speed of the parachutists is approximately 67.22 m/s and his coordinates are (558, 3766).

The parachutist fall down around 234 m from the launching point till the moment he/she deploys the parachute.

The parachutist has deviated, in horizontal direction, around 560 m from the launching point.

Example 5. Use the results obtained in example 2 to find the dynamic deceleration force exerted on the parachutist as a result of its deceleration:

(a) if the parachutist deploys the parachute when he reaches the minimum speed, $v_m = 66.99m/s$;
(b) if the parachutist deploys his parachute when his speed is approximately equal to the limit speed, $v_l = 79.55m/s$.

Assume that the parachute is fully deployed after 3 sec. and the speed of parachutist at the time of full deployment of parachute is $10m/s$.

Solution
We will use the impulse-momentum theorem to find an average force F_{ave}

$$F_{ave} \cdot \Delta t = mv_2 - mv_1$$

Substituting, we find that

(a) $F_{ave} = \dfrac{mv_2 - mv_1}{\Delta t} = \dfrac{103(10-67)}{3} = -1,957\,N$

(b) $F_{ave} = \dfrac{mv_2 - mv_1}{\Delta t} = \dfrac{103(10-79.55)}{3} = -2,388\,N$

Dividing both results by the mass of the parachutist, we find that the acceleration of the parachutist is respectively

(a) $a_1 = -1957/103 = -19m/s^2 = -1.94g$
(b) $a_{21} = -2388/103 = -23.18m/s^2 = -2.36g$,

where $g = 9.80665m/s^2$.

9.5 The Siacci Solution Method

We use the Siacci pseudo-speed to integrate the equations of motion of the parachutist.

The system of differential equations (9.2.1),

$$
\begin{cases}
\dfrac{dv_x}{dp}=\dfrac{b}{mg}\cdot h(y)\cdot v_x^3 (1+p^2)^{1/2} \\[2mm]
\dfrac{dx}{dp}=-\dfrac{v_x^2}{g} \\[2mm]
\dfrac{dy}{dp}=-\dfrac{p\cdot v_x^2}{g} \\[2mm]
\dfrac{dt}{dp}=-\dfrac{v_x}{g}
\end{cases}
\tag{9.5.1}
$$

where

$$p=\tan\alpha$$

and

$$b=c_D\frac{\rho_0\pi\cdot d^2}{8}=c_D\frac{\rho_0 A}{2} \tag{9.5.2}$$

can be expressed through the pseudo-speed,

$$u=v\frac{\cos\alpha}{\cos\alpha_0}, \tag{9.5.3}$$

defined in section 5.1.

Thus, it can be shown that (9.5.1) can be written in the form

$$
\begin{cases}
\dfrac{du}{dp}=Bu^{3}\cos^{2}\alpha_{0} \\[2mm]
\dfrac{dt}{dp}=-\dfrac{u}{g}\cos\alpha_{0} \\[2mm]
\dfrac{dx}{dp}=-\dfrac{u^{2}}{g}\cos^{2}\alpha_{0} \\[2mm]
\dfrac{dy}{dp}=-p\dfrac{u^{2}}{g}\cos^{2}\alpha_{0}
\end{cases}
\tag{9.5.4}
$$

where

$$
B=\beta\frac{b}{m\cdot g}, \quad \beta=h(\overline{y})/\cos\alpha_{0}=h(\overline{y}), \quad b=c_{D}\frac{\rho_{0}A}{2}
\tag{9.5.5}
$$

and

$$
h(\overline{y})=(\frac{289.08-0.006328\cdot\overline{y}}{289.08})^{4.4}
\tag{9.56}
$$

The system of equations (9.5.4) is similar to system (5.3.3).

Adapting the solution obtained in section 5.3, we find that the Siacci solution of system (9.5.4) that describes the parachutist fall is given by the set of equations

$$
p=-\frac{1}{2B}(\frac{1}{u^{2}}-\frac{1}{v_{0}^{2}})
\tag{9.5.7}
$$

$$
t=\frac{1}{Bg}(\frac{1}{u}-\frac{1}{v_{0}})
\tag{9.5.8}
$$

and

$$
x=-\frac{1}{Bg}(\ln u-\ln v_{0})
\tag{9.5.9}
$$

The trajectory equation is

$$y=y_0+\frac{x}{2B}[\frac{1}{2}(\frac{1}{u^2}-\frac{1}{v_0^2})\cdot(\frac{1}{\ln u-\ln v_0})+\frac{1}{v_0^2}]$$

(9.5.10)

since $p_0=\tan\alpha_0=\tan 0°=0$.

Example 1. Use the Siacci method to describe the fall of the parachutist with closed parachute if his mass is $m=103$ kg, the airplane speed is $v_0=115$ m/s, and the jumping altitude is $h=4,000$ m. Find the elements of the parachutist trajectory 10 sec. after he leaves the airplane. (See example 2 and example 3 of section 9.4.)

Solution
We consider as a first approach

$$h(y)=(\frac{289.08-0.006328\cdot y}{289.08})^{4.4}=(\frac{289.08-0.006328\cdot(4,000)}{289.08})^{4.4}=0.66839$$
.

Since

$$b=c\frac{\rho_0 A}{2}=(0.8)\frac{(1.205)(0.5)}{2}=0.241$$
,

for B, we obtain the following value:

$$B=\frac{b}{m\cdot g}h(y)=\frac{(0.241)}{(103)(9.80665)}\cdot(0.66839)=1.5947\cdot10^{-4}$$
.

Solving (9.5.8) for u, we obtain the pseudo-speed as a function of the time of flight

$$u=\frac{v_0}{1+Bgv_0t}$$
.

Substituting $t=10$ sec., we find that the pseudo-speed is

$$u=\frac{v_0}{1+Bgv_0t}=\frac{115}{1+0.179788256\cdot(10)}=41.10m/s$$

Employing (9.5.9) and (9.5.10), we find that the coordinates of the parachutist at $t=10$ sec. are respectively

$$x=-\frac{1}{Bg}(\ln u-\ln v_0)=-\frac{1}{1.5942\cdot10^{-4}g}[\ln(41.103)-\ln(115)]=658.10m$$

and

$$y=y_0+\frac{x}{2B}[\frac{1}{2}(\frac{1}{u^2}-\frac{1}{v_0^2})\cdot(\frac{1}{\ln u-\ln v_0})+\frac{1}{v_0^2}]$$

$$=4000+\frac{658.10}{2\cdot(1.5942\cdot10^{-4})}[\frac{1}{2}(\frac{1}{41.103^2}-\frac{1}{115^2})\frac{1}{\ln(41.103)-\ln(115)}+\frac{1}{115^2}]=3638.20m$$

Improving Results

Considering an average altitude of

$$\bar{y}=(1/2)(y_{max}+y_{10})=(1/2)(4000+3638.20)=3819m,$$

we can write

$$h(\bar{y})=(\frac{289.06-0.006328\cdot\bar{y}}{289.06})^{4.4}=(\frac{289.06-0.006328\cdot(3819)}{289.06})^{4.4}=0.681029$$

and

$$B=\frac{b}{m\cdot g}h(y)=\frac{(0.241)}{(103)(9.80665)}\cdot(0.668167)=1.624888\cdot10^{-4}.$$

The pseudo-speed at $t=10$ sec. is

$$u=\frac{v_0}{1+Bgv_0t}=\frac{115}{1+0.183249\cdot(10)}=40.60m/s.$$

Employing (9.5.9) and (9.5.10), we find that the coordinates of the parachutist at that moment are respectively

$$x=-\frac{1}{Bg}(\ln u-\ln v_0)=-\frac{1}{1.624888\cdot10^{-4}\,g}[\ln(40.60)-\ln(115)]=653.40m$$

and

$$y=y_0+\frac{x}{2B}[\frac{1}{2}(\frac{1}{u^2}-\frac{1}{v_0^2})\cdot(\frac{1}{\ln u-\ln v_0})+\frac{1}{v_0^2}]$$

$$=4000+\frac{653.40}{2\cdot(1.624888\cdot10^{-4})}[\frac{1}{2}(\frac{1}{40.60^2}-\frac{1}{115^2})\frac{1}{\ln(40.60)-\ln(115)}+\frac{1}{115^2}]=3639.30m$$

Employing (9.5.7), we find that

$$p=-\frac{1}{2B}(\frac{1}{u^2}-\frac{1}{v_0^2})=-\frac{1}{2\cdot(1.624888\cdot10^{-4})}(\frac{1}{40.6^2}-\frac{1}{115^2})=-1.6341104$$

Hence, since $p=\tan\alpha$, we find that the angle the velocity of parachutist forms with the x-axis is

$$\alpha=\tan^{-1}(-1.6341104)=-58.54°$$

Employing (9.5.3), we find that the parachutist speed at time $t=10sec.$ is

$$v=u\frac{\cos\alpha_0}{\cos\alpha}=(40.60)\frac{\cos(0°)}{\cos(-58.54°)}=77.78m/s$$

Note: Comparing the results obtained in the above example using the Siacci method with the data in table 1 of example 3 in section 9.4 (for $t=10sec.$), we see that the outcomes obtained for the coordinates of the parachute deployment are a little different.

In the following table are displayed for comparison the results obtained using respectively the approximate method of integration (table 1 of exercise 3 in section 9.4), the Siacci method, and the numerical integration of differential equations of parachute fall (PC program Parach.Bas).

Table 1

Method	x meter	y meter	v	Angle Degree	t
Approximate	635 m	3654	68.71 m/s	-60.00	10 s
Siacci	653 m	3639	77.78 m/s	-58.54	10 s
Parach.Bas	637 m	3654	68.48 m/s	-59.96	10 s

9.6 The Numerical Integration of Differential Equations of Parachute Flight

Consider the differential equations of parachute flight in standard atmosphere, (9.1.19),

$$\begin{cases} \dfrac{dv_x}{dx} = -\dfrac{b}{m}h(y) \cdot v^2 \dfrac{1}{v} \\[2mm] \dfrac{dp}{dx} = -\dfrac{g}{v_x^2} \\[2mm] \dfrac{dt}{dx} = \dfrac{1}{v_x} \\[2mm] \dfrac{dy}{dx} = p \end{cases}$$

(9.6.1)

where

$$b = c_D \frac{\rho_0 \pi \cdot d^2}{8} = c_D \frac{\rho_0 A}{2}$$

(9.6.2)

and

$$h(y) = \left(\frac{289.08 - 0.006328y}{289.08}\right)^{4.4}$$

(9.6.3)

is the density function.

For the parachutist jumping from the plane with closed parachute

$$b=c\frac{\rho_0 A}{2}=(0.8)\frac{(1.205)(0.5)}{2}=0.241$$

(9.6.4)

To integrate numerically (9.6.1), we consider the following initial conditions:

$$x_0=0, \; y(0)=y_0, \; z(0)=0, \; t(0)=0, \; \alpha_0=0, \; p(0)=\tan\alpha_0=\tan(0)=0$$

$$v_{0x}=v_0\cos\alpha_0=v\cos(0)=v_0, \; v_{0y}=v_0\sin\alpha_0=v_0\sin(0)=0, \; v_{0z}=0$$

(9.6.5)

The QBasic PC program Parach.Bas is compiled using numerical methods integrating (9.6.1). It is valid as far as the falling angle is not less than (−90°).

The following example demonstrates the use of Parach.Bas to estimate the elements of the trajectory of a parachutist jumping from a horizontally flying airplane in standard atmosphere.

Example 1. A parachutist of mass $m=103kg$ jumps from an airplane that flies horizontally with a speed $v_0=115m/s$ at an altitude $y(0)=4000m$.

Find the elements of the trajectory when the parachutist is at the altitude $y(0)=3730m$ using Parach.Bas. Use

$$b=c\frac{\rho_0 A}{2}=(0.8)\frac{(1.205)(0.5)}{2}=0.241$$

.

Solution
  Parach.Bas PC Program

Input data:
Jumping altitude $y(0)=4000$; Altitude of parachute deployment $y=3790m$; Launching speed $v_0=115$; Launching angle $\alpha_0=0$; Initial x-coordinate $x_0=0$; Initial time $t_0=0$; Drag coefficient b = 0.241

Mass of parachutist (including equipment and jumping suit) $m=103$; Integration step $h_0=0.01$

Executing the program, we obtain the following data:

Data Output
At the altitude $y=3790m$, the horizontal distance from the launching point is $x=540m$; the speed, the falling angle, and the time of flight to the point are respectively $v=67.02m/s$, $\alpha_0=-48.04°$, $t=7.56s$.

Thus, the coordinates of parachutist at the moment he deploys the parachute are $x=540m$, $y=3790m$.

Note: In table 1 of example 3 in section 9.4, for the same altitude $y=3790m$, we have the following values: $x=538m$, $v=67.09m/s$, $\alpha_0=-48°$, $t=7.5s$.
We see that the results obtained using the approximate integration (example 2 of section 9.4) are quite identical to the results obtained by numerical integration of differential equations of flight.

Example 2. (Parachutist fails to deploy the parachute). A parachutist of mass $m=100kg$ jumps from an airplane that flies horizontally at an altitude $y(0)=1200m$, with speed $v_0=80m/s$.

Use PC program to find the following:

(a) The elements of the trajectory when the parachutist is at the altitude $y(0)=300m$
(b) The elements of the trajectory if the parachute fails to deploy at $y(0)=300m$

$$b=c\frac{\rho_0 A}{2}=(0.8)\frac{(1.205)(0.5)}{2}=0.241$$
.

Solution
   Parach.Bas PC Program

(a) Input data:

   Jumping altitude $y(0)=1200$; Altitude of parachute deployment $y=300m$
   Launching speed $v_0=80$; Launching angle $\alpha_0=0$;
   Initial x-coordinate $x_0=0$; Initial time $t_0=0$;
   Drag coefficient $= 0.241$
   Mass of parachutist (including equipment and jumping suit) $m=100$; Integration step $h_0=0.01$

Executing the program, we obtain the following data:

Data Output
At the altitude $y=300m$, the horizontal distance from the launching point is $x=553m$; the speed, the falling angle, and the time of flight to the point are respectively $v=64.50m/s$, $\alpha_0=-84.7°$, $t=19s$.

   Thus, the coordinates of parachutist at the moment he deploys the parachute are $x=553m$, $y=300m$.

(b) Input Data

   The input data are the same as in (a), but instead of altitude of deployment $y=300m$, the input is $y=0m$.

Data Output
On the ground $y=0m$, the horizontal distance from the launching point is $x=573m$; the speed, the falling angle, and the time of flight to the point are respectively $v=64.20m/s$, $\alpha_0=-87°$, $t=24s$.

   Thus, the parachutist at the moment he falls on the ground has a relatively high speed, $v=64.20m/s$.

Comment. The speed of impact on the ground is the same as the speed the parachutist will have if he falls in absence of air resistance from a high building of

$$h=v^2/2g=(64.20)^2/(2\cdot9.80665)=210m$$.

9.7 Skydiving in a Nonstandard Atmosphere and in Presence of Wind

To study the fall of the parachutist in nonstandard atmosphere, we will adapt the results obtained in chapter 8.

Considering the differential equations of parachute flight, (9.5.1), and the differential equations of projectile flight in nonstandard atmosphere, (8.5.26), we can write the differential equations of a parachutist's fall in nonstandard atmosphere as

$$\begin{cases} \dfrac{dv_x}{dx}=-\dfrac{b}{m}h_\tau(y)\cdot v^2\dfrac{1}{v} \\[2mm] \dfrac{dp}{dx}=-\dfrac{g}{v_x^2} \\[2mm] \dfrac{dt}{dx}=\dfrac{1}{v_x} \\[2mm] \dfrac{dy}{dx}=p \end{cases} \qquad (9.7.1)$$

where

$$h_\tau(y)=\frac{\rho}{\rho_0}=(\frac{\tau_0-0.006328y}{\tau_0})^{4.4}, \quad \tau_{0N}=289.08K,$$

$$\tau_0=\frac{T_0}{1-0.3785(e_0/p_0)} \qquad (9.7.2)$$

and

$$b=c\frac{\rho_0 A}{2}=(0.8)\frac{(1.205)(0.5)}{2}=0.241.$$

The only changes (with respect to nonstandard atmosphere) we consider are the changes in temperature, pressure, and wind.

To take into account the influence of the range wind, we substitute in (9.7.1)

$$v = [(\vec{v} - \vec{w}_x)^2]^{1/2} = [(v^2 + w_x^2 - 2v \cdot w_x \cdot \cos \alpha)]^{1/2}, \qquad (9.7.3)$$

since the drag force depends on the relative velocity of air.

The QBasic PC program Paramet.Bas is compiled using numerical methods integrating (9.7.1).

The following example demonstrates the use of Paramet.Bas to estimate the elements of the parachutist trajectory jumped from a horizontally flying airplane in standard atmosphere.

Example 1. A parachutist of mass $m=100kg$ jump from an airplane that flies horizontally with a speed $v_0=100m/s$ at an altitude $y(0)=2500m$. The parachutist deploys his parachute when he/she is $y=1200m$ above sea level.

Find the following:

(a) The elements of the trajectory of flight at the altitude $y=1200m$
(b) The elements of the trajectory after $t=8s$

Input Data
Input: Altitude of Launching point [m] = 2,500
Input: Altitude of parachute deployment [m] = 1,200
Launching Speed [m/s] = 100,
Launching Angle [Degree]: 0,
Launching x-coordinate: 0,
Input: Initial Time = 0;
Input: Time of Interest [s] = 8
Input: Temperature of Air = 5
Input: Atmospheric Pressure = 760
Input: Range Wind [m/s] = 10
Input: Coefficient b = 0.241

Input: Mass of Parachutist [kg] = 100
Input: Integration Step h0 = 0.01

Results
(a) x-coordinate of Parachutist = 895m; y-coordinate of Parachutist= 1200; Speed of Parachutist = 66m/s; Falling Angle = -87.5 degree; Time of flight = 24.24.
(b) At time $t = 8$; Parachutist Location is (512, 2269); Corresponding Speed = 58m/s; Corresponding Angle = -57.77.

The coefficient (b/m) is = 0.00241.

9.8 The Influence of Crosswind

Estimation of the deviation of the parachutist trajectory due to the crosswind $\vec{w}_z$ requires the solution of the vector differential equation of flight in three dimensions

$$\frac{d\vec{v}}{dt}=\vec{g}-c{\cdot}h(y)K_D(V)\frac{\vec{v}}{v}$$
(9.8.1)

where V , given by the equation

$$V^2=(\vec{v}-\vec{w}_z)^2$$
(9.8.2)

is the relative speed of projectile with respect to range wind.

To simplify the solution, we will adapt the results obtained in section 6.6 for the crosswind.

In other words, we will solve numerically the differential equations with respect to an observer moving along the z-axis with velocity of wind, i.e., the equations

$$
\begin{cases}
\dfrac{dv_x}{dx}=-\dfrac{b}{m}h_\tau(y)\cdot\dfrac{K(v)}{v} \\[2mm]
\dfrac{dp}{dx}=-\dfrac{g}{v_x^2} \\[2mm]
\dfrac{dt}{dx}=\dfrac{1}{v_x} \\[2mm]
\dfrac{dy}{dx}=p
\end{cases}
$$

(9.8.3)

with initial conditions

$$
v'_0=(v_0^2+w_z^2)^{1/2}, \quad \tan\alpha'_0=\dfrac{v_0\sin\alpha_0}{v_0\cos\alpha_0(1+w^2/(v_0^2\cos^2\alpha_0))^{1/2}}
$$

(9.8.4)

where

$$
b=c\dfrac{\rho_0 A}{2}=(0.8)\dfrac{(1.205)(0.5)}{2}=0.241 \ .
$$

To find the actual coordinates of the parachutist, we will use the formulas results of section 6.7:

$$
\tan\varphi=-\dfrac{w_z}{v_0\cos\alpha_0} \ .
$$

(9.8.5)

Note: The angle φ is negative since the rotation of the firing plane is clockwise.

In figure 1, we see that the relative deviation of the projectile in the direction of the z-axis is

$$
z_2(t)=x\cdot\tan\varphi=-\dfrac{w_z}{v_0\cos\alpha_0}x \ .
$$

(9.8.6)

$$z_2(t)=z(t)-w_z \cdot t \ , \tag{9.8.7}$$

the z-coordinate of the projectile with respect to the fixed system of coordinates is

$$z(t)=w_z t-\frac{w_z}{v_0 \cos\alpha_0}x \ . \tag{9.8.8}$$

Example 1. A parachutist of mass $m=100kg$ jumps from an airplane that flies horizontally with a speed $v_0=100m/s$ at an altitude $y(0)=2500m$. The parachutist deploys his parachute when he/she is $y=1200m$ above sea level.

Find the coordinates of parachutist at the moment of parachute deployment if there is only a crosswind that blows with velocity $w_z=10m/s$. Use the PC program Parach.Bas and the value

$$b=c\frac{\rho_0 A}{2}=(0.8)\frac{(1.205)(0.5)}{2}=0.241 \ .$$

Solution

Initial conditions:
    Initial speed is

$$v'_0=(v_0^2+w_z^2)^{1/2}=(100^2+10^2)^{1/2}=100.50$$

while the initial angle is $\alpha'_0=0$, since

$$\tan\alpha'_0=\frac{v_0 \sin\alpha_0}{v_0 \cos\alpha_0 (1+w^2/(v_0^2 \cos^2\alpha_0)^{1/2}}=0 \ .$$

Using the PC program Parach.Bas:

Input Data
Altitude of Launching point [m] = 2,500
Input: Initial Time = 0; Input: Launching Angle [Degree]: 0,
Launching Speed [m/s] = 100.5
Input: Altitude of parachute deployment [m] = 1,200
Input: Time of Interest [s] = 8

Results: In relative coordinate system, moving with the velocity of wind.

x-coordinate of Parachutist =719; y-coordinate of Parachutist=1200; Speed of Parachutist =68.08; Falling Angle = -86.89; Time of Flight = 24.64.

The z-coordinate of the parachutist with respect to the launching plane is

$$z(t)=w_z t-\frac{w_z}{v_0\cos\alpha_0}x=(10)\cdot(24.64)-\frac{10}{100}\cdot(719.2)=174.28m$$

Thus, the coordinates of parachutist at the moment of deployment are $x=719m$, $y=1200m$, $z=174m$.

Example 2. For the parachutist of example 1, find the coordinates at the moment $t=7s$

Use the PC program Parach.Bas.

Solution
The initial conditions are the same:
Initial speed is $v'_0=100.50$ and the initial angle is $\alpha'_0=0$

Using Parach.Bas:

Input Data
Altitude of Launching point [m] = 2,500
Input: Initial Time = 0
Input: Launching Angle [Degree]: 0
Launching Speed [m/s] = 100.5
Input: Altitude of parachute deployment [m] = 1,200
Input: Time of Interest [s] = 7

Results: (In relative coordinate system) moving with the velocity of wind

   At time $t=7s$:

   x-coordinate of Parachutist = 444; y-coordinate of Parachutist = 2314; Speed of Parachutist = 60.86; Falling Angle = -50.05.

   The z-coordinate of the parachutist with respect to the launching plane is

$$z(t)=w_z t-\frac{w_z}{v_0 \cos\alpha_0}x=(10)\cdot(7)-\frac{10}{100}(444)=55.6m$$

.

   Thus, the coordinates of parachutist at $t=7s$ are $x=414m$, $y=2314m$, $z=55.6m$.

Example 3. Find the coordinates of parachutist of example 1 at $t=7s$ if there is a wind with speed $w=12m/s$ that blows under an angle $a=30°$ with respect to the positive direction of the flying plane. The temperature of air and the atmospheric pressure are respectively $T=0°celsius$ and $P=765mm\ Hg$.

   Use the PC program Paramet.bas.

Solution
   The range wind and the crosswind are respectively

$w_x=w\cos(a)=12\cdot\cos(30°)=10.39m/s$, $w_z=w\cos(a)=12\cdot c\sin(30°)=6m/s$

Initial speed due to the crosswind is

$$v'_0 = (v_0^2 + w_z^2)^{1/2} = (100^2 + 6^2)^{1/2} = 100.18 \approx 100,$$

while the initial angle is $\alpha'_0 = 0$.

Input Data
Input: Altitude of Launching point [m] = 2,500
Input: Altitude of parachute deployment [m] = 1,200
Input: Launching Speed [m/s] = 100
Input: Launching Angle [Degree]: 0
Input: Initial x-coordinate = 0
Input: Initial Time = 0
Input: Time of Interest [s] = 7
Input: Temperature of Air = 0
Input: Atmospheric Pressure = 765
Input: Mass of Parachutist [kg] = 100
Input: Pressure of Vapor = 6.35
Input: Range Wind [m/s] = 10
Input: Integration Step h0 = 0.01

Results: (In the relative coordinate system) moving with the velocity of crosswind
  At time $t=7s$:

x-coordinate of Parachutist = 467; y-coordinate of Parachutist= 2319; Speed of Parachutist = 57; Falling Angle = -52.77

At time $t=7s$, the z-coordinate of the parachutist with respect to the launching plane is

$$z(t) = w_z t - \frac{w_z}{v_0 \cos \alpha_0} x = (6).(7) - \frac{6}{100}.(476) = 14m.$$

Thus, the coordinates of parachutist at $t=7s$ are $x=467m$, $y=2311m$, $z=13.44m$.

The deviation of parachutist from launching plane (*xoy*): Employing (9.8.5), we find that

$$\tan\varphi = -\frac{w_z}{v_0\cos\alpha_0} = -\frac{6}{100} = -0.06 \ .$$

Hence,

$$\varphi = \tan^{-1}(-0.06) = -3.43° \ .$$

9.9 The Vertical Fall of the Parachutist

The vertical fall of a parachutist is a common activity in military and sports training.

Such scenario we encounter in the following instances:

- The parachutist jumps from the parachute tower or from a helicopter hovering above the ground. The vertical fall from a fixed location (tower) cannot be studied using the systems of differential equations we have seen in preceding sections.
- The parachutist falling with closed parachute forms an angle close to $\alpha=-90°$.

The vertical fall of the parachutist is performed assuming that the initial velocity $\vec{v}_0$ is directed vertically downward, in the opposite direction of the *y*-axis, in standard weather conditions and in absence of wind.

We will study the vertical fall of the parachutist for the parachutist falling vertically with fully deployed parachute. We apply the results obtained for the parachutist falling vertically with closed parachute.

Consider the differential equation of parachutist flight (9.1.13) that for our scenarios can be written in the form

$$\frac{d\vec{v}}{dt} = \vec{g} - B_1 v^2 \frac{\vec{v}}{v} \ ,$$

$$(9.9.1)$$

where

$$B_1 = \frac{b}{m}h(y) = c\frac{\rho_0\pi\cdot d^2}{8m}h(y) = c\frac{\rho_0 A}{2m}h(y) \qquad b = c\frac{\rho_0 A}{2}$$

The forces exerted on the parachutist have the direction of the y-axis since no force is acting along x- and z-axes and since at the time the parachutist jumps from helicopter at rest or from a tower the components of initial velocity along x- and z-axes are zero. So the motion of the parachutist is vertically downward along the y-axis. The velocity of the parachutist at any moment forms an angle $\alpha = -\pi$ with the y-axis

Projecting the above vector differential equation, we have

$$\begin{cases} \dfrac{dv_x}{dt} = 0 \\ \dfrac{dv}{dt} = -g + B_1 v^2 \\ \dfrac{dx}{dt} = 0 \\ \dfrac{dy}{dt} = -v \end{cases} \qquad (9.9.2)$$

Expressing the above system through variable y, we get

$$\begin{cases} \dfrac{dv}{dy} = \dfrac{-g + B_1 \cdot v^2}{v} \\ \dfrac{dt}{dy} = -\dfrac{1}{v} \end{cases} \qquad (9.9.3)$$

where

$$B_1 = \frac{b}{m}h(y), \text{ and } b = c\frac{\rho_0\pi\cdot d^2}{8} = c\frac{\rho_0 A}{2}.$$

The system of equations (9.9.3) can be integrated considering the initial conditions

$$v(y)=v_0 , \quad t(y)=t_0.$$

We consider $h(y)$ to be a constant equal to an average value

$$h(y)=h(\bar{y}),$$

where $\bar{y}=(y_0+y)/2$.

Thus, B is also constant

$$B_1=\frac{b}{m}h(\bar{y})=c\frac{\rho_0 \pi \cdot d^2}{8m}h(\bar{y})=c\frac{\rho_0 A}{2m}h(\bar{y}).$$

(9.9.4)

We assume the acceleration of the parachutist along the y-axis to be zero at a certain altitude y_e during the flight, i.e.,

$$\frac{dv}{dy}=(-g+B_1 \cdot v^2)/v=0$$

(9.9.5)

At that point (y_e), the motion of the parachutist will have a theoretical maximum value $v_e=v(y_e)$ determined by equation (9.9.5), i.e.,

$$g-B_1 \cdot v^2=0.$$

Hence, since the speed is positive, we find that

$$v_e=(g/B_1)^{1/2},$$

(9.9.6)

where

$$B_1=\frac{b}{m}h(y), \text{ and } b=c\frac{\rho_0 \pi \cdot d^2}{8}=c\frac{\rho_0 A}{2}.$$

Studying the sign of the second derivative of the speed v,

$$\frac{d^2v}{dy^2}=(B_1 \cdot v^2 - g)/v^2$$

$$\text{(9.9.7)}$$

we find that v_e is a relative minimum. Integrating the first equation of (9.9.3), we find the speed of the parachutist as a function of altitude y

$$v^2 = \frac{g}{B_1}+(v_0^2 - \frac{g}{B_1}) \cdot e^{2 \cdot B(y-y_0)}$$

$$\text{(9.9.8)}$$

Using (9.9.6), we can write (9.9.8) in the following form:

$$v^2 = v_e^2 +(v_0^2 - v_e^2) \cdot e^{2 \cdot g(y-y_0)/v_e^2}$$

$$\text{(9.9.9)}$$

Equation (9.9.8) shows that when y decreases ($y \rightarrow -\infty$) the speed of the parachutist becomes again equal to v_e, given in (9.9.6).

Taking into account the relative minimum velocity and the fact that again the velocity tends asymptotically to the same value, we can conclude the following:

- After the parachutist passes through the point where the speed reaches the minimum absolute value v_e, it seems that the parachutist's speed increases and takes a relative maximum value. After that, falling down, the speed of the parachutist practically becomes again v_e.

The following example 1 shows that practically after falling a few meters from the launching point, the speed of the parachutist takes the minimum value. At that point, the gravity and drag force become equal. Because of inertia, the speed the parachutist's fall tends to increase. A slight increase of speed causes an immediate increase in drag function that tends to bring the drag equal to gravity.

Practically we can say that after the parachute is deployed, the fall of parachutist is a motion with speed equal to

$$v_e = (g/B_1)^{1/2} = (\frac{mg}{b \cdot h(y)})^{1/2} .$$
(9.9.10)

Integrating the second equation of (9.9.3), considering the parachutist speed constant and equal to the speed given in (9.9.6), we find the time the parachutist will be at an altitude y in absence of wind

$$t = (y_0 - y) \cdot (g/\overline{B}_1)^{-1/2} = (y_0 - y)/\overline{v}_e ,$$
(9.9.11)

where

$$v_e = (g/\overline{B}_1)^{1/2} = (\frac{mg}{b \cdot h(\overline{y})})^{1/2} .$$
(9.9.12)

Note 1: The obtained results can be applied to the parachutist falling vertically without parachute.

Note 2: Consider a parachutist that jumps from a tower (or hovering helicopter) with initial speed $v_0 = 0$. Substituting $v_0 = 0$ in (9.9.9), we find that the speed of parachutist at an altitude y (in absence of wind) is

$$v^2 = v_e^2 (1 - e^{2 \cdot g(y - y_0)/v_e^2}) .$$
(9.9.13)

In this case, the speed given in (9.9.12) is a relative maximum speed of the parachutist.

Note 3: The following example shows that we can assume that, from the moment that the parachute is fully deployed, the parachutist falls down with an instantaneous speed estimated by the formula

$$v_e = (\frac{mg}{b \cdot h(y)})^{1/2} , \text{ or } v_e = \sqrt{2mg/(\rho_0 c \cdot A \cdot h(y))} .$$

The speed of parachutist at the deployment stage becomes equal to the minimum (limiting) speed v_e after he/she falls down a few meters from the altitude where he/she starts to deploy his/her parachute.

Example 1. The parachutist falling down instantaneously deploys the parachute when she is at an altitude $y_0=2000m$ and her speed is $v_0=70m/s$.

Find the speed of the parachutist at the following altitudes: (a) $y=1999m$, (b) $y=1998m$; (c) $y=1997m$, (d) $y=1996m$; (e) $y=1995m$, and (f) $y=1990m$.

The diameter of the parachute is $d=7.5m$. Mass of the parachutist is $m=100m$.

Solution

We estimate

$$b=c\frac{\rho_0 \pi \cdot d^2}{8}=(1.33)\frac{(1.205)\cdot\pi(7.5)^2}{8}=35.40$$

and

$$h(\bar{y})=(\frac{\tau_0-0.006328\bar{y}}{\tau_0})^{4.4}=(\frac{289.08-0.006328(2000)}{289.08})^{4.4}=0.8212.$$

We have

$$B_1=\frac{b}{m}h(\bar{y})=\frac{35.40}{100}0.8212=0.2907$$

and

$$v_e^2=g/B_1=(9.80665)/(0.2907)=33.734.$$

The limiting speed is

$$v_e=\sqrt{33.734}=5.808m/s.$$

Substituting the value of $v_e^2 = 33.734$ in (9.9.9), we have

(a) For $y=1999m$,

$$v^2 = v_e^2 + (v_0^2 - v_e^2) \cdot e^{2 \cdot g(y-y_0)/v_e^2} = 33.734 + (70^2 - 33.734) \cdot e^{2 \cdot 9.80665 \cdot (1999-2000)/33.734} = 49.98$$

Hence, for the parachutist speed, we have

$$v = (49.98)^{1/2} = 7.069 m/s$$

In the same way, we find

(b) For $y=1998m$: $v=5.813$
(c) For $y=1997m$: $v=5.808$
(d) For $y=1996m$: $v=5.808$
(e) For $y=1995m$: $v=5.808$
(f) For $y=1990m$: $v=5.808$

Time of Flight
(c) The time the parachutist will be at the altitude $y=1997m$ is

$$t = (y_0 - y)/\bar{v} = (2000 - 1997)/(37.94) = 0.08s,$$

considering that the parachutist falls with an average speed

$$v = (v_0 + v_e)/2 = (70 + 5.808)/2 = 37.94 m/s$$

Note: This example shows that falling down 2–3 m from the initial altitude $y_0 = 2000m$, the speed of parachutist becomes equal to the minimum (limiting) speed

$$v_e = 5.808 m/s$$

We can assume that at the moment that the parachute is fully deployed the parachutist falls down with an instantaneous speed estimated by the formula,

$$v_e = (\frac{mg}{b \cdot h(y)})^{1/2}, \text{ or } v_e = \sqrt{2mg / (\rho_0 c \cdot A \cdot h(\overline{y}))}$$

Example 2. The parachutist of example 1 jumps from a tower $y_0 = 50m$ high. Find the speed of parachutist at altitudes: (a) $y=48m$, (b), $y=46m$ (c) $y=44m$, (d), $y=42m$ (e) $y=40m$, (f) $y=38m$, and (g) $y=36m$.

Use

$$b = c\frac{\rho_0 \pi \cdot d^2}{8} = (1.33)\frac{(1.205) \cdot \pi (7.5)^2}{8} = 35.40$$

Solution

The maximum (limit) speed is

$$v_e = (\frac{mg}{b \cdot h(y)})^{1/2} = [\frac{100 \cdot (9.80665)}{35.4 \cdot (1)}]^{1/2} = 5.263$$

(a) For $y=48m$, employing (9.9.13), we have

$$v^2 = v_e^2 (1 - e^{2 \cdot g(y-y_0)/v_e^2}) = 27.70(1 - e^{2(9.80665) \cdot (48-50)/27.70}) = 20.97$$

The speed is

$$v = 4.58 m/s.$$

In the same way, we find

b) For $y=48m$: $v=4.58$

(c) For $y=46m$: $v=5.106$

(d) For $y=42m$: $v=5.23$

(e) For $y=40m$: $v=5.25$

(f) For $y=38m$: $v=5.26$ g. For $y=36m$: $v=5.26$

Note: The parachutist reaches the maximum speed falling approximately 12 m down from the launching point.

From that point, the motion of the parachutist can be considered uniform, with a speed $v=5.26$.

9.10 The Parachute Deployment

A parachutist jumping from an airplane deploys the parachute at a safe distance, at a certain altitude, after he has fallen for a few seconds with closed parachute.

To study the parachutist's fall, we assume that in average, the time needed to fully deploy the parachute is 3 sec.

We assume as well that the trajectory of the parachutist during the process of deployment is a vertical line.

Indeed, during the deployment period, the form of parachute varies; and as a result, the drag coefficient of the parachutist increases from the value it had when the parachute is closed to the value it has when the parachute is fully deployed.

The drag force increases as well, while the parachute speed decreases at least ten times.

Since the mass of parachutist remains unchanged, the gravity force remains the same as well. The normal component of the gravity force causes the normal acceleration of projectile,

$$a_N = v^2/r ,$$

where r is the radius of curvature.

In such conditions, the above formula shows that the radius of curvature is proportional to the square of the parachutist speed. That means that the radius of curvature of the trajectory decreases and the trajectory tends to become a vertical line.

To illustrate the parachutist fall to the ground with the parachute fully deployed, we consider the following example:

Example 1. A parachutist of mass 100 kg jumps from a horizontal airplane flying at an altitude of 2,340 m, with speed 100 m/s. He deploys the parachute after 10 sec.

Find the position of the parachutist

(a) at   the   moment   he   opens   the   parachute

$$(b=c\frac{\rho_0 A}{2}=(0.8)\frac{(1.205)(0.5)}{2}=0.241),$$

(b) at the moment the parachute is fully deployed, and

(c) at the time the parachutist lands on the ground, $y=0$.

Assume an atmosphere in standard conditions. No presence of wind.

The parachutist b value with parachute fully deployed is

$$b=c\frac{\rho_0 \pi \cdot d^2}{8}=(1.33)\frac{(1.205)\cdot\pi(7.5)^2}{8}=35.40$$

Solution

(a) Using PC program Parach.Bas ($b=0.241$).

Input Data

Initial x-coordinate = 0; Initial Altitude = 2,340 m; Initial time = 0; Initial Angle = 0; Initial Speed = 100; Altitude of Deployment (guessed altitude) = 1,800; Time of Interest = 10; Mass of Parachutist = 100; Integration Step = 0.01

Results

At time t=10 sec., the parachutist is at the point with coordinates (535 m, 2,003 m). The speed and the falling angle are respectively 63 m/s and -64.445°.

(b) We will consider that after 3 sec., the parachute is fully deployed and that the speed of parachutist drops from $v=63m/s$ to the minimum value given by (9.8.12), i.e., to the value

$$v_e=(\frac{mg}{b\cdot h(\bar{y})})^{1/2}=(\frac{100\cdot(9.80665)}{35.40(0.82095)})^{1/2}=(33.7443)^{1/2}=5.809m/s,$$

since

$$h(y)=(\frac{\tau_0-0.006328y}{\tau_0})^{4.4}=(\frac{289.08-0.006328(2003)}{289.08})^{4.4}=0.82096.$$

We assume as well that the trajectory of parachutist is a vertical line.

During 3 sec., the average acceleration of parachutist is

$$\bar{a}=(v-v_0)/dt=(5.809-63)/3=-19\cdot064m/s^2.$$

Thus, the parachutist has fallen down approximately

$$dy=v_0t+at^2/2=(63)\cdot(3)-19.0648(3)^2/2=103.2m.$$

The y-coordinate of parachutist at the time the parachute is fully deployed is

$$y=2003-103=1900m.$$

(b) The coordinates of parachutist at time $t=13$ sec., when his parachute is fully deployed, are (535, 1900).
(c) The instantaneous speed of the parachutist,

$$v=(\frac{mg}{b\cdot h(y)})^{1/2},$$

is a function of altitude.

We can write

$$v_e=(\frac{mg}{b\cdot h(y)})^{1/2}=[\frac{100\cdot(9.80665)}{35.4}]^{1/2}\cdot\frac{1}{h^{1/2}(y)}=5.263\cdot(\frac{289.08}{289.08-0.006328\cdot(y)})^{2.2}.$$

Substituting in the second equation of (9.8.2), we have

$$\frac{dt}{dy} = \frac{1}{v_e} - 0.19(\frac{289.088 - 0.006328 \cdot y}{289.08})^{2.2} .$$

Integrating for the time of impact to the ground, we have

$$\Delta t = -0.19 \int_{1900}^{0} (\frac{289.08 - 0.006328 y}{289.08})^{2.2} dy .$$

Using a TI-83 Plus graphing calculator, we find that

$$\Delta t = -0.19 \int_{1900}^{0} (\frac{289.08 - 0.006328 y}{289.08})^{2.2} dy = (0.19) \cdot (1814.5) = 345s .$$

Thus, the total time of flight is about

$$t = 10 + 3 + 345 = 391.26 \text{sec} = 358s = 5.96 \text{minutes} .$$

Note: At the time of landing, the parachutist has deviated in x-direction 535 m from the point where he jumped from the airplane.

9.11 The Influence of Wind in Vertical Fall

The influence of wind (range wind or crosswind) in vertical fall of parachutist can be studied using a Cartesian coordinate system that moves with the velocity of wind w_x (in direction of the x-axis).

For an observer that moves with the wind, the fall of parachutist is vertical and is described by system (9.8.3) as

$$\begin{cases} \dfrac{dv}{dy} = \dfrac{-g + B_1 \cdot v^2}{v} \\ \dfrac{dt}{dy} = -\dfrac{1}{v} \end{cases} \tag{9.11.1}$$

Suppose that at a time t the y-coordinate of parachutist is $y = y(t)$.

At time t, the parachutist (with respect to a fixed system of coordinates) is at the point with coordinates $x=w_x t$, $y=y(t)$.

Example 1 illustrates the way to solve the problem.

Example 1. For the parachutist of example 1, section 9.9, find the coordinates of the parachutist if there is a range wind of $w_x=10m/s$.

The parachutist with the mass of $m=100$ kg jumps from the plane flying horizontally at an altitude of 2,340 m with a speed of $v=100m/s$. He pulls the cord to open the parachute after 10 sec. The parachute is fully deployed after 3 sec.

Assume the following b values of the parachutist: $b=0.241$ and $b=35.40$, respectively, when falling with closed parachute and with deployed parachute.

Find as well the speed of parachutist at the moment he lands on the ground (sea level), considering that the speed of wind on the ground is $w_g=3.5m/s$.

Note: As a matter of fact, the speed of wind near the ground is much lower than the average speed of the wind (ballistics wind).

To estimate the speed of impact, we need to measure the speed of wind near the ground.

Solution
(a) Using the PC program Paramet.bas ($b=0.241$), we have:

Input Data
Input: Launching Altitude [m] = 2,340
Input: Altitude of Parachute Deployment [m] (guessed)= 1,500
Input: Time of Interest [s] = 10
Input: Ballistics Constant [cd] = 0.241
Input: Initial x-coordinate
Input: Initial Time = 0
Input: Launching Angle [Degree]: 0,
Input: Launching Speed [m/s] = 100
Input: x-coordinate of a Point on Trajectory [m] = 400
Input: Range Wind [m/s] = 10
Input: Temperature of Air = 15

Input: Atmospheric Pressure = 750
Input: Pressure of Air Vapor = 6.35
Input: Mass of Parachutist [kg] = 100
Input: Integration Step h_0 = 0.01

Results:
    At time $t=10s$:

x-coordinate of parachutist = 599; y-coordinate of parachutist, 1999;
Speed of Parachutist, 63; Falling Angle, -65.76

(b) We will consider that after 3 sec., the parachute is fully deployed
    and the speed of parachutist drops to the limiting (minimum) value
    given in (9.8.12),

$$v_e=(\frac{mg}{b\cdot h(\bar{y})})^{1/2}=(\frac{100\cdot(9.80665)}{35.40(0.82227)})^{1/2}=5.804m/s$$

since

$$h(y)=(\frac{\tau_0-0.006328y}{\tau_0})^{4.4}=(\frac{289.08-0.006328(1987)}{289.08})^{4.4}=0.822$$

We consider as well that the trajectory of the parachutist is a
vertical line.
    During 3 sec., the average acceleration of the parachutist is

$$\bar{a}=(v-v_0)/dt=(5.804-67.89)/3=-20.70m/s^2$$

Thus, the parachutist has fallen down approximately

$$dy=v_0t+at^2/2=(67.89)\cdot(3)-20.70(3)^2/2=110.5m$$

Thus, the y-coordinate of parachutist at the time the parachute is
fully deployed is

$$y=1987-110.5=1876.5m$$

Because of the range wind, the parachutist has moved in the direction of airplane flight for

$$dx=(10)\cdot(3)=30m$$.

Thus, the parachutist, at the end of parachute deployment process, is at the point with coordinates

$$x = 599+30 = 629m, \quad y = 1999-110 = 1889m.$$

In absence of wind, the trajectory of parachutist from the moment the parachute is fully deployed till he lands on the ground can be considered vertical.

Since there is a range wind, the trajectory is not vertical, and we have to estimate the influence of wind in the falling of parachutist with parachute.

(c) Time of flight from the moment the parachute is fully deployed

Considering the instantaneous speed of parachutist to be a function of altitude,

$$v_e = (\frac{mg}{b \cdot h(y)})^{1/2} = 5.263 \cdot h(y)^{-1/2} = 5.263 \cdot (\frac{289.08}{289.08 - 0.006328 \cdot y})^{0.5},$$

since

$$(mg/b)^{1/2} = [(100)\cdot(9.80665)/(35.40)] = 5.263$$.

Substituting v_e in the second equation of (9.8.2), we have

$$\frac{dt}{dy} = 0.19 \cdot (\frac{289.08 - 0.006328 \cdot y}{289.08})^{2.2}$$.

Integrating and using a TI-83 Plus, we find

$$\Delta t = -0.19 \int_{1877}^{0} (\frac{289.08 - 0.006328 y}{289.08})^{2.2} dy = (0.19)\cdot(1793.6) = 341s$$.

Because of the range wind, the parachutist deviates in the direction of wind by a quantity

$$\Delta x = w_x t = (10) \cdot (341) = 3410 m$$
.

Thus, the total time of flight is

$$t = 10 + 3 + 341 = 354 \text{sec} = 5.9 \text{minutes}$$
.

The coordinates of parachutist at the time he lands on the ground are respectively

$$x = 629 + 3410 = 4,039 mm.$$

Thus, the parachutist has fallen from the altitude 2,340 m above sea level and has deviated in the direction of airplane flight for $x = 4029 m$.

(c) The speed of parachutist at the moment of impact

The vertical speed of parachutist is

$$v_e = (\frac{mg}{b \cdot h(y)})^{1/2} = (\frac{100 \cdot (9.80665)}{35.40 \cdot (1)})^{1/2} = 5.26 m/s$$
.

The velocity of wind near the point of impact in the direction of the x-axis is $w_g = 3.5 m/s$.

The landing speed of parachutist is

$$v = (v_e^2 + w_g^2)^{1/2} = 6.16 m/s$$
.

9.12 The Free-Fall Drag Coefficient of a Skydiver. The Limiting Speed

The practice of parachute falling is related to a three-stage process:

(1) Jumping from airplane with closed parachute (the so-called free-fall period)
(2) Parachute deployment (the intermediate period)
(3) Falling and landing with parachute fully deployed (parachute landing)

The free-fall period usually is related with a stable face-to-earth body position (known also as the box position).

Parachute deployment stage lasts normally 2–3 sec. till the parachute is fully deployed. At that instant the parachutist trajectory (in absence of wind) can be considered vertical, and the speed can be considered equal to the limiting speed.

The optimum and limiting speed are basic parameters to determine the flight of parachutist at any stage.

For the free fall stage the minimum speed and the limiting speed (see 9.4) are determined respectively by the equations

$$v_m^2 = \frac{v_0^2(1+p_m^2)}{1-Bv_0^2 J(p_m)}$$

$$(9.12.1)$$

and

$$v_l^2 = \frac{mg}{b \cdot h(y)},$$

$$(9.12.2)$$

where

$$J(p_m) = p_m\sqrt{1+p_m^2} + \ln(p_m + \sqrt{1+p_m^2}),$$

$$(9.12.3)$$

$$B = \frac{b}{mg} h(y),$$

$$(9.12.4)$$

and

$$b = c_D \frac{\rho_0 A}{2}.$$

(9.12.5)

An important conclusion, useful for the practice of flight, is the fact that the limiting speed of the parachutist is independent from the launching speed v_0, while the minimum (optimum) speed, determined in (9.12.1) depends on the launching speed (that practically is the speed of airplane).

As the parachutist speed approaches the limiting speed, the falling angle approaches to $90°$. For such angles, the systems of equations of flight (9.12.16), (9.12.18), (9.12.19), and (9.12.20) are not more valid.

To study the parachutist fall beyond the point where the launching angle is too close to $90°$, we have to use system (9.8.2) or system (9.8.3).

Substituting $v_l = [mg/(bh(y))]^{1/2}$ in (9.12.2), we can write the limiting speed in the following form

$$v_l = \left(\frac{2mg}{\rho_0 \cdot c_D A \cdot h(y)} \right)^{1/2},$$

(9.12.6)

where

$$h(y) = \frac{\rho}{\rho_0} = \left(\frac{289.08 - 0.006328 y}{289.08} \right)^{4.4}.$$

(9.12.7)

The limiting speed (9.12.6) depends on the parachutist's mass m, cross-section area A, coefficient of air resistance c_D, and the altitude of parachutist y (density function $h(y)$, in standard atmosphere).

For the parachutist falling with closed parachute, let's say in face-to-earth position, we can determine experimentally the product $c_D A$. For example, measuring the limiting speed of a parachutist by measuring the fall distance in a short time interval, we can find the unknown product $c_D A$ and then the parameter (5), i.e.,

$$b=\rho_0(c_D A)/2 .$$

Example 1. Experiments show that the limiting speed of a parachutist with mass $m=80kg$ at 1,220 m above sea level is 57.5 m/s. Find the product $c_D \cdot A$ of the given parachutist and the parameter $b=\rho_0(c_D A)/2$.

Solution

To use the formula (9.12.6), we estimate

$$h(y)=(\frac{289.08-0.006328(1220)}{289.08})^{4.4}=0.8877 .$$

Substituting in (9.12.4), we have

$$57.5=(\frac{2\cdot(80)\cdot(9.80665)}{1.205\cdot(c_d A)(0.8877)})^{1/2} .$$

Hence, we find that

$$c_D A=0.443 .$$

The parameter b for the given parachutist is

$$b=\rho_0(c_D A)/2=1.205(0.443)/2=0.2669 .$$

Example 2. Use the results of example 1 to find the limiting speed of the same parachutist at the altitude 2,240 m.

Solution

We have

$$h(y)=(\frac{289.08-0.006328(2240)}{289.08})^{4.4}=0.8015 .$$

The limiting speed is

$$v_l = (\frac{2mg}{\rho_0 \cdot c_d A \cdot h(y)})^{1/2} = (\frac{2 \cdot (80) \cdot (9.80665)}{1.205 \cdot (0.443)(0.8015)})^{1/2} = 60.56 m/s$$

Example 3. A parachutist of mass 100 kg jumps from an airplane flying horizontally with a speed $v_0 = 100 m/s$ at an altitude of 3,200 m above sea level.

Experiments show that the limiting speed at 1,500 m above sea level is 65 m/s.

Consider a standard atmosphere and no wind.

Find the product $c_D A$ and the parameter b.

Solution

Estimating $h(y)$ for $y = 1800 m$,

$$h(1800) = (\frac{289.08 - 0.006328(1500)}{289.08})^{4.4} = 0.86338$$

Substituting in (9.12.4), we have

$$65 = (\frac{2 \cdot (100) \cdot (9.80665)}{1.205 \cdot (c_d A)(0.86338)})^{1/2}$$

Solving for $c_D A$, we find

$$c_D A = 0.4462$$

The parameter b is

$$b = \rho_0 (c_D A)/2 = 1.205(0.4462)/2 = 0.26884$$

Note: In a similar way, we can find experimentally the product $c_D A$ for a parachutist falling and landing with parachute fully deployed (parachute landing).

Example 4. The parachutist of example 4 (mass 100 kg, $c_D A = 0.4462$, $b = 0.26884$) jumps from an airplane flying horizontally with a speed of

$v_0=100m/s$ at an altitude of 4,000 m above sea level. Assume a standard atmosphere and no wind.

(a) Find the elements of parachutist trajectory when he deploys his parachute at an altitude of 3,300 m above sea level.
(b) Find the coordinates of the parachutist when he lands on the ground that is 100 m above sea level. The parameter b of the parachutist with fully deployed parachute is $b=35.40$ (example 1 of section 9.10).

Solution
(a) Using PC QBasic program Parach.Bas

Input Data
Input: Altitude of Launching point [m] = 4,000
Input: Altitude of Parachute Deployment [m] = 3,300
Input: Launching Speed [m/s] = 100,
Input: Launching Angle [Degree]: 0,
Input: Initial Time = 0;
Input: Time of Interest [s] = 10
Input: Coefficient b= 0.26884
Input: Mass of Parachutist [kg] = 100
Input: Integration Step ho = 0.01

Results:
The elements of trajectory at altitude 3,300 m are the following: x-coordinate of parachutist = 665 m; Time = 15.66 sec., Speed of Parachutist = 69; Falling Angle = -78.31°

(b) We will use the same method we used in example 1 of section 9.10.
    We will consider that after 3 sec., the parachute is fully deployed and the speed of the parachutist drops to the limiting (minimum) value given in (9.8.12),

$$v_e=(\frac{mg}{b \cdot h(\bar{y})})^{1/2}=(\frac{100 \cdot (9.80665)}{35.40(0.82227)})^{1/2}=6.2m/s$$
,

since

$$h(y)=(\frac{\tau_0-0.006328y}{\tau_0})^{4.4}=(\frac{289.08-0.006328(3300)}{289.08})^{4.4}=0.71899$$

We consider as well that the trajectory of the parachutist is a vertical line.

During 3 sec., the average acceleration of the parachutist is

$$\bar{a}=(v-v_0)/dt=(6.2-69)/3=-31.40m/s^2$$

Thus, the parachutist has fallen down approximately

$$dy=v_0t+at^2/2=(69)\cdot(3)-31.40(3)^2/2=75.565m$$

The y-coordinate of parachutist at the time the parachute is fully deployed is

$$y=3300-76=3224m$$

So the parachutist, at the end of parachute deployment process, is at the point with coordinates $x=665m$, $y=3224m$.

(c) Considering the instantaneous speed of parachutist to be a function of altitude

$$v_e=(\frac{mg}{b\cdot h(y)})^{1/2}=5.263\cdot h(y)^{-1/2}=5.263\cdot(\frac{289.08}{289.08-0.006328\cdot y})^{0.5},$$

since $(mg/b)^{1/2}=[(100)\cdot(9.80665)/(35.40)]=5.263$.

Substituting v_e in the second equation of (9.8.2), we have

$$\frac{dt}{dy}=0.19\cdot(\frac{289.08-0.006328\cdot y}{289.08})^{2.2}$$

Integrating using a TI-83 Plus, we find

$$\Delta t = -0.19 \int_{3224}^{0} (\frac{289.08-0.006328y}{289.08})^{2.2} dy = (0.19)\cdot(2980.74) = 566.34s$$

Thus, the total time of flight is

$$t = 15.66+3+566.34 = 585sec = 9.75 minutes.$$

The deviation of the parachutist from the direction of the airplane flight is $x=665m$.

Note: To fall at particular location, the parachutist should live the airplane when around $x=665m$ from the landing location.

9.13 Unguided Airplane Bombing (Application of PC Programs)

The trajectory of bombing from slow-flying bombardiers can be studied using the PC program Parach.Bas.
The following example illustrates the solution of an airplane bombing problem.

Example 1. (*Application*: Airplane Bombing) An airplane flying 2,200 m above sea level with a speed of 140 m/s drops an unguided 250 kg bomb on a target located 100 m above sea level. The product $c_D A$ for the given bomb is 0.05.

(a) How far from the target should the airplane drop the bomb to hit the target? Assume a horizontal flight?
(b) How far from the target should the airplane drop the bomb if the direction of flight forms a depression angle $-10°$?
Consider a standard atmosphere with no wind.

Solution
We can use the PC program Parach.Bas (instead of the parachutist, we have the bomb).

First, we estimate

$$b=\rho_0(c_D A)/2=1.205(0.05)/2=0.03$$.

(a) Input Data
    Input: Altitude of Launching Point [m] = 2,200
    Input: Altitude of Parachute Deployment [m] = 0
    Input: Launching Speed [m/s] = 120
    Input: Launching Angle [Degree]: 0
    Input: Initial Time = 0
    Input: Time of Interest [s] = 10
    Input: Drag Coefficient b= 0.03
    Input: Mass of Bomb [kg] = 250
    Input: Integration Step h0 = 0.01

Results:
    x-coordinate of impact is $x=2321m$, Impact Speed $v=199m/s$;
    Time of Flight $t=22.21s$; Angle of Impact $\alpha=-65°$.

To hit the target, the airplane must drop the bomb when it is located at the point with coordinates at $x=2321m$, $y=2200m$.

(b) Imputing the same data as in (a), but launching angle [degree]: $\alpha_0=-10°$, we obtain the following results: x-coordinate of Impact is $x=2095m$; Impact Speed $v=200m/s$; Time of Flight $t=20.40s$; Angle of Impact $\alpha=-65°$.

The airplane should drop the bomb when it is at the point with coordinates $x=2095m$, $y=2200m$.

10

The Ballistics of Fragments of Antipersonnel Ammunitions

Introduction

A ntipersonnel ammunitions (hand grenades, land mines, mortar artillery shells, etc.) are constructed to incapacitate a large number of army personnel.

In general, any antipersonnel ammunition has a metallic body that includes an explosive charge (TNT, RDX, C-4, etc.). The detonation of the explosive charge produces a large number of fast-flying metallic fragments that can hit and incapacitate the military personnel located in relatively great distances from the center of explosion.

In general, the shape of projectile fragments is irregular and is too far from the aerodynamic form of bullets and artillery projectiles. Only some ammunitions and antipersonnel mines distribute around the center of explosion regular fragments in the form of metallic spheres.

Though the projectile fragments are launched with relatively high speeds during the detonation of the explosive charge, their speed decreases very quickly with the distance due to the enormous drag forces. The fast-launched fragments loose very soon the incapacitating effect.

The flight of fragments to the target, as well as the problems related with the construction of fragmentation ammunitions, and their efficacy can be studied using the differential equations of projectile flight in presence of drag.

Exterior ballistics can be applied to estimate the penetration of projectile fragments in liquids and in soft targets as well.

10.1 The Incapacitation Criteria

The incapacitation action (injury or lethal action) of any antipersonnel ammunition is a result of the kinetic energy that the metallic fragment transmits to the human body during the collision.

The experience of the wars and experiments have shown that a projectile fragment that hits a vital part of a person would be lethal if the projectile fragment will have a kinetic energy not less than 78 J.

That means that a projectile fragment that is launched during the detonation of an ammunition conserves its lethal effect at a distance where the speed of the metallic fragment of mass m will not be reduced (as a result of drag and gravity) under the speed limit value v_l determined by the equation

$$\frac{m \cdot v_l^2}{2} = 78 .$$

(10.1.1)

Hence, we find that the lowest speed that a projectile fragment must have in order to be lethal is

$$v_l = \sqrt{\frac{156}{m}} = \frac{12.5}{\sqrt{m}} .$$

(10.1.2)

The limit value given in (10.1.2) restricts the lethal range of the incapacitation of the personnel located in open space or behind light shields and might serve as a "criteria" for the construction of antipersonnel ammunitions.

Not all lethal fragments are dangerous for the personnel because part of them fly over the personnel's head while another part hits the ground before reaching the persons around the detonation point.

Antiaircraft fragmentation projectiles are used to incapacitate airplanes, helicopters, and parachutists. Their incapacitating effect is related with fragments launched with high speeds when the projectile is detonated.

The distance R from the center of detonation where a projectile fragment of mass m preserves the incapacitation effect on some airplanes is determined by the condition that the kinetic energy of the fragment at the time of impact with the vital parts of the aircraft must be at least 1,500 J.

Thus, the lowest incapacitation speed of a fragment of mass m that hits an aircraft (helicopter) can be determined by the equation

$$mv_T^2/2=1,500 \qquad (10.1.3)$$

Hence, we find that the lowest incapacitation speed of the given fragment is

$$v_T=\sqrt{\frac{3000}{m}}=\frac{54.77}{\sqrt{m}} \qquad (10.1.4)$$

The above speed is the speed relative to the flying target.

Example 1. Find the lethal speed limit value of a fragment of mass 3 g launched during the detonation of an antipersonnel grenade.

Solution
Employing (10.1.2), we find that the speed limit value is

$$v_l=\frac{12.5}{\sqrt{m}}=\frac{12.5}{\sqrt{0.003}}=228m/s$$

Example 2. The M18 Claymore fragmentation mine contains steel spheres each of mass 0.68 g. Find the lethal speed limit value of the given spheres.

Solution

Employing (10.1.2), we find that the speed of the given sphere needed to incapacitate a person will be at least

$$v_i = \frac{12.5}{\sqrt{m}} = \frac{12.5}{\sqrt{0.00068}} = 479 m/s$$

Note: The following tableau shows the lethal speed limit value (in meter per second) for some metallic fragments of different masses (1–10 g) estimated using (10.1.2).

Table 1

m	1 g	2 g	3 g	4 g	5 g	6 g	7 g	8 g	9 g	10 g
v_i	395	280	228	198	177	161	149	140	132	125

The table shows that for relatively large fragments, the lethal speed limit is less than 256 m/s.

Example 3. Find the lowest incapacitation airplane speed for projectile fragments of masses: 5 g, 10 g, 15 g, 20 g, 25 g, 30 g, 35 g, 40 g, 45 g, and 50 g.

Solution

For a fragment of mass 5 g (0.005 kg) substituting in (10.1.4), we have

$$v_I = \frac{54.77}{\sqrt{m}} = \frac{54.77}{\sqrt{0.005}} = 775$$

m/s.

Using (10.1.4) and substituting the given masses, we can easily estimate the lowest incapacitation speed for the given masses of other projectile fragments.

The results obtained using (10.1.4) are given in table 2.

Table 2

m	5 g	10 g	15 g	20 g	25 g	30 g	35 g	40 g	45 g	10 g
v_I	775	547	447	387	346	316	293	274	258	245

10.2 The Ballistics of Projectile Fragments

The projectile fragments produced during the detonation of fragmentation ammunitions are launched with relatively high speeds that are determined by the mass of explosive charge and the mass of the metallic body. The speed of the projectile fragments decreases during the flight because of the resistance of air and gravity. For that reason, we need to estimate the distance (from the detonation point) where a metallic fragment of a given mass will preserve the incapacitation effect on the personnel.

It is obvious that effective projectiles are those that will hit and incapacitate an average person located within a certain distance from the center of detonation of the ammunition. A part of projectile fragments produced during the detonation will not hit the target (figure 1).

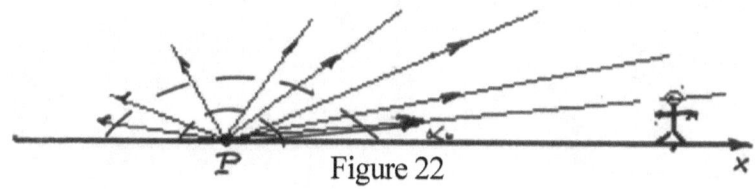

Figure 22

A lethal projectile will incapacitate a person if it hits any point of his/her body, injuring or killing that person. An injured person cannot continue to perform his military duties; he/she is out of action.

The lowest point of the body are the feet, located at ground level ($y=0$), while the highest point is the head, located on average $y=1.75$ m above the ground.

The antipersonnel fragmentation ammunitions are designed to be effective in relatively short distances. Since a person is relatively not tall (at most 1.9–2.0 m), effective fragments are fragments launched with relatively narrow angles.

Thus, to find approximate solutions it is obvious to assume the following:

- The projectile fragment is launched at narrow angles.
- The trajectory is a straight line, i.e., the vertical motion (the drop) of the projectile fragment is ignored.

To study the flight of such projectiles, we consider the resistance of air (2.1.1)

$$D=A\frac{\rho v^2}{2}C(\frac{v}{a})$$

(10.2.1)

where A is the cross-section area of the projectile fragment, ρ is the density of air, and $C(v/a)$ is the drag coefficient.

The direction of the resistance of air is opposite to the projectile velocity since we assume that the projectile trajectory is a line. Thus, ignoring the effect of gravity, the differential equation of motion of the center of mass of a projectile fragment of mass m (equation [2.3.1]) along the straight line can be written in the form

$$\frac{dv}{dt}=-c{\cdot}h(y)K_D(v)$$

(10.2.2)

where

$$c=\frac{id^2}{m}1000$$

(10.2.3)

$$h(y)=(\frac{289.08-0.006328y}{289.08})^{4.4}$$

;

(10.2.4)

and

$$K_D(v)=\begin{cases}1.212{\cdot}10^{-4}v^2 & for & v{\le}256m/s \\ (v-240)/3 & for & v>256m/s\end{cases}$$

(10.2.5)

is the Siacci function of resistance.

To solve the problems related with the ballistics of projectile fragments, we can use as well the results obtained in chapters 2, 3, and 4.

Example 1. The value of the ballistics parameter,[63]

$$B=\frac{\rho \cdot \pi \cdot d^2}{8 \cdot m}C(v/a) \qquad \qquad (1)$$

for a 4.5 mm steel sphere is

(a) 0.00921 m^{-1} when the speed of sphere is up to 250 m/s;
(b) 0.02501 m^{-1} when the speed of sphere is in the interval 700–1,000 m/s.

The drag coefficient $C(v/a)$ is considered constant in each of the given intervals. The density of air is ρ=1.225 kg/m^3.

Using the above data, find the Siacci coefficient of form i of the given steel sphere.

The density of steel sphere can be considered equal to 7,850 kg/m^3.

Solution

The mass of the given steel sphere is

$$m=\frac{4}{3}\pi(\frac{d}{2})^3 \rho_{steel}=\frac{4}{3}\cdot \pi \cdot(\frac{0.0045}{2})^3 \cdot(7,850)=3.7455\times10^{-4} \text{ kg.}$$

Using (1), for the drag coefficient, we can write

$$C(v/a)=\frac{8m \cdot B}{\rho \cdot \pi \cdot d^2}.$$

---

[63]   Sellier, K.G., and Kneubuehel, B. P *Wound Ballistics and the Scientific Background,* page 116.
Elsevier, 1994.

Substituting

(a) For $B=0.00921$, we find that the drag coefficient is

$$C(v/a)=\frac{8m \cdot B}{\rho \cdot \pi \cdot d^2}=\frac{8 \cdot (3.7455 \cdot 10^{-4}) \cdot (0.00921)}{(1.225) \cdot \pi \cdot (0.0045)^2}=0.3541$$

(b) For $B=0.02501$, we find the drag coefficient is

$$C(v/a)=\frac{8m \cdot B}{\rho \cdot \pi \cdot d^2}=\frac{8 \cdot (3.7455 \cdot 10^{-4}) \cdot (0.02501)}{(1.225) \cdot \pi \cdot (0.0045)^2}=0.96161$$

To find the coefficient of form, we will use formulas (2.6.5) and (2.6.10), respectively, for each interval of projectile speeds,

$$i=3.90431 \cdot C(\frac{v}{a})$$

and

$$i=0.0014196\frac{v^2}{v-240} \cdot C(\frac{v}{a})$$

(a) The coefficient of resistance, when the speed of the steel sphere is up to 256 m/s, is

$$i=3.90431 \cdot C(\frac{v}{a})=3.90431 \cdot (0.3541)=1.3825$$

(b) To find an average coefficient of the given steel sphere for speeds 700–1,000 m/s, we consider the average velocity of steel sphere $v=(700+1000)/2=850$ m/s.

The average coefficient of resistance for the steel sphere is

$$i=0.0014196\frac{v^2}{v-240} \cdot C(\frac{v}{a})=0,0014196\frac{850^2}{(850-240)}(0.9616)=1.61686$$

The coefficient of steel sphere changes from

$$i=0.0014196\frac{v^2}{v-240}\cdot C(\frac{v}{a})=0,0014196\frac{700^2}{(700-240)}(0.9616)=1.4541$$

when the speed of projectile is $v=700m/s$ to

$$i=0.0014196\frac{v^2}{v-240}\cdot C(\frac{v}{a})=0,0014196\frac{1000^2}{(1000-240)}(0.9616)=1.7962$$

when the speed of projectile (sphere) is $v=1000m/s$.

Note: The form factor of a spherical ball is[64]

(a) $i=2$ for speeds less than 300 m/s,
(b) $i=1.5$ for speeds greater than 300 m/s.

The Siacci form coefficient $i=1.61686$ estimated in the above example for speeds in the interval 700–1,000 m/s is compatible with the value $i=1.5$ given by Ugolini for speeds greater than 300 m/s.

For speeds less than 300 m/s, there is a large difference between the value estimated in this example and the Ugolini's value.

For speeds lower than 256 m/s, we will use the Siacci value given by Ugolini $i=2$.

Thus, it is reasonable to accept as average values for Siacci form factor the values given by A. Ugolini:

(a) $i=2$ for speeds less than 256 m/s,
(b) $i=1.5$ for speeds greater than 256 m/s.

---

[64] Ugolini, A. L'Esperto Balistico, p. 322. Vol. I. Editoriale Olimpia, 1983.

10.3 The Trajectory of Fast-Flying Fragments

We suppose that the launching angle of the fragment is relatively narrow.

Substituting the second relation of (10.2.5) in (10.2.2), we obtain the equation of motion of the projectile fragment when the fragment speed is greater than 256 m/s,

$$\frac{dv}{dt}=-ch(y)\cdot\frac{(v-240)}{3}.$$

(10.3.1)

Denoting

$$b=c\cdot h(y)/3,$$

(10.3.2)

we write (10.3.1) as

$$\frac{dv}{dt}=-b\cdot(v-240).$$

(10.3.3)

Expressing the above equation through the distance r from the center of detonation, we obtain

$$\frac{dv}{dr}\cdot\frac{dr}{dt}=-b\cdot(v-240).$$

Hence, since $dr/dt=v$, we get the differential equation

$$\frac{dv}{dr}=-b\cdot\frac{v-240}{v}.$$

(10.3.4)

Integrating the above differential equation for the initial conditions $v=v_0$ when, $r=0$ we obtain the distance from the center of detonation where the projectile fragment have the speed v,

$$r=\frac{240}{b}\ln(\frac{v_0-240}{v-240})+\frac{v_0-v}{b},$$

(10.3.5)

Example 1. The Ballistics Coefficient of Irregular Projectile Fragments

The ballistics coefficient for the fragments of artillery ammunitions has a value in the interval 60–90. Find the coefficient of the form of a projectile fragment of mass m=20 g whose ballistics coefficient is c=90.[65]

Solution

The forms of fragments are irregular cylinders with unknown base and height equal to the width of the metallic body of the projectile that is fragmented during the detonation.

To find the coefficient of the form, we employ

$$c=\frac{id^2}{m}1000$$
                                                    (1)

To find c, we will consider the projectile fragment as a steel sphere of diameter D. For the mass of this sphere, we can write

$$m=(volume)\cdot(density)=\frac{4}{3}\pi(\frac{d}{2})^3\cdot(7,800)$$

Substituting m=0.020kg and solving for d, we find d=0.01698 m. Substituting in (1), we can write

$$90=\frac{i(0.01698)^2}{0.020}1000$$

Hence, we find approximately that

$$i=6.24$$

---

[65] **Vreto, Zhivko.** *Mundesite taktiko teknike te kompleksit 100 mm per qitje kunder objektivave ajrore*, **p. 106-134. MMP, Tirana.**

Note: We will use the value $i=6.24$ to estimate the ballistics coefficient of irregular projectile fragments.

Example 2. The projectile of the antiaircraft gun 100 mm explodes 6,000 m above sea level. A projectile fragment of mass 20 g is thrown with a speed of 968 m/s. Find the distance r where the projectile speed have the lowest incapacitation value of 387 m/s.

The ballistics coefficient of the fragment is 90.

Solution

To estimate b, we first find $h(y)$ at altitude $y=6,000m$.
Substituting in

$$h(y)=(\frac{289.08-0.006328y}{289.08})^{4.4},$$

we have

$$h(y)=[(289.08-0.006328y)/289.08]^{4.4}=$$
$$= [(289.08-0.006328 \cdot (6,000))/289.08]^{4.4}=0.538.$$

For b, we have

$$b=c \cdot h(y)/3=(90) \cdot (0.538) \cdot /3)=16.1398.$$

The distance from the detonation center where the projectile fragment has the lowest incapacitation value is

$$r=\frac{240}{b}\ln(\frac{v_0-240}{v-240})+\frac{v_0-v}{b}=$$
$$=\frac{240}{16.1398}\ln(\frac{986-240}{387-240})+\frac{986-387}{16.1398}=61.27m.$$

Example 3. The M18 Claymore directional fragmentation mine contains 700 steel balls each of mass 0.68 g. [66] The casualty radius is 100 m, while the killing radius is 50 m.

Assuming that each fragment on average is lethal at a distance of 50 m from the center of detonation, estimate the initial velocity of each steel ball.

Use the coefficient of form $i=1.61686$ for a steel sphere found in example 1 in section 10.2.

Solution

Using (10.1.2), we find that the lowest lethal speed of spherical fragment of M18 is

$$v_l=\frac{12.5}{\sqrt{m}}=\frac{12.5}{\sqrt{0.00068}}=479m/s$$.

The lowest lethal speed is at a distance of 50 m from the M18 Claymore mine.

Let's find the diameter of a spherical fragment. Substituting the mass of the spherical fragment in

$$m=(volume)\cdot(density)=\frac{4}{3}\pi(\frac{d}{2})^3\cdot(7,850)$$,

we find that the diameter of the sphere is

$$d=0.0055\,m.$$

The ballistics coefficient is

$$c=\frac{id^2}{m}1000=\frac{(1.612)\cdot(0.0055)^2}{0.00068}\cdot1000=71.71$$.

---

[66] M18 Claymore, *http://www.fas.org/man/dod-101/sys/land/m18-claymore.htm* [Web site]

For b, we have

$$b = c \cdot h(y)/3 = (71.71) \cdot (1) \cdot /3 = 23.90332 .$$

Substituting in (10.3.5), we can write

$$r = \frac{240}{b} \ln(\frac{v-240}{v_0-240}) + \frac{v_0-v}{b}$$

$$50 = \frac{240}{(23.90332)} \ln(\frac{v_0-240}{479-240}) + \frac{v_0-479}{(23.90332)} .$$

Hence,

$$6.975691064 - 0.00416667 v_0 = \ln(\frac{v_0-240}{239}) .$$

Using a graphing calculator TI-83 Plus, we solve the above equation for v_0 as intersection of the following curves

$$y_1 = 6.975691064 - 0.00416667 v_0, \quad y_2 = \ln(\frac{v_0-240}{239}).$$

We find that the initial speed is

$$v_0 = 1314 m/s .$$

We see that the speed of the projectile fragment drops very quickly.

Note: The initial speed obtained here is approximate. It can be corrected through the experiments.

Example 4. For the M18 Claymore mine of example 2, find a more accurate solution and the departure angle that assures that all fragments will hit a vertical panel 2 m above the ground at a distance of 50 m considering that all 700 fragments will fly horizontally to cover a sector 60° wide centered at the M18 mine. Initial speed is $v_0 = 1314 m/s$.

Solution

Since the M18 projectiles are launched with a speed greater than 256 m/s, we use the results obtained in section 3.3. The equation of the trajectory (3.3.4) is

$$y = p_0 \cdot x - \frac{g}{2 \cdot v_0^2} x^2 - \frac{gb(v_0 - 240)}{3v_0^4} x^3 - gb^2(v_0 - 240)\frac{v_0 - 320}{4v_0^6} x^4.$$

We suppose that the M18 Claymore mine is constructed in such a way to ensure that at the distance 50 m from the mine the fragments will be distributed vertically over the ground within a height of 2 m.

Let's find the vertical launching angle.

(a) First, we estimate the launching angle of those fragments that will hit the target at the point that is 2 m above the ground. (We neglect the height of the mine.)

We have found in example 2 the value $b = 23.90332$.
Substituting in the equation of projectile trajectory, we can write

$$p_0(50) - \frac{g}{2(1{,}314)^2}(50)^2 - \frac{g(23.90332)(1{,}314 - 240)}{3(1{,}314)^4}(50)^4$$

$$-g(23.90332)^2(1{,}314 - 240)\frac{(1{,}314 - 320)}{4(1{,}314)^6}(50)^4 = 2.$$

Solving the last equation for p_0, we have

$$p_0 = 0.04024862.$$

The launching angle is

$$\alpha_0 = \tan^{-1}(0.04024862) = 2.305°.$$

(b) Let's find the launching angle of those fragments that will hit the target at the point that is on the ground at the distance 50 m from the mine, i.e., $y=0$. (We neglect the height of mine.)

The solution is similar to case (a), but the initial coordinate is $y=0$.

It is obvious that we have to solve the same equation with the left side equal to zero, i.e.,

$$p_0(50)-\frac{g}{2(1,314)^2}(50)^2-\frac{g(23.90332)(1,314-240)}{3(1,314)^4}(50)^4$$

$$-g(23.90332)^2(1,314-240)\frac{(1,314-320)}{4(1,314)^6}(50)^4=0 \qquad .$$

Solving the above equation, we find

$$p_0=0.01424865 \qquad .$$

The launching angle is

$$\alpha_0=\tan^{-1}(0.001424865)=0.8163^\circ \qquad .$$

Summary: The sphere fragments will be distributed within (0.8163°, 2.305°). The distribution of fragments is supposed to be assured by the construction of M18 mine.

Time of Flight

$$(\alpha_0=0.8163^\circ)$$

$$v_{x0}=v_0\cos\alpha_0=1314\cdot\cos(0.8163^\circ)=1313.87$$

Substituting in (3.2.8),

$$x=240t+\frac{(v_{x0}-240)}{b}(1-e^{-b\cdot t}) \qquad ,$$

we have

$$240t+\frac{(v_{x0}-240)}{b}(1-e^{-b\cdot t})=240t+\frac{(1313.87-240)}{(23..90332)}\cdot(1-e^{-23.9033\cdot t})=50$$

or

$$5.34217t-1.112952=e^{-23.9033\cdot t}-1.$$

Solving the above equation (using, for example, a graphing calculator), we find

$$t=0.06283s.$$

Terminal Speed
    The main component of the terminal velocity (formula [3.2.9]) is

$$v_x=240+(v_{x0}-240)\cdot e^{-b\cdot t}=$$
$$=240+(1313.87-240)e^{-(1.501844)}=479.17m/s.$$

The value of p (see [3.2.10]) is

$$p=p_0-\frac{g}{240\cdot b}\ln(\frac{240\cdot e^{b\cdot t}+v_{x0}-240}{v_{x0}})=$$

$$=\tan(0.8163°)-\frac{g}{240(23.9033)}\ln(\frac{240e^{1.501127}+1313.87-240}{1313.87})=0.014195$$

The y-component of velocity is

$$v_Y=pv_X=(0.014195)(1213.87)=6.8m/s.$$

The terminal speed is

$$v=(v_x^2+v_y^2)^{0.5}=[479.17^2+(6.8)^2]^{1/2}=479.22m/s.$$

Example 5. An antipersonnel mine have an explosive charge of $q=0.075$ kg TNT. The metallic body of the mine is composed of 400 steel balls each with a mass of 1 g. Find the greatest distance from the center of detonation of the mine where a steel ball is still lethal. The initial speed of each fragment can be estimated using the following Gurney's formula:

$$v_0=\frac{D}{\sqrt{8}}\cdot(M/q+3/5)^{-1/2}$$

,

where M is the mass of the metallic body that encloses an explosive charge with mass q and D is the speed of detonation of explosive charge. For explosive charges of TNT, the detonation speed is $D=7,000m/s$.

Solution

The mass of all 400 steel balls is $M=0.4kg$. Substituting in the above formula, we find that the initial speed of each ball is

$$v_0=\frac{D}{\sqrt{8}}\cdot(M/q+3/5)^{-1/2}=\frac{7,000}{\sqrt{8}}(0.4/0.10+3/5)^{-1/2}=1154m/s$$

.

The lethal speed of a projectile of mass 1 g is 395 m/s (see table 1 of section 10.1).

To find the distance, we use (10.3.5),

$$r=\frac{240}{b}\ln(\frac{v_0-240}{v-240})+\frac{v_0-v}{b}$$

,

where $b=c\cdot h(y)/3$, $c=\frac{id^2}{m}-1000$.

The coefficient of form of a spherical steel ball is $i=1.61686$.

To find the diameter d of the spherical ball, we will use the same method we used in example 1, i.e., employing the formula

$$m=\frac{4}{3}\pi(\frac{d}{2})^3\cdot(7,800)$$

.

Substituting the mass of the ball $m=0.001kg$ and solving for the diameter d of the ball, we find

$d=0.006256$ m.

The ballistics coefficient is

$$c=\frac{id^2}{m}1000=\frac{(1.61686)\cdot(0.006256)^2}{0.001}\cdot1{,}000=63.279916$$.

For b, we have

$$b=c\cdot h(y)/(3g)=(63.2799)\cdot(1)\cdot/3=13.0458$$.

The greatest lethal distance is

$$r=\frac{240}{b}\ln(\frac{v_0-240}{v-240})+\frac{v_0-v}{b}=$$
$$=\frac{240}{(13.0458}\ln(\frac{1154-240}{387-240})+\frac{1154-387}{(13.0458)}=60.62m$$.

Example 6. Consider a fragmentation mine that is similar to the mine of example 5, but the metallic body of the mine is a compact steel shell of 400 g that is fragmented during the detonation of the explosive charge in projectile fragments of different shapes and masses.

Find the greatest distance where a fragment of mass 1 g will still be lethal. Assume a coefficient of the form $i=6.24$ (estimated in example 1).

Solution

Taking into consideration the data obtained in example 1 and example 5, it is obvious that the coefficient $i=6.24$ of the form of an irregular projectile fragment of mass 1 g is the only parameter that is different from the respective parameter of the steel ball of mass 1 g.

Thus for c and b, we have respectively

$$c=\frac{id^2}{m}1000=\frac{(6.24)\cdot(0.006256)^2}{0.001}\cdot1{,}000=244.22$$

and

$$b=c{\cdot}h(y)/3=(244.22){\cdot}(1){\cdot}/3=81.407 .$$

The greatest distance from the detonation center where the fragment is still lethal is approximately

$$r=\frac{240}{b}\ln(\frac{v_0-240}{v-240})+\frac{v_0-v}{b}=$$

$$=\frac{240}{(81.407)}\ln\frac{1154-240}{387-240}+\frac{1154-387}{(81.407)}=16.62m .$$

10.4 The Ballistics of Slow-Speed Fragments

We can find approximate solutions for the motion of a projectile fragment substituting the first relation of (10.2.5) in (10.2.2).
We can write

$$\frac{dv}{dt}=-1.212{\cdot}10^{-4}c{\cdot}h(y)v^2 . \tag{10.4.1}$$

The density function $h(y)$ can be considered constant, and we can write (10.4.1) in the form

$$\frac{dv}{dr}{\cdot}\frac{dr}{dt}=-1.212{\cdot}10^{-4}ch(y)v^2 .$$

Hence, since $dr/dt=v$, we obtain

$$\frac{dv}{dr}=-1.212{\cdot}10^{-4}ch(y)v . \tag{10.4.2}$$

Denoting

$$b=1.212{\cdot}10^{-4}c{\cdot}h(y) , \tag{10.4.3}$$

we write (10.4.2)

$$\frac{dv}{dr}=-b{\cdot}v . \tag{10.4.4}$$

Integrating (10.4.4), for initial conditions $v=v_0$ when $r=0$, we obtain the fragment speed as a function of the distance r along the linear trajectory

$$v=v_0 \cdot e^{-b \cdot r} \qquad (10.4.5)$$

Hence, for the distance from the center of detonation where the projectile fragment has a given speed v, we have

$$r=\frac{1}{b}\ln(\frac{v_0}{v}) \qquad (10.4.6)$$

For projectile speed less than $v_k=256m/s$, exterior ballistics considers the drag

$$\vec{D}=-A\frac{\rho v^2}{2}c_D\frac{\vec{v}}{v} \qquad (10.4.7)$$

where c_D is the drag coefficient that is assumed constant for all projectile velocities under $v_k=256m/s$, ρ is the density of air, and A is the cross-section area of the projectile.

If we denote

$$k=\rho \cdot A \cdot c_D /2 \qquad (10.4.8)$$

then

$$\vec{D}=-k \cdot v^2 \frac{\vec{v}}{v} \qquad (10.4.9)$$

The equation of projectile flight, similar to (2.2.1), in Cartesian coordinate system, can be written in the form

$$\frac{d^2\vec{r}}{dt^2}=\vec{g}-A\frac{\rho v^2}{2m}C_D\frac{\vec{v}}{v} \qquad (10.4.10)$$

or

$$\frac{d\vec{v}}{dt}=\vec{g}-\frac{k}{m}v^2\cdot\frac{\vec{v}}{v}.$$

(10.4.11)

If we consider a linear trajectory for fragments, i.e., neglecting the gravity, from (10.4.11) we obtain an equation, similar to (10.4.4),

$$\frac{dv}{dr}=-\frac{k}{m}v.$$

(10.4.12)

From (10.4.12), we obtain two equations, similar to (10.4.5) and (10.4.6), i.e.,

$$v=v_0\cdot e^{-k\cdot r/m}$$

(10.4.13)

and

$$r=\frac{m}{k}\ln(\frac{v_0}{v}).$$

(10.4.14)

For a more accurate solution of the problems related with the projectile fragments, we can adapt the formulas (4.2.20), i.e.

$$\left\{\begin{array}{l} v_x=v_0\cos\alpha_0\cdot e^{-b\cdot x} \\[2ex] \tan\alpha=\tan\alpha_0+\dfrac{g(1-e^{2bx})}{2bv_0^2\cos^2\alpha_0} \\[2ex] v_y=(\tan\alpha_0+\dfrac{g(1-e^{2b\cdot x})}{2bv_0^2\cos^2\alpha_0})\cdot(v_0\cos\alpha_0 e^{-b\cdot x}) \\[2ex] y=(\tan\alpha_0+\dfrac{g}{2bv_0^2\cos^2\alpha_0})x+\dfrac{g(1-e^{2b\cdot x})}{4b^2v_0^2\cos^2\alpha_0} \\[2ex] t=\dfrac{e^{b\cdot x}-1}{bv_0\cos\alpha_0} \end{array}\right.$$

(10.4.15)

by substituting

$$b=k/m, \quad k=\rho A \cdot c_D /2 . \tag{10.4.16}$$

Example 1. A directional fragmentation land mine contains steel balls each of mass 4 g. Estimate the initial speed v_0 the spherical fragment must have to be lethal for a person located 10 m from the center of detonation of the mine.

(We assume that launching angle is such that the projectile will hit an average person 8 m away from the mine. The coefficient of form for steel sphere for speeds up to 256 m/s, see the note in example 1 in section 10.2, is $i=2$).

Solution

To estimate the initial speed of the fragment, we solve equation (10.4.5) for v_0. We obtain

$$v_0 = v \cdot e^{b \cdot r} . \tag{1}$$

The lethal limit speed of a fragment (see table 1 in section 10.1) of mass 4 g is $v=198$. We need to estimate the ballistics coefficient c and the value of b as well.

The mass of the sphere is

$$m = (volume) \cdot (density) = \frac{4}{3}\pi(\frac{d}{2})^3 \cdot \rho_f .$$

Hence,

$$d = (6m/\pi\rho_f)^{1/3} .$$

The density of steel is $\rho_f = 7{,}850 kg/m^3$. Substituting in the above formula $m=4g=0.004 kg$ and $\rho_f = 7{,}850 kg/m^3$, we find

$$d = (6m/\pi\rho_f)^{1/3} = [6 \cdot (0.004)]/\pi(7850)]^{1/3} = 0.00991 .$$

The ballistics coefficient is

$$c=\frac{id^2}{m}1000=\frac{(2)\cdot(0.00991)^2}{(0.004)}1{,}000=49.10$$.

For b, we have

$$b=1.212\cdot10^{-4}c\cdot h(y)=1.212\cdot10^{-4}(49.10)\cdot(1)=0.0059514$$.

Substituting in (1), we find that the initial speed of the spherical fragment is

$$v_0=(198)\cdot e^{(0.00595)\cdot(10)}=210m/s$$.

Example 2. A directional fragmentation antipersonnel mine contains steel balls each with a mass of 2 g. The initial speed given to a spherical fragment at the moment of detonation is $v_0=180m/s$. Find the speed of the projectile on a wood board located 20 m from the center of detonation.

The form coefficient estimated in example 1 in section 10.2 is $i=2$. The density of air is $\rho=1.205kg/m^3$. (Employ formula (10.4.13), $v=v_0\cdot e^{-k\cdot r/m}$, to solve the problem).

Solution

For the mass of the spherical fragment, we can write

$$m=\frac{4}{3}\pi(\frac{d}{2})^3\rho_{steel}$$,

where d is the diameter of the sphere and ρ_{steel} is the density of steel, $\rho_{steel}=7850kg/m^3$.

Substituting the mass of the sphere $m=0.002kg$ and the density of steel $\rho_{steel}=7850$, then solving the obtained equation for the diameter of sphere, we find $d=0.00787m$.

We find that

$$c=\frac{id^2}{m}1000=\frac{(2)\cdot(0.00787)^2}{0.002}\cdot1{,}000=61.9369$$

and

$$b=1.212 \cdot 10^{-4} c \cdot h(y)=1.212 \cdot 10^{-4}(61.94) \cdot (1)=0.007507$$.

Employing (10.4.5), we find that the speed of the given fragment is

$$v=v_0 \cdot e^{-b \cdot r}=(180) \cdot e^{-(0.007507) \cdot (20)}=155 m/s$$.

Example 3. Consider the spherical projectile of example 2. Find the launching angle needed to direct the projectiles in order that they will hit the wooden board located 20 m from the mine if the board is 2 m high. Ignore the height of the mine.

Solution

The projectiles should be directed from the surface of the mine in such a way that they will be distributed along the height of the wooden board. None of the spherical fragments should hit the ground before reaching the board or pass over it.

Let's find the narrowest and the widest launching angle.

(a) The narrowest launching angle: Since the initial speed is less than 256 m/s and the angle of fire is relatively narrow (mine and the bottom of wooden board are in horizontal plane), to find the launching angle α_0, we can use the fourth equation of (15), considering $b=0.007507$.

The y-coordinate of the point of impact is zero, i.e., $y=0$.
Thus, to find the launching angle α_0, we have to solve for α_0

$$(\tan\alpha_0 + \frac{g}{2bv_0^2 \cos^2\alpha_0})x + \frac{g(1-e^{2bx})}{4b^2 v_0^2 \cos^2\alpha_0}=0 \tag{1}$$.

Substituting the data of example 2 and $b=k/m=0.00518$, we have

$$\left(\tan\alpha_0+\frac{g}{2(0.007507)(180)^2\cos^2\alpha_0}\right)\cdot(20)+\frac{g(1-e^{2(0.007507)(20)})}{4(0.007507)^2(180)^2\cos^2\alpha_0}=0 \quad . \text{(2)}$$

Hence, we have

$$\tan\alpha_0-\frac{0.003354}{\cos^2\alpha_0}=0 \quad .$$

We can write

$$\frac{\sin(2\alpha_0)-0.00608}{2\cos^2\alpha_0}=0$$

or

$$\sin(2\alpha_0)=0.006708 \quad .$$

Solving the above equation, we find that

$$\alpha_0=\frac{1}{2}\sin^{-1}(0.00608)=0.192° \quad .$$

(a) The widest launching angle: The widest launching angle is determined by the condition that the target is the upper part of the board, $y=2m$ high.

In a similar way, like in (a), we obtain the following equation

$$\left(\tan\alpha_0+\frac{g}{2(0.007507)(180)^2\cos^2\alpha_0}\right)\cdot(20)+\frac{g(1-e^{2(0.007507)(20)})}{4(0.007507)^2(180)^2\cos^2\alpha_0}=2$$

$$\tan\alpha_0-\frac{0.003354}{\cos^2\alpha_0}=0.1 \quad .$$

Substituting $\cos^2\alpha_0=(1+\tan^2\alpha_0)$, we can write

$$\tan\alpha_0-0.003754\tan^2(\alpha_0)-0.1003354=0 \quad .$$

Solving the last equation with a TI-83 Plus graphing calculator, we obtain

$$\tan\alpha_0 = 0.100373$$

Hence, we find that the widest angle is

$$\alpha_0 = \tan^{-1}(0.100373) = 5.732°$$

10.5 The Square Law of Air Resistance for Spherical Projectiles

For spherical projectiles, the acceleration due to the drag for relatively short distances from the launching point is[67]

$$a_D = i\frac{1}{2}\frac{\pi \cdot d^2}{4m}\rho_0 h(y) \cdot v^2 C_D(\frac{v}{a})$$ (10.5.1)

where i is the form coefficient relative to the square law, while $C_D(v/a)$ is the drag coefficient. Substituting $\rho_0 = 1.205$ in (10.5.1), we find that

$$a_D = 0.4732024i\frac{d^2}{m}h(y) \cdot v^2 C_D(\frac{v}{a})$$ (10.5.2)

If we denote

$$k = i\frac{d^2}{m}0.4732024$$ (10.5.3)

then (10.5.2) can be written as

$$a_D = k \cdot h(y) \cdot v^2 C_D(\frac{v}{a})$$ (10.5.4)

---

[67] Sellier, K. G. and Kneubuehel, B. P. *Wound Ballistics and the Scientific Background*, p. 112-114.
Elsevier, 1994,

To study the flight of projectile spheres, we need to know the form coefficient i related with the square law of drag. For spheres, when the shooting distances are relatively short, we assume the following:

- The function $C_D(v/a)$ can be considered constant. Since $C_D(v/a)$ is constant, we can assume $C_D(v/a)=1$; otherwise, we can include the value of $C_D(v/a)$ in the form coefficient i.
- For a steel sphere of diameter $d=0.0045m$ (mass $m=3.7455 \cdot 10^{-4}$ kg), the coefficient k, when the density of air is $\rho_1=1.225 kg/m$, has the following experimental value:

$k=0.02501 m^{-1}$, for speeds in the interval 700–1,000 m/s.

Using (10.5.2), (10.5.3), and (10.5.4) for the form coefficient of a spherical projectile that corresponds to the square law, we can write

$$k=0.4732024i\frac{d^2}{m}\frac{\rho_1}{\rho_0}.$$

(10.5.5)

From (10.5.5), we find that the form factor related with the quadratic function of resistance of the spherical projectile is

$$i=(0.473204)^{-1}k\frac{\rho_0 m d}{\rho_1 d^2}.$$

(10.5.6)

Thus the form coefficient of the spherical projectile for projectile speed greater than 250 m/s is

$$i=(0.473204)^{-1}k\frac{\rho_0 m}{\rho_1 d^2}=(0.4732024)^{-1}(0.02501)\frac{(1.205)(3.7455 \cdot 10^{-4})}{(1.225)(0.0045)^2}=0.961617781$$

$$i=0.961617781.$$

(10.5.7)

We can write the drag acceleration (10.5.2) in the following form:

$$a_D=(0.4732024)i\frac{d^2}{m}h(y)\cdot v^2,$$

(10.5.8)

where i is given in (10.5.7). Thus, the acceleration of the spherical projectile is

$$a_D = k \cdot h(y) \cdot v^2 C_D\left(\frac{v}{a}\right),$$

(10.5.9)

where

$$k = 0.4732024i\frac{d^2}{m}.$$

For short-flying distances, considering $h(y)$ to be constant and equal to the value the density function has at the launching points, we can denote

$$b = k \cdot h(y).$$

(10.5.10)

The differential equation of flight of a spherical projectile in a Cartesian coordinate system is similar to equation (10.4.11),

$$\frac{d\vec{v}}{dt} = \vec{g} - bv^2 \cdot \frac{\vec{v}}{v}.$$

(10.5.11)

To study the flight of a spherical projectile not far from the launching point, we adapt the results obtained in section 10.4.

I. Approximate solutions of the equations of projectile flight

Assuming a linear trajectory (the absence of the gravity) from (10.5.11), we obtain the following equations for the projectile speed and the range of the spherical projectile, we obtain respectively

$$v = v_0 \cdot e^{-b \cdot r}$$

(10.5.12)

and

$$r = \frac{1}{b}\ln\left(\frac{v_0}{v}\right).$$

(10.5.13)

II. The analytical solutions of the differential equation (10.5.11) for a spherical projectile launched under a given angle obtained in the same way as in chapter 4 are

$$v_x = v_0 \cos\alpha_0 \cdot e^{-b \cdot x}$$

$$\tan\alpha = \tan\alpha_0 + \frac{g(1 - e^{2bx})}{2bv_0^2 \cos^2\alpha_0}$$

$$v_y = (\tan\alpha_0 + \frac{g(1 - e^{2b \cdot x})}{2bv_0^2 \cos^2\alpha_0}) \cdot (v_0 \cos\alpha_0 e^{-b \cdot x})$$

$$\qquad\qquad (10.5.14)$$

$$y = (\tan\alpha_0 + \frac{g}{2bv_0^2 \cos^2\alpha_0})x + \frac{g(1 - e^{2b \cdot x})}{4b^2 v_0^2 \cos^2\alpha_0}$$

$$t = \frac{e^{b \cdot x} - 1}{bv_0 \cos\alpha_0}$$

where

$$b = k \cdot h(y) , \quad k = 0.4732024 i \frac{d^2}{m} , \text{ and } i = 0.961617781 . \quad (10.5.15)$$

Example 1. Use the results obtained in this section to solve the problem of example 5 in section 10.3:

The antipersonnel mine has an explosive charge of $q = 0.075$ kg TNT. The metallic body of the mine is composed of 400 steel balls each with a mass of 1 g. Find the greatest distance from the center of detonation of the mine where a steel ball is still lethal. The initial speed of each fragment can be estimated using the following Gurney's formula:

$$v_0 = D \cdot (M/q + 3/5)^{-1/2} / \sqrt{8}$$

where M is the mass of the metallic body that encloses an explosive charge with mass q and D is the speed of detonation of explosive charge. For explosive charges of TNT, the detonation speed is $D=7,000m/s$.

Solution

In example 5 of section 10.3, we have found the following:

(a) The launching speed of a steel ball of diameter $d=0.006256m$ and mass $m=0.001kg$ is $v_0=1154m/s$. The lowest lethal speed of the given projectile is $v=395m/s$.

We find that

$$k=i\frac{d^2}{m}0.4732024=(0.9616178)\frac{(0.006256)^2}{(0.001)}0.4732024=0.0178902$$.

Considering $h(y)=1$, we find that $. b=kh(y)=0.0178902(1)=0.0178902$
The lethal range of the given spherical projectile is

$$r=\frac{1}{b}\ln(\frac{v_0}{v})=\frac{1}{(0.0178902)}\ln(\frac{1154}{395})=60.20m$$.

The result obtained using the square law is the same as the result we obtained in example 5 of section 10.3 using the Siacci law of resistance.

APPENDIX A

Siacci's Functions In SI Units
(Projectile Speed $v \geq 256 m/s$)

Tabulated Siacci's Functions

$$D(u)=[u+240 \cdot \ln(u-240)], \quad T(u)=\ln(u-240),$$

$$J(u)=\ln(\frac{u-240}{u}), \quad A(u)=\int_{1000}^{u} \frac{u}{u-240}\ln\frac{u-240}{u}du$$

Elements of Projectile Trajectory

The parameter $p=\tan\alpha$

$$p=p_0+\frac{1}{240B\cos^2\alpha_0}[J(u)-J(v_0)],$$

The coordinates (x, y) of projectile

$$x=-\frac{1}{Bg}[D(u)-D(v_0)],$$

$$y = p_0 x + \frac{x}{240B\cos^2\alpha_0} \cdot [\frac{A(u)-A(v_0)}{D(u)-D(v_0)} - J(v_0)]$$

The time of flight

$$t = -\frac{1}{Bg\cos\alpha_0}[T(u)-T(v_0)]$$

Launching Angle α_0:

For field artillery fire

$$\sin(2\alpha_0) = -\frac{1}{120B} \cdot [\frac{A(u)-A(v_0)}{D(u)-D(v_0)} - J(v_0)]$$

u	D(u)	J(u)	T(u)	A(u)	u	D(u)	J(u)	T(u)	A(u)
256	921.4213	-2.77259	2.77259	0	301	1287.6097	-1.59624	4.11087	-788.51392
257	936.9712	-2.71586	2.83321	-42.67189	302	1292.5123	-1.58329	4.12713	-796.30777
258	951.6892	-2.66259	2.89037	-82.25145	303	1297.3523	-1.5706	4.14313	-803.94031
259	965.6654	-2.61239	2.94444	-119.11289	304	1302.1319	-1.55814	4.15888	-811.41738
260	978.9757	-2.56495	2.99573	-153.56868	305	1306.8529	-1.54592	4.17439	-818.74454
261	991.6854	-2.52	3.04452	-185.88224	306	1311.5171	-1.53393	4.18965	-825.92706
262	1003.8502	-2.4773	3.09104	-216.2775	307	1316.1262	-1.52216	4.20469	-832.96995
263	1015.5186	-2.43666	3.13549	-244.94629	308	1320.6818	-1.51059	4.21951	-839.87797
264	1026.7329	-2.3979	3.17805	-272.05412	309	1325.1856	-1.49923	4.23411	-846.65567
265	1037.5302	-2.36085	3.21888	-297.74464	310	1329.6389	-1.48808	4.2485	-853.30737
266	1047.9432	-2.3254	3.2581	-322.14332	311	1334.0432	-1.47711	4.26268	-859.83719
267	1058.0008	-2.29141	3.29584	-345.36032	312	1338.3999	-1.46634	4.27667	-866.24906
268	1067.7291	-2.25878	3.3322	-367.49281	313	1342.7103	-1.45574	4.29046	-872.54673
269	1077.151	-2.22742	3.3673	-388.62692	314	1346.9756	-1.44533	4.30407	-878.73379
270	1086.2874	-2.19722	3.4012	-408.83934	315	1351.1971	-1.43508	4.31749	-884.81367
271	1095.1569	-2.16813	3.43399	-428.19858	316	1355.376	-1.42501	4.33073	-890.78964
272	1103.7766	-2.14007	3.46574	-446.76609	317	1359.5133	-1.4151	4.34381	-896.66484
273	1112.1618	-2.11296	3.49651	-464.59722	318	1363.6101	-1.40534	4.35671	-902.44226
274	1120.3265	-2.08677	3.52636	-481.74189	319	1367.6675	-1.39574	4.36945	-908.1248
275	1128.2835	-2.06142	3.55535	-498.24538	320	1371.6864	-1.38629	4.38203	-913.71519
276	1136.0445	-2.03688	3.58352	-514.14877	321	1375.6678	-1.37699	4.39445	-919.21609
277	1143.6203	-2.0131	3.61092	-529.4895	322	1379.6126	-1.36783	4.40672	-924.63003
278	1151.0207	-1.99003	3.63759	-544.30177	323	1383.5217	-1.35881	4.41884	-929.95945
279	1158.2548	-1.96765	3.66356	-558.61686	324	1387.396	-1.34993	4.43082	-935.20669
280	1165.3311	-1.94591	3.68888	-572.46349	325	1391.2363	-1.34117	4.44265	-940.37399
281	1172.2573	-1.92478	3.71357	-585.86807	326	1395.0434	-1.33255	4.45435	-945.46351
282	1179.0407	-1.90424	3.73767	-598.85491	327	1398.8179	-1.32405	4.46591	-950.47734
283	1185.688	-1.88425	3.7612	-611.44648	328	1402.5608	-1.31568	4.47734	-955.41747
284	1192.2055	-1.86478	3.78419	-623.66355	329	1406.2727	-1.30742	4.48864	-960.28582
285	1198.599	-1.84583	3.80666	-635.52536	330	1409.9543	-1.29928	4.49981	-965.08426
286	1204.8739	-1.82735	3.82864	-647.04979	331	1413.6063	-1.29126	4.51086	-969.81457
287	1211.0354	-1.80933	3.85015	-658.25344	332	1417.2293	-1.28335	4.52179	-974.47847
288	1217.0882	-1.79176	3.8712	-669.15178	333	1420.8239	-1.27554	4.5326	-979.07762

u	D(u)	J(u)	T(u)	A(u)	u	D(u)	J(u)	T(u)	A(u)
289	1223.0369	-1.77461	3.89182	-679.75924	334	1424.3907	-1.26785	4.54329	-983.61362
290	1228.8855	-1.75786	3.91202	-690.08927	335	1427.9305	-1.26025	4.55388	-988.08801
291	1234.6382	-1.7415	3.93183	-700.15449	336	1431.4436	-1.25276	4.56435	-992.5023
292	1240.2985	-1.72551	3.95124	-709.96668	337	1434.9306	-1.24537	4.57471	-996.85792
293	1245.8701	-1.70988	3.97029	-719.53691	338	1438.3922	-1.23808	4.58497	-1001.1563
294	1251.3562	-1.6946	3.98898	-728.87555	339	1441.8288	-1.23088	4.59512	-1005.3987
295	1256.76	-1.67964	4.00733	-737.99237	340	1445.2408	-1.22378	4.60517	-1009.5865
296	1262.0844	-1.66501	4.02535	-746.89654	341	1448.6289	-1.21676	4.61512	-1013.7209
297	1267.3323	-1.65068	4.04305	-755.59672	342	1451.9935	-1.20984	4.62497	-1017.8031
298	1272.5063	-1.63665	4.06044	-764.10106	343	1455.335	-1.203	4.63473	-1021.8343
299	1277.609	-1.62291	4.07754	-772.41726	344	1458.6538	-1.19625	4.64439	-1025.8158
300	1282.6427	-1.60944	4.09434	-780.55259	345	1461.9505	-1.18958	4.65396	-1029.7485
346	1465.2254	-1.183	4.66344	-1033.63353	391	1595.1472	-0.95143	5.01728	-1171.67828
347	1468.4789	-1.1765	4.67283	-1037.47193	392	1597.7313	-0.94738	5.02388	-1174.13177
348	1471.7115	-1.17007	4.68213	-1041.2647	393	1600.3051	-0.94337	5.03044	-1176.56503
349	1474.9235	-1.16372	4.69135	-1045.01281	394	1602.8686	-0.9394	5.03695	-1178.97837
350	1478.1153	-1.15745	4.70048	-1048.71722	395	1605.422	-0.93546	5.04343	-1181.37208
351	1481.2872	-1.15126	4.70953	-1052.37883	396	1607.9654	-0.93156	5.04986	-1183.74646
352	1484.4397	-1.14513	4.7185	-1055.99854	397	1610.499	-0.92769	5.05625	-1186.10179
353	1487.5731	-1.13908	4.72739	-1059.5772	398	1613.0228	-0.92386	5.0626	-1188.43835
354	1490.6876	-1.1331	4.7362	-1063.11566	399	1615.537	-0.92006	5.0689	-1190.75642
355	1493.7837	-1.12719	4.74493	-1066.61472	400	1618.0417	-0.91629	5.07517	-1193.05626
356	1496.8616	-1.12134	4.75359	-1070.07518	401	1620.537	-0.91256	5.0814	-1195.33813
357	1499.9217	-1.11556	4.76217	-1073.4978	402	1623.0231	-0.90886	5.0876	-1197.60229
358	1502.9643	-1.10985	4.77068	-1076.88333	403	1625.5	-0.90519	5.09375	-1199.849
359	1505.9896	-1.1042	4.77912	-1080.2325	404	1627.9679	-0.90155	5.09987	-1202.0785
360	1508.998	-1.09861	4.78749	-1083.546	405	1630.4269	-0.89794	5.10595	-1204.29103
361	1511.9897	-1.09309	4.79579	-1086.82452	406	1632.8771	-0.89437	5.11199	-1206.48683
362	1514.9651	-1.08762	4.80402	-1090.06873	407	1635.3185	-0.89082	5.11799	-1208.66613
363	1517.9242	-1.08222	4.81218	-1093.27928	408	1637.7514	-0.8873	5.12396	-1210.82916
364	1520.8676	-1.07687	4.82028	-1096.45679	409	1640.1757	-0.88382	5.1299	-1212.97614
365	1523.7953	-1.07158	4.82831	-1099.60189	410	1642.5916	-0.88036	5.1358	-1215.1073
366	1526.7077	-1.06635	4.83628	-1102.71516	411	1644.9993	-0.87693	5.14166	-1217.22283
367	1529.6049	-1.06117	4.84419	-1105.7972	412	1647.3987	-0.87353	5.14749	-1219.32296

u	D(u)	J(u)	T(u)	A(u)	u	D(u)	J(u)	T(u)	A(u)
368	1532.4873	-1.05605	4.85203	-1108.84857	413	1649.79	-0.87016	5.15329	-1221.4079
369	1535.355	-1.05098	4.85981	-1111.86982	414	1652.1733	-0.86681	5.15906	-1223.47783
370	1538.2083	-1.04597	4.86753	-1114.86149	415	1654.5486	-0.86349	5.16479	-1225.53297
371	1541.0474	-1.041	4.8752	-1117.8241	416	1656.9162	-0.8602	5.17048	-1227.57351
372	1543.8725	-1.03609	4.8828	-1120.75817	417	1659.2759	-0.85694	5.17615	-1229.59963
373	1546.6838	-1.03123	4.89035	-1123.66419	418	1661.6281	-0.8537	5.18178	-1231.61153
374	1549.4816	-1.02642	4.89784	-1126.54266	419	1663.9726	-0.85049	5.18739	-1233.60938
375	1552.2659	-1.02165	4.90527	-1129.39403	420	1666.3096	-0.8473	5.19296	-1235.59338
376	1555.0372	-1.01693	4.91265	-1132.21879	421	1668.6393	-0.84414	5.1985	-1237.56369
377	1557.7954	-1.01226	4.91998	-1135.01737	422	1670.9616	-0.841	5.20401	-1239.5205
378	1560.5409	-1.00764	4.92725	-1137.79023	423	1673.2767	-0.83789	5.20949	-1241.46396
379	1563.2737	-1.00306	4.93447	-1140.53778	424	1675.5846	-0.8348	5.21494	-1243.39425
380	1565.9942	-0.99853	4.94164	-1143.26045	425	1677.8854	-0.83173	5.22036	-1245.31154
381	1568.7024	-0.99404	4.94876	-1145.95864	426	1680.1792	-0.82869	5.22575	-1247.21598
382	1571.3985	-0.98959	4.95583	-1148.63277	427	1682.4661	-0.82568	5.23111	-1249.10774
383	1574.0827	-0.98519	4.96284	-1151.28321	428	1684.7461	-0.82268	5.23644	-1250.98697
384	1576.7552	-0.98083	4.96981	-1153.91036	429	1687.0193	-0.81971	5.24175	-1252.85382
385	1579.4161	-0.97651	4.97673	-1156.51458	430	1689.2858	-0.81676	5.24702	-1254.70845
386	1582.0656	-0.97223	4.98361	-1159.09623	431	1691.5456	-0.81383	5.25227	-1256.55099
387	1584.7038	-0.96799	4.99043	-1161.65568	432	1693.7989	-0.81093	5.2575	-1258.38161
388	1587.3309	-0.96379	4.99721	-1164.19328	433	1696.0456	-0.80805	5.26269	-1260.20043
389	1589.9471	-0.95963	5.00395	-1166.70935	434	1698.286	-0.80519	5.26786	-1262.00761
390	1592.5525	-0.95551	5.01064	-1169.20424	435	1700.5199	-0.80235	5.273	-1263.80328
436	1702.7475	-0.79953	5.27811	-1265.58757	481	1797.3513	-0.69107	5.4848	-1335.9331
437	1704.9689	-0.79673	5.2832	-1267.36061	482	1799.3451	-0.68901	5.48894	-1337.30902
438	1707.1841	-0.79395	5.28827	-1269.12255	483	1801.3347	-0.68696	5.49306	-1338.67802
439	1709.3932	-0.79119	5.2933	-1270.87351	484	1803.3204	-0.68492	5.49717	-1340.04017
440	1711.5962	-0.78846	5.29832	-1272.61361	485	1805.302	-0.68289	5.50126	-1341.39552
441	1713.7932	-0.78574	5.3033	-1274.34298	486	1807.2796	-0.68088	5.50533	-1342.74415
442	1715.9842	-0.78304	5.30827	-1276.06174	487	1809.2532	-0.67888	5.50939	-1344.08611
443	1718.1694	-0.78036	5.31321	-1277.77001	488	1811.2229	-0.67689	5.51343	-1345.42147
444	1720.3488	-0.7777	5.31812	-1279.46792	489	1813.1887	-0.67491	5.51745	-1346.75028
445	1722.5224	-0.77506	5.32301	-1281.15558	490	1815.1506	-0.67294	5.52146	-1348.07261
446	1724.6903	-0.77244	5.32788	-1282.8331	491	1817.1087	-0.67099	5.52545	-1349.38852

u	D(u)	J(u)	T(u)	A(u)	u	D(u)	J(u)	T(u)	A(u)
447	1726.8525	-0.76984	5.33272	-1284.50059	492	1819.063	-0.66905	5.52943	-1350.69806
448	1729.0091	-0.76726	5.33754	-1286.15817	493	1821.0135	-0.66712	5.53339	-1352.00129
449	1731.1602	-0.76469	5.34233	-1287.80595	494	1822.9602	-0.6652	5.53733	-1353.29828
450	1733.3058	-0.76214	5.34711	-1289.44404	495	1824.9033	-0.66329	5.54126	-1354.58907
451	1735.446	-0.75961	5.35186	-1291.07253	496	1826.8426	-0.6614	5.54518	-1355.87372
452	1737.5807	-0.7571	5.35659	-1292.69154	497	1828.7783	-0.65951	5.54908	-1357.15229
453	1739.7101	-0.7546	5.36129	-1294.30117	498	1830.7103	-0.65764	5.55296	-1358.42483
454	1741.8342	-0.75212	5.36598	-1295.90151	499	1832.6387	-0.65578	5.55683	-1359.69139
455	1743.9531	-0.74966	5.37064	-1297.49268	500	1834.5636	-0.65393	5.56068	-1360.95203
456	1746.0668	-0.74721	5.37528	-1299.07476	501	1836.4849	-0.65209	5.56452	-1362.2068
457	1748.1754	-0.74479	5.3799	-1300.64785	502	1838.4027	-0.65026	5.56834	-1363.45574
458	1750.2788	-0.74237	5.3845	-1302.21205	503	1840.317	-0.64844	5.57215	-1364.69892
459	1752.3772	-0.73998	5.38907	-1303.76745	504	1842.2278	-0.64663	5.57595	-1365.93638
460	1754.4706	-0.7376	5.39363	-1305.31415	505	1844.1352	-0.64483	5.57973	-1367.16817
461	1756.559	-0.73524	5.39816	-1306.85223	506	1846.0391	-0.64304	5.5835	-1368.39434
462	1758.6426	-0.73289	5.40268	-1308.38178	507	1847.9397	-0.64126	5.58725	-1369.61494
463	1760.7212	-0.73056	5.40717	-1309.9029	508	1849.8369	-0.63949	5.59099	-1370.83001
464	1762.7951	-0.72824	5.41165	-1311.41566	509	1851.7307	-0.63774	5.59471	-1372.0396
465	1764.8641	-0.72594	5.4161	-1312.92016	510	1853.6213	-0.63599	5.59842	-1373.24376
466	1766.9284	-0.72365	5.42053	-1314.41647	511	1855.5085	-0.63425	5.60212	-1374.44254
467	1768.988	-0.72138	5.42495	-1315.90469	512	1857.3925	-0.63252	5.6058	-1375.63598
468	1771.043	-0.71912	5.42935	-1317.38489	513	1859.2732	-0.6308	5.60947	-1376.82412
469	1773.0933	-0.71688	5.43372	-1318.85716	514	1861.1507	-0.6291	5.61313	-1378.00701
470	1775.139	-0.71465	5.43808	-1320.32156	515	1863.0251	-0.6274	5.61677	-1379.18469
471	1777.1803	-0.71244	5.44242	-1321.77819	516	1864.8962	-0.62571	5.6204	-1380.35721
472	1779.217	-0.71024	5.44674	-1323.22712	517	1866.7642	-0.62403	5.62402	-1381.52461
473	1781.2492	-0.70806	5.45104	-1324.66842	518	1868.6291	-0.62235	5.62762	-1382.68693
474	1783.2771	-0.70589	5.45532	-1326.10217	519	1870.4908	-0.62069	5.63121	-1383.84421
475	1785.3005	-0.70373	5.45959	-1327.52845	520	1872.3495	-0.61904	5.63479	-1384.99649
476	1787.3196	-0.70159	5.46383	-1328.94732	521	1874.2051	-0.6174	5.63835	-1386.14382
477	1789.3344	-0.69946	5.46806	-1330.35886	522	1876.0577	-0.61576	5.64191	-1387.28623
478	1791.345	-0.69734	5.47227	-1331.76313	523	1877.9073	-0.61413	5.64545	-1388.42377
479	1793.3513	-0.69524	5.47646	-1333.16022	524	1879.7538	-0.61252	5.64897	-1389.55647
480	1795.3533	-0.69315	5.48064	-1334.55019	525	1881.5974	-0.61091	5.65249	-1390.68438

u	D(u)	J(u)	T(u)	A(u)	u	D(u)	J(u)	T(u)	A(u)
526	1883.438	-0.60931	5.65599	-1391.80752	571	1963.5084	-0.54527	5.80212	-1437.96898
527	1885.2757	-0.60772	5.65948	-1392.92594	572	1965.2324	-0.544	5.80513	-1438.90811
528	1887.1105	-0.60614	5.66296	-1394.03968	573	1966.9542	-0.54274	5.80814	-1439.84388
529	1888.9424	-0.60456	5.66643	-1395.14877	574	1968.6738	-0.54149	5.81114	-1440.77631
530	1890.7714	-0.603	5.66988	-1396.25326	575	1970.3913	-0.54024	5.81413	-1441.70542
531	1892.5976	-0.60144	5.67332	-1397.35316	576	1972.1067	-0.539	5.81711	-1442.63125
532	1894.4209	-0.59989	5.67675	-1398.44853	577	1973.8199	-0.53776	5.82008	-1443.5538
533	1896.2414	-0.59835	5.68017	-1399.5394	578	1975.531	-0.53653	5.82305	-1444.4731
534	1898.0591	-0.59682	5.68358	-1400.6258	579	1977.24	-0.5353	5.826	-1445.38917
535	1899.8741	-0.59529	5.68698	-1401.70777	580	1978.9469	-0.53408	5.82895	-1446.30204
536	1901.6863	-0.59377	5.69036	-1402.78533	581	1980.6518	-0.53287	5.83188	-1447.21173
537	1903.4957	-0.59227	5.69373	-1403.85853	582	1982.3546	-0.53166	5.83481	-1448.11825
538	1905.3024	-0.59077	5.69709	-1404.9274	583	1984.0553	-0.53046	5.83773	-1449.02163
539	1907.1065	-0.58927	5.70044	-1405.99198	584	1985.754	-0.52926	5.84064	-1449.92188
540	1908.9078	-0.58779	5.70378	-1407.05228	585	1987.4507	-0.52807	5.84354	-1450.81904
541	1910.7065	-0.58631	5.70711	-1408.10835	586	1989.1453	-0.52688	5.84644	-1451.71311
542	1912.5025	-0.58484	5.71043	-1409.16022	587	1990.8379	-0.5257	5.84932	-1452.60413
543	1914.2959	-0.58338	5.71373	-1410.20792	588	1992.5286	-0.52452	5.8522	-1453.4921
544	1916.0866	-0.58192	5.71703	-1411.25149	589	1994.2173	-0.52335	5.85507	-1454.37706
545	1917.8748	-0.58047	5.72031	-1412.29094	590	1995.904	-0.52219	5.85793	-1455.25901
546	1919.6604	-0.57903	5.72359	-1413.32632	591	1997.5887	-0.52103	5.86079	-1456.13798
547	1921.4435	-0.5776	5.72685	-1414.35765	592	1999.2715	-0.51988	5.86363	-1457.01399
548	1923.2239	-0.57618	5.7301	-1415.38496	593	2000.9523	-0.51873	5.86647	-1457.88706
549	1925.0019	-0.57476	5.73334	-1416.40829	594	2002.6313	-0.51758	5.8693	-1458.7572
550	1926.7774	-0.57335	5.73657	-1417.42766	595	2004.3083	-0.51644	5.87212	-1459.62443
551	1928.5503	-0.57194	5.73979	-1418.4431	596	2005.9834	-0.51531	5.87493	-1460.48878
552	1930.3208	-0.57054	5.743	-1419.45464	597	2007.6566	-0.51418	5.87774	-1461.35026
553	1932.0888	-0.56915	5.7462	-1420.46231	598	2009.3279	-0.51306	5.88053	-1462.20889
554	1933.8543	-0.56777	5.74939	-1421.46613	599	2010.9974	-0.51194	5.88332	-1463.06468
555	1935.6174	-0.5664	5.75257	-1422.46614	600	2012.665	-0.51083	5.8861	-1463.91766
556	1937.3781	-0.56503	5.75574	-1423.46237	601	2014.3307	-0.50972	5.88888	-1464.76785
557	1939.1364	-0.56366	5.7589	-1424.45483	602	2015.9946	-0.50861	5.89164	-1465.61525
558	1940.8923	-0.56231	5.76205	-1425.44355	603	2017.6567	-0.50751	5.8944	-1466.45989
559	1942.6459	-0.56096	5.76519	-1426.42857	604	2019.3169	-0.50642	5.89715	-1467.30179
560	1944.397	-0.55962	5.76832	-1427.40991	605	2020.9754	-0.50533	5.8999	-1468.14096

u	D(u)	J(u)	T(u)	A(u)	u	D(u)	J(u)	T(u)	A(u)
561	1946.1459	-0.55828	5.77144	-1428.38759	606	2022.632	-0.50425	5.90263	-1468.97741
562	1947.8924	-0.55695	5.77455	-1429.36165	607	2024.2868	-0.50317	5.90536	-1469.81117
563	1949.6366	-0.55563	5.77765	-1430.3321	608	2025.9399	-0.50209	5.90808	-1470.64226
564	1951.3784	-0.55431	5.78074	-1431.29897	609	2027.5912	-0.50102	5.9108	-1471.47068
565	1953.118	-0.553	5.78383	-1432.26229	610	2029.2407	-0.49996	5.9135	-1472.29646
566	1954.8554	-0.5517	5.7869	-1433.22209	611	2030.8885	-0.49889	5.9162	-1473.11961
567	1956.5904	-0.5504	5.78996	-1434.17838	612	2032.5345	-0.49784	5.91889	-1473.94014
568	1958.3233	-0.54911	5.79301	-1435.13119	613	2034.1788	-0.49679	5.92158	-1474.75808
569	1960.0539	-0.54782	5.79606	-1436.08054	614	2035.8214	-0.49574	5.92426	-1475.57344
570	1961.7822	-0.54654	5.79909	-1437.02646	615	2037.4622	-0.4947	5.92693	-1476.38623
616	2039.1014	-0.49366	5.92959	-1477.19647	661	2111.2319	-0.45112	6.04263	-1511.24424
617	2040.7388	-0.49262	5.93225	-1478.00418	662	2112.8013	-0.45026	6.04501	-1511.95179
618	2042.3746	-0.49159	5.93489	-1478.80936	663	2114.3693	-0.4494	6.04737	-1512.65739
619	2044.0087	-0.49057	5.93754	-1479.61204	664	2115.936	-0.44855	6.04973	-1513.36104
620	2045.6411	-0.48955	5.94017	-1480.41223	665	2117.5014	-0.4477	6.05209	-1514.06276
621	2047.2719	-0.48853	5.9428	-1481.20995	666	2119.0654	-0.44685	6.05444	-1514.76256
622	2048.9009	-0.48752	5.94542	-1482.00521	667	2120.6282	-0.44601	6.05678	-1515.46044
623	2050.5284	-0.48651	5.94803	-1482.79801	668	2122.1896	-0.44516	6.05912	-1516.15642
624	2052.1542	-0.48551	5.95064	-1483.58839	669	2123.7497	-0.44433	6.06146	-1516.85051
625	2053.7784	-0.48451	5.95324	-1484.37635	670	2125.3085	-0.44349	6.06379	-1517.54272
626	2055.401	-0.48351	5.95584	-1485.16191	671	2126.8659	-0.44266	6.06611	-1518.23305
627	2057.0219	-0.48252	5.95842	-1485.94508	672	2128.4221	-0.44183	6.06843	-1518.92152
628	2058.6413	-0.48153	5.96101	-1486.72587	673	2129.9771	-0.44101	6.07074	-1519.60814
629	2060.259	-0.48055	5.96358	-1487.50431	674	2131.5307	-0.44019	6.07304	-1520.29291
630	2061.8752	-0.47957	5.96615	-1488.28039	675	2133.083	-0.43937	6.07535	-1520.97584
631	2063.4898	-0.4786	5.96871	-1489.05414	676	2134.6341	-0.43855	6.07764	-1521.65696
632	2065.1028	-0.47763	5.97126	-1489.82557	677	2136.184	-0.43774	6.07993	-1522.33625
633	2066.7143	-0.47666	5.97381	-1490.59469	678	2137.7325	-0.43693	6.08222	-1523.01374
634	2068.3242	-0.4757	5.97635	-1491.36152	679	2139.2799	-0.43612	6.0845	-1523.68943
635	2069.9326	-0.47474	5.97889	-1492.12607	680	2140.8259	-0.43532	6.08677	-1524.36334
636	2071.5394	-0.47378	5.98141	-1492.88835	681	2142.3708	-0.43452	6.08904	-1525.03547
637	2073.1447	-0.47283	5.98394	-1493.64838	682	2143.9144	-0.43372	6.09131	-1525.70582
638	2074.7485	-0.47189	5.98645	-1494.40616	683	2145.4567	-0.43293	6.09357	-1526.37442
639	2076.3507	-0.47094	5.98896	-1495.16171	684	2146.9979	-0.43213	6.09582	-1527.04126
640	2077.9515	-0.47	5.99146	-1495.91505	685	2148.5378	-0.43134	6.09807	-1527.70636

u	D(u)	J(u)	T(u)	A(u)	u	D(u)	J(u)	T(u)	A(u)
641	2079.5507	-0.46907	5.99396	-1496.66618	686	2150.0765	-0.43056	6.10032	-1528.36973
642	2081.1485	-0.46814	5.99645	-1497.41512	687	2151.6141	-0.42978	6.10256	-1529.03137
643	2082.7448	-0.46721	5.99894	-1498.16188	688	2153.1504	-0.429	6.10479	-1529.6913
644	2084.3396	-0.46628	6.00141	-1498.90647	689	2154.6855	-0.42822	6.10702	-1530.34951
645	2085.9329	-0.46536	6.00389	-1499.64891	690	2156.2194	-0.42744	6.10925	-1531.00603
646	2087.5248	-0.46445	6.00635	-1500.3892	691	2157.7522	-0.42667	6.11147	-1531.66086
647	2089.1152	-0.46353	6.00881	-1501.12737	692	2159.2837	-0.4259	6.11368	-1532.314
648	2090.7041	-0.46262	6.01127	-1501.86341	693	2160.8141	-0.42514	6.11589	-1532.96547
649	2092.2916	-0.46172	6.01372	-1502.59734	694	2162.3433	-0.42437	6.1181	-1533.61527
650	2093.8777	-0.46082	6.01616	-1503.32918	695	2163.8714	-0.42361	6.1203	-1534.26341
651	2095.4624	-0.45992	6.01859	-1504.05893	696	2165.3983	-0.42286	6.12249	-1534.90991
652	2097.0456	-0.45902	6.02102	-1504.78661	697	2166.924	-0.4221	6.12468	-1535.55476
653	2098.6274	-0.45813	6.02345	-1505.51223	698	2168.4486	-0.42135	6.12687	-1536.19798
654	2100.2078	-0.45724	6.02587	-1506.23579	699	2169.9721	-0.4206	6.12905	-1536.83958
655	2101.7868	-0.45636	6.02828	-1506.95732	700	2171.4944	-0.41985	6.13123	-1537.47955
656	2103.3645	-0.45548	6.03069	-1507.67681	701	2173.0155	-0.41911	6.1334	-1538.11792
657	2104.9407	-0.4546	6.03309	-1508.39429	702	2174.5356	-0.41837	6.13556	-1538.75468
658	2106.5155	-0.45372	6.03548	-1509.10976	703	2176.0545	-0.41763	6.13773	-1539.38985
659	2108.089	-0.45285	6.03787	-1509.82324	704	2177.5723	-0.41689	6.13988	-1540.02343
660	2109.6611	-0.45199	6.04025	-1510.53473	705	2179.089	-0.41616	6.14204	-1540.65544
706	2180.6046	-0.41543	6.14419	-1541.28587	751	2247.7287	-0.38504	6.23637	-1568.14268
707	2182.119	-0.4147	6.14633	-1541.91474	752	2249.1979	-0.38441	6.23832	-1568.70822
708	2183.6324	-0.41398	6.14847	-1542.54205	753	2250.6662	-0.38379	6.24028	-1569.27248
709	2185.1447	-0.41325	6.1506	-1543.16782	754	2252.1336	-0.38317	6.24222	-1569.83548
710	2186.6558	-0.41253	6.15273	-1543.79204	755	2253.6001	-0.38255	6.24417	-1570.39723
711	2188.1659	-0.41181	6.15486	-1544.41473	756	2255.0656	-0.38193	6.24611	-1570.95773
712	2189.675	-0.4111	6.15698	-1545.03589	757	2256.5303	-0.38132	6.24804	-1571.51698
713	2191.1829	-0.41039	6.1591	-1545.65553	758	2257.9941	-0.38071	6.24998	-1572.07499
714	2192.6898	-0.40968	6.16121	-1546.27366	759	2259.4569	-0.3801	6.2519	-1572.63177
715	2194.1956	-0.40897	6.16331	-1546.89029	760	2260.9189	-0.37949	6.25383	-1573.18732
716	2195.7003	-0.40826	6.16542	-1547.50542	761	2262.38	-0.37888	6.25575	-1573.74164
717	2197.204	-0.40756	6.16752	-1548.11905	762	2263.8402	-0.37828	6.25767	-1574.29475
718	2198.7066	-0.40686	6.16961	-1548.7312	763	2265.2996	-0.37768	6.25958	-1574.84664
719	2200.2081	-0.40616	6.1717	-1549.34188	764	2266.758	-0.37708	6.26149	-1575.39732

u	D(u)	J(u)	T(u)	A(u)	u	D(u)	J(u)	T(u)	A(u)
720	2201.7087	-0.40547	6.17379	-1549.95108	765	2268.2156	-0.37648	6.2634	-1575.94681
721	2203.2081	-0.40477	6.17587	-1550.55882	766	2269.6723	-0.37588	6.2653	-1576.49509
722	2204.7066	-0.40408	6.17794	-1551.16511	767	2271.1281	-0.37529	6.2672	-1577.04218
723	2206.204	-0.40339	6.18002	-1551.76994	768	2272.5831	-0.37469	6.2691	-1577.58808
724	2207.7004	-0.40271	6.18208	-1552.37333	769	2274.0372	-0.3741	6.27099	-1578.1328
725	2209.1957	-0.40202	6.18415	-1552.97529	770	2275.4905	-0.37351	6.27288	-1578.67634
726	2210.6901	-0.40134	6.18621	-1553.57581	771	2276.9429	-0.37293	6.27476	-1579.21871
727	2212.1834	-0.40066	6.18826	-1554.17491	772	2278.3944	-0.37234	6.27664	-1579.75991
728	2213.6757	-0.39999	6.19032	-1554.7726	773	2279.8451	-0.37176	6.27852	-1580.29995
729	2215.167	-0.39931	6.19236	-1555.36887	774	2281.295	-0.37118	6.2804	-1580.83883
730	2216.6573	-0.39864	6.19441	-1555.96374	775	2282.744	-0.3706	6.28227	-1581.37656
731	2218.1466	-0.39797	6.19644	-1556.55721	776	2284.1922	-0.37002	6.28413	-1581.91313
732	2219.6349	-0.3973	6.19848	-1557.14929	777	2285.6395	-0.36944	6.286	-1582.44857
733	2221.1222	-0.39664	6.20051	-1557.73999	778	2287.0861	-0.36887	6.28786	-1582.98286
734	2222.6085	-0.39597	6.20254	-1558.32931	779	2288.5317	-0.3683	6.28972	-1583.51602
735	2224.0939	-0.39531	6.20456	-1558.91725	780	2289.9766	-0.36772	6.29157	-1584.04805
736	2225.5782	-0.39465	6.20658	-1559.50383	781	2291.4206	-0.36716	6.29342	-1584.57896
737	2227.0616	-0.394	6.20859	-1560.08905	782	2292.8638	-0.36659	6.29527	-1585.10874
738	2228.544	-0.39334	6.2106	-1560.67292	783	2294.3062	-0.36602	6.29711	-1585.63741
739	2230.0255	-0.39269	6.21261	-1561.25544	784	2295.7478	-0.36546	6.29895	-1586.16497
740	2231.5059	-0.39204	6.21461	-1561.83661	785	2297.1886	-0.3649	6.30079	-1586.69142
741	2232.9855	-0.39139	6.21661	-1562.41646	786	2298.6286	-0.36434	6.30262	-1587.21677
742	2234.464	-0.39075	6.2186	-1562.99497	787	2300.0677	-0.36378	6.30445	-1587.74102
743	2235.9416	-0.39011	6.22059	-1563.57215	788	2301.5061	-0.36322	6.30628	-1588.26417
744	2237.4183	-0.38946	6.22258	-1564.14802	789	2302.9436	-0.36267	6.3081	-1588.78624
745	2238.894	-0.38883	6.22456	-1564.72258	790	2304.3804	-0.36211	6.30992	-1589.30723
746	2240.3688	-0.38819	6.22654	-1565.29583	791	2305.8164	-0.36156	6.31173	-1589.82713
747	2241.8426	-0.38755	6.22851	-1565.86778	792	2307.2515	-0.36101	6.31355	-1590.34596
748	2243.3155	-0.38692	6.23048	-1566.43843	793	2308.6859	-0.36047	6.31536	-1590.86372
749	2244.7875	-0.38629	6.23245	-1567.0078	794	2310.1195	-0.35992	6.31716	-1591.3804
750	2246.2586	-0.38566	6.23441	-1567.57588	795	2311.5523	-0.35937	6.31897	-1591.89603
796	2312.9844	-0.35883	6.32077	-1592.4106	841	2376.6628	-0.336	6.39859	-1614.53494
797	2314.4157	-0.35829	6.32257	-1592.92411	842	2378.0618	-0.33552	6.40026	-1615.00502

u	D(u)	J(u)	T(u)	A(u)	u	D(u)	J(u)	T(u)	A(u)
798	2315.8462	-0.35775	6.32436	-1593.43658	843	2379.4601	-0.33505	6.40192	-1615.47421
799	2317.2759	-0.35721	6.32615	-1593.94799	844	2380.8578	-0.33458	6.40357	-1615.94252
800	2318.7048	-0.35667	6.32794	-1594.45837	845	2382.2548	-0.33411	6.40523	-1616.40995
801	2320.133	-0.35614	6.32972	-1594.96771	846	2383.6512	-0.33364	6.40688	-1616.8765
802	2321.5604	-0.35561	6.3315	-1595.47601	847	2385.0469	-0.33317	6.40853	-1617.34219
803	2322.9871	-0.35508	6.33328	-1595.98329	848	2386.442	-0.33271	6.41017	-1617.80701
804	2324.413	-0.35455	6.33505	-1596.48954	849	2387.8364	-0.33224	6.41182	-1618.27096
805	2325.8382	-0.35402	6.33683	-1596.99477	850	2389.2301	-0.33178	6.41346	-1618.73406
806	2327.2626	-0.35349	6.33859	-1597.49898	851	2390.6233	-0.33132	6.4151	-1619.19629
807	2328.6862	-0.35296	6.34036	-1598.00217	852	2392.0157	-0.33085	6.41673	-1619.65767
808	2330.1091	-0.35244	6.34212	-1598.50436	853	2393.4076	-0.33039	6.41836	-1620.1182
809	2331.5313	-0.35192	6.34388	-1599.00554	854	2394.7988	-0.32994	6.41999	-1620.57787
810	2332.9527	-0.3514	6.34564	-1599.50572	855	2396.1893	-0.32948	6.42162	-1621.03671
811	2334.3734	-0.35088	6.34739	-1600.00491	856	2397.5793	-0.32902	6.42325	-1621.49469
812	2335.7934	-0.35036	6.34914	-1600.5031	857	2398.9686	-0.32857	6.42487	-1621.95184
813	2337.2126	-0.34985	6.35089	-1601.0003	858	2400.3572	-0.32812	6.42649	-1622.40815
814	2338.6311	-0.34933	6.35263	-1601.49651	859	2401.7453	-0.32766	6.42811	-1622.86363
815	2340.0488	-0.34882	6.35437	-1601.99174	860	2403.1327	-0.32721	6.42972	-1623.31828
816	2341.4658	-0.34831	6.35611	-1602.48599	861	2404.5195	-0.32676	6.43133	-1623.7721
817	2342.8821	-0.3478	6.35784	-1602.97927	862	2405.9056	-0.32632	6.43294	-1624.22509
818	2344.2977	-0.34729	6.35957	-1603.47158	863	2407.2912	-0.32587	6.43455	-1624.67726
819	2345.7126	-0.34678	6.3613	-1603.96292	864	2408.6761	-0.32542	6.43615	-1625.12861
820	2347.1267	-0.34628	6.36303	-1604.45329	865	2410.0604	-0.32498	6.43775	-1625.57915
821	2348.5402	-0.34577	6.36475	-1604.94271	866	2411.4441	-0.32453	6.43935	-1626.02887
822	2349.9529	-0.34527	6.36647	-1605.43117	867	2412.8272	-0.32409	6.44095	-1626.47778
823	2351.3649	-0.34477	6.36819	-1605.91868	868	2414.2096	-0.32365	6.44254	-1626.92588
824	2352.7762	-0.34427	6.3699	-1606.40523	869	2415.5915	-0.32321	6.44413	-1627.37318
825	2354.1868	-0.34377	6.37161	-1606.89085	870	2416.9728	-0.32277	6.44572	-1627.81968
826	2355.5967	-0.34327	6.37332	-1607.37552	871	2418.3534	-0.32234	6.44731	-1628.26538
827	2357.006	-0.34278	6.37502	-1607.85925	872	2419.7335	-0.3219	6.44889	-1628.71028
828	2358.4145	-0.34229	6.37673	-1608.34205	873	2421.1129	-0.32147	6.45047	-1629.15438
829	2359.8223	-0.34179	6.37843	-1608.82392	874	2422.4917	-0.32103	6.45205	-1629.5977
830	2361.2294	-0.3413	6.38012	-1609.30486	875	2423.87	-0.3206	6.45362	-1630.04023
831	2362.6358	-0.34081	6.38182	-1609.78487	876	2425.2477	-0.32017	6.4552	-1630.48197
832	2364.0416	-0.34033	6.38351	-1610.26396	877	2426.6247	-0.31974	6.45677	-1630.92293

u	D(u)	J(u)	T(u)	A(u)	u	D(u)	J(u)	T(u)	A(u)
833	2365.4467	-0.33984	6.38519	-1610.74214	878	2428.0012	-0.31931	6.45834	-1631.36311
834	2366.851	-0.33935	6.38688	-1611.2194	879	2429.3771	-0.31888	6.4599	-1631.80251
835	2368.2547	-0.33887	6.38856	-1611.69576	880	2430.7524	-0.31845	6.46147	-1632.24114
836	2369.6578	-0.33839	6.39024	-1612.1712	881	2432.1271	-0.31803	6.46303	-1632.679
837	2371.0601	-0.33791	6.39192	-1612.64575	882	2433.5012	-0.3176	6.46459	-1633.11608
838	2372.4618	-0.33743	6.39359	-1613.11939	883	2434.8747	-0.31718	6.46614	-1633.55241
839	2373.8628	-0.33695	6.39526	-1613.59213	884	2436.2477	-0.31676	6.4677	-1633.98796
840	2375.2631	-0.33647	6.39693	-1614.06398	885	2437.6201	-0.31634	6.46925	-1634.42276
886	2438.9919	-0.31592	6.4708	-1634.85679	931	2500.1536	-0.29812	6.53814	-1653.64287
887	2440.3631	-0.3155	6.47235	-1635.29007	932	2501.5006	-0.29775	6.53959	-1654.0446
888	2441.7338	-0.31508	6.47389	-1635.7226	933	2502.8472	-0.29738	6.54103	-1654.44569
889	2443.1039	-0.31466	6.47543	-1636.15437	934	2504.1933	-0.297	6.54247	-1654.84613
890	2444.4734	-0.31425	6.47697	-1636.5854	935	2505.5388	-0.29663	6.54391	-1655.24592
891	2445.8423	-0.31383	6.47851	-1637.01568	936	2506.8839	-0.29627	6.54535	-1655.64507
892	2447.2107	-0.31342	6.48004	-1637.44522	937	2508.2285	-0.2959	6.54679	-1656.04357
893	2448.5785	-0.31301	6.48158	-1637.87401	938	2509.5726	-0.29553	6.54822	-1656.44144
894	2449.9458	-0.3126	6.48311	-1638.30207	939	2510.9162	-0.29516	6.54965	-1656.83867
895	2451.3125	-0.31219	6.48464	-1638.72939	940	2512.2593	-0.2948	6.55108	-1657.23526
896	2452.6786	-0.31178	6.48616	-1639.15598	941	2513.6019	-0.29444	6.55251	-1657.63122
897	2454.0442	-0.31137	6.48768	-1639.58184	942	2514.944	-0.29407	6.55393	-1658.02655
898	2455.4092	-0.31097	6.4892	-1640.00697	943	2516.2857	-0.29371	6.55536	-1658.42125
899	2456.7736	-0.31056	6.49072	-1640.43137	944	2517.6268	-0.29335	6.55678	-1658.81532
900	2458.1376	-0.31015	6.49224	-1640.85505	945	2518.9675	-0.29299	6.5582	-1659.20877
901	2459.5009	-0.30975	6.49375	-1641.27801	946	2520.3077	-0.29263	6.55962	-1659.60159
902	2460.8637	-0.30935	6.49527	-1641.70025	947	2521.6474	-0.29227	6.56103	-1659.99379
903	2462.226	-0.30895	6.49677	-1642.12177	948	2522.9866	-0.29191	6.56244	-1660.38537
904	2463.5877	-0.30855	6.49828	-1642.54258	949	2524.3253	-0.29155	6.56386	-1660.77633
905	2464.9489	-0.30815	6.49979	-1642.96267	950	2525.6636	-0.2912	6.56526	-1661.16668
906	2466.3095	-0.30775	6.50129	-1643.38206	951	2527.0014	-0.29084	6.56667	-1661.55641
907	2467.6696	-0.30735	6.50279	-1643.80074	952	2528.3387	-0.29049	6.56808	-1661.94553
908	2469.0292	-0.30696	6.50429	-1644.21872	953	2529.6755	-0.29013	6.56948	-1662.33404
909	2470.3882	-0.30656	6.50578	-1644.63599	954	2531.0119	-0.28978	6.57088	-1662.72194
910	2471.7467	-0.30617	6.50728	-1645.05256	955	2532.3478	-0.28943	6.57228	-1663.10924
911	2473.1046	-0.30577	6.50877	-1645.46844	956	2533.6832	-0.28908	6.57368	-1663.49592

u	D(u)	J(u)	T(u)	A(u)
912	2474.462	-0.30538	6.51026	-1645.88362
913	2475.8189	-0.30499	6.51175	-1646.29811
914	2477.1752	-0.3046	6.51323	-1646.7119
915	2478.531	-0.30421	6.51471	-1647.12501
916	2479.8863	-0.30382	6.51619	-1647.53743
917	2481.2411	-0.30344	6.51767	-1647.94917
918	2482.5953	-0.30305	6.51915	-1648.36022
919	2483.9491	-0.30266	6.52062	-1648.7706
920	2485.3023	-0.30228	6.52209	-1649.18029
957	2535.0182	-0.28873	6.57508	-1663.88201
958	2536.3527	-0.28838	6.57647	-1664.2675
959	2537.6867	-0.28803	6.57786	-1664.65238
960	2539.0203	-0.28768	6.57925	-1665.03667
961	2540.3534	-0.28734	6.58064	-1665.42037
962	2541.686	-0.28699	6.58203	-1665.80347
963	2543.0182	-0.28664	6.58341	-1666.18597
964	2544.3499	-0.2863	6.58479	-1666.56789
965	2545.6812	-0.28596	6.58617	-1666.94922
921	2486.655	-0.3019	6.52356	-1649.58932
922	2488.0071	-0.30152	6.52503	-1649.99766
923	2489.3588	-0.30113	6.52649	-1650.40534
924	2490.7099	-0.30075	6.52796	-1650.81235
925	2492.0605	-0.30037	6.52942	-1651.21869
926	2493.4106	-0.3	6.53088	-1651.62437
927	2494.7602	-0.29962	6.53233	-1652.02939
928	2496.1093	-0.29924	6.53379	-1652.43374
966	2547.012	-0.28561	6.58755	-1667.32996
967	2548.3424	-0.28527	6.58893	-1667.71012
968	2549.6723	-0.28493	6.5903	-1668.0897
969	2551.0017	-0.28459	6.59167	-1668.46869
970	2552.3307	-0.28425	6.59304	-1668.8471
971	2553.6592	-0.28391	6.59441	-1669.22494
972	2554.9873	-0.28358	6.59578	-1669.6022
973	2556.315	-0.28324	6.59715	-1669.97888
929	2497.4579	-0.29887	6.53524	-1652.83744
930	2498.806	-0.29849	6.53669	-1653.24048
974	2557.6422	-0.2829	6.59851	-1670.355
975	2558.9689	-0.28257	6.59987	-1670.73054
976	2560.2952	-0.28223	6.60123	-1671.10551
991	2580.1374	-0.27731	6.62141	-1676.66294
977	2561.6211	-0.2819	6.60259	-1671.47992
992	2581.4567	-0.27699	6.62274	-1677.02903
978	2562.9465	-0.28157	6.60394	-1671.85376
993	2582.7757	-0.27667	6.62407	-1677.39458
979	2564.2715	-0.28123	6.6053	-1672.22703
994	2584.0942	-0.27634	6.62539	-1677.75959
980	2565.596	-0.2809	6.60665	-1672.59975
995	2585.4123	-0.27602	6.62672	-1678.12406
981	2566.9202	-0.28057	6.608	-1672.9719
996	2586.7299	-0.27571	6.62804	-1678.488
982	2568.2438	-0.28024	6.60935	-1673.34349
997	2588.0472	-0.27539	6.62936	-1678.8514
983	2569.5671	-0.27991	6.6107	-1673.71453
998	2589.364	-0.27507	6.63068	-1679.21427
984	2570.8898	-0.27958	6.61204	-1674.08501
999	2590.6804	-0.27475	6.632	-1679.5766
985	2572.2122	-0.27926	6.61338	-1674.45493
1000	2591.9964	-0.27444	6.63332	-1679.9384
986	2573.5341	-0.27893	6.61473	-1674.82431
987	2574.8556	-0.2786	6.61607	-1675.19313
988	2576.1767	-0.27828	6.6174	-1675.5614
989	2577.4974	-0.27796	6.61874	-1675.92913
990	2578.8176	-0.27763	6.62007	-1676.29631

APPENDIX B

```
'                        QBasic PC Program
'                           ANGLEC.BAS

' FIND:     LAUNCHING Angle, Time of Flight, y-
'           coordinate of a point
' GIVEN:    RANGE, Ballistics Coefficient, Launching
'           Speed (Standard Atmosphere)
'----------------------------------------------------
' CONTROL DATA
' INPUT:    Range = 10,000: Launching Speed = 885:
'           Ballistics Coefficient c = 0.2389,
'           x-coordinate of a point on Trajectory = 5500
'
' RESULTS: Launching Angle = 6.20 [Degree], Time of
'           Flight = 16.75 [Sec.]
' Terminal Angle = -10.196 [Degree]
'           Coordinates of Trajectory vertex (5634, 347)
'           For x =5500m: y = 346.62m, Time = 7.62s,
'           Angle = 0.21171 Degree, Speed = 593.20m/s
'----------------------------------------------------
' Note: Round the Input RANGE to the nearest TEN:
'  Example 10036 = 10040
'----------------------------------------------------
' FUNCTIONS, SUBS.

DECLARE SUB y1z1v1w1 (x, y, z, v, w, y1, z1, v1, w1,
    koef)
DECLARE SUB InfHyres (xx, koef, Speed, xc1, n)
```

620

```
DECLARE SUB InfDales (x0, y0, z0, v0, w0, xc, yc, tc,
    ac, vc, xm, ym, a)
DECLARE SUB NPxyzvw (nk, x, x0, y, y0, z, z0, v, v0, w,
    w0, h, h0, k, l, r, q)
DECLARE SUB NPkoef (k, l, r, q, h, y1, z1, v1, w1)
DECLARE SUB menu (cog, cof, xf, yf, xfu, yfu, t$)
DECLARE SUB y0z0 (y0, z0, Speed, a)
DECLARE SUB c (koef, a)

'Variables

SCREEN 0
1 :
DIM m(4, 4), v(4)
rendi = 4
cog = 7: cof = 0
cikli = 0
a = 23               'Initial Guessed Angle 23
kendi = 22
kov = 1
gab = 1
tt = 1

'Solution
CLS
'Initial Data
      menu cog, cof, 3, 10, 8, 70, "INPUT"
      InfHyres xx, koef, Speed, xc1, n
      hap = 1
'Start
f:
      x0 = 0: v0 = 0: w0 = 0
      y0z0 y0, z0, Speed, a: h0 = hap
      c koef, a
ff:
      FOR nk = 1 TO rendi
          NPxyzvw nk, x, x0, y, y0, z, z0, v, v0, w,
      w0, h, h0, k, l, r, q
          y1z1v1w1 x, y, z, v, w, y1, z1, v1, w1, koef
          NPkoef k, l, r, q, h, y1, z1, v1, w1
          m(nk, 1) = k: m(nk, 2) = l
          m(nk, 3) = r: m(nk, 4) = q
      NEXT nk

'Calculation
      FOR i = 1 TO rendi
          v(i) = 1 / 6 * (m(1, i) + 2 * m(2, i) + 2 *
              m(3, i) + m(4, i))
      NEXT i
```

```
'New Point
      x0 = x0 + h: y0 = y0 + v(1): z0 = z0 + v(2)
      v0 = v0 + v(3): w0 = w0 + v(4)

IF ABS(x0 - xc1) <= .5 THEN
xc = x0
yc = v0
tc = w0
ac = (180 / 3.141592654#) * ATN(z0)
vc = y0 / COS(ATN(z0))
END IF


IF ABS(z0) <= .0001 THEN
xm = x0
ym = v0
END IF


'Test if the y-coordinate passes here
IF kov = 1 THEN kov = -1: GOTO ff:
IF v0 <= (gab * TAN(a * 3.1415954# / 180)) AND v0 >= ((-1
* gab) * TAN(a * 3.1415954# / 180)) AND ABS(x0 - xx) <
.01 THEN
c:
                'Display of results
                CLS
                PLAY "a8a16a32b8"
                menu 12, 0, 5, 10, 11, 70, "RESULTS:"
                COLOR 7
                InfDales x0, y0, z0, v0, w0, xc, yc, tc, ac,
                   vc, xm, ym, a
                CLS
                GOTO 1:
END IF

IF v0 > (gab * TAN(a * 3.1415954# / 180)) AND ABS(x0 -
xx) < .01 THEN
                t$ = "   * ? *"
                menu 18, 0, 10, 20, 14, 60, t$
                COLOR 14
                LOCATE 12, 30: PRINT "Wait a Moment Please
                   (+)";
                LOCATE 12, 53: PRINT tt
                tt = tt + 1
                COLOR 7
                a = a - kendi
                GOTO fff:
END IF
```

```
IF v0 < ((-1 * gab) * TAN(a * 3.1415954# / 180)) THEN
          t$ = "    * ? *"
          menu 18, 0, 10, 20, 14, 60, t$
          COLOR 14
          LOCATE 12, 30: PRINT "Wait a Moment Please
             (-)";
          LOCATE 12, 53: PRINT tt
          tt = tt + 1
          COLOR 7
          a = a + kendi
          GOTO fff:
END IF
GOTO ff:

fff:
'Restart Cycle
cikli = cikli + 1
IF cikli = 20 THEN GOTO c:
kendi = kendi / 2
kov = 1
GOTO f:
SUB c (koef, a)
koef = koef
      END SUB

SUB InfDales (x0, y0, z0, v0, w0, xc, yc, tc, ac, vc, xm,
ym, a)
      g = (a - INT(a)) * .6 + INT(a)
      zg = y0 / COS(ATN(z0))
      z0 = (180 / 3.1415169542#) * ATN(z0)
      sh = a / .06
      LOCATE 6, 16: PRINT "RANGE                   :"; x0;
          " meter"
      LOCATE 7, 16: PRINT "LAUNCHING ANGLE     :"; a;
          " Degree"
      LOCATE 8, 16: PRINT "TIME OF FLIGHT      :"; w0;
          " seconds"
      LOCATE 9, 16: PRINT "TERMINAL SPEED      :"; zg;
          "m/s"
      LOCATE 10, 16: PRINT "TERMINAL ANGLE      :"; z0;
          " Degree"
      LOCATE 12, 14: PRINT "Coordinates of Vertex
          (x,y) :"; "("; xm; ","; ym; ")"
      LOCATE 14, 18: PRINT "x-coordinate of a point[m]
          :"; xc
      LOCATE 15, 18: PRINT "Corresponding y [m]
          :"; yc
      LOCATE 16, 18: PRINT "Corresponding Time [sec]
          :"; tc
```

```
        LOCATE 17, 18: PRINT "Corresponding Speed [m/s]
            :"; vc
        LOCATE 18, 18: PRINT "Corresponding Angle [Deg]
            :"; ac
COLOR 7
LOCATE 22, 15: PRINT "_ Press [ P ] to Repeat and [ Esc ]
to End _";
COLOR 10
LOCATE 22, 25: PRINT "P";
LOCATE 22, 45: PRINT "Esc";
COLOR 7
        cc$ = INPUT$(1)
'Variables

SCREEN 0
1 :
DIM m(4, 4), v(4)
rendi = 4
cog = 7: cof = 0
cikli = 0
a = 23              'Initial Guessed Angle 23
kendi = 22
kov = 1
gab = 1
tt = 1

'Solution
CLS
'Initial Data
        menu cog, cof, 3, 10, 8, 70, "INPUT"
        InfHyres xx, koef, Speed, xc1, n
        hap = 1
'Start
f:
        x0 = 0: v0 = 0: w0 = 0
        y0z0 y0, z0, Speed, a: h0 = hap
        c koef, a
ff:
        FOR nk = 1 TO rendi
                NPxyzvw nk, x, x0, y, y0, z, z0, v, v0, w,
                    w0, h, h0, k, l, r, q
                y1z1v1w1 x, y, z, v, w, y1, z1, v1, w1,
                    koef
                NPkoef k, l, r, q, h, y1, z1, v1, w1
                m(nk, 1) = k: m(nk, 2) = l
                m(nk, 3) = r: m(nk, 4) = q
        NEXT nk

'Calculation
```

```
      FOR i = 1 TO rendi
            v(i) = 1 / 6 * (m(1, i) + 2 * m(2, i) + 2
                * m(3, i) + m(4, i))
      NEXT i

'New Point
      x0 = x0 + h: y0 = y0 + v(1): z0 = z0 + v(2)
      v0 = v0 + v(3): w0 = w0 + v(4)

IF ABS(x0 - xc1) <= .5 THEN
xc = x0
yc = v0
tc = w0
ac = (180 / 3.141592654#) * ATN(z0)
vc = y0 / COS(ATN(z0))
END IF


IF ABS(z0) <= .0001 THEN
xm = x0
ym = v0
END IF


'Test if the y-coordinate passes here
IF kov = 1 THEN kov = -1: GOTO ff:
IF v0 <= (gab * TAN(a * 3.1415954# / 180)) AND v0 >=
   ((-1 * gab) * TAN(a * 3.1415954# / 180)) AND ABS(x0
   - xx) < .01 THEN
c:
                    'Display of results
                    CLS
                    PLAY "a8a16a32b8"
                    menu 12, 0, 5, 10, 11, 70,
                       "RESULTS:"
                    COLOR 7
                    InfDales x0, y0, z0, v0, w0, xc, yc,
                       tc, ac, vc, xm, ym, a
                    CLS
                    GOTO 1:
END IF

IF v0 > (gab * TAN(a * 3.1415954# / 180)) AND ABS(x0 -
xx) < .01 THEN
                    t$ = "   * ? *"
                    menu 18, 0, 10, 20, 14, 60, t$
                    COLOR 14
                    LOCATE 12, 30: PRINT "Wait a Moment
                       Please (+)";
                    LOCATE 12, 53: PRINT tt
                    tt = tt + 1
```

```
                    COLOR 7
                    a = a - kendi
                    GOTO fff:
END IF

IF v0 < ((-1 * gab) * TAN(a * 3.1415954# / 180)) THEN
                    t$ = "   * ? *"
                    menu 18, 0, 10, 20, 14, 60, t$
                    COLOR 14
                    LOCATE 12, 30: PRINT "Wait a Moment
                        Please (-)";
                    LOCATE 12, 53: PRINT tt
                    tt = tt + 1
                    COLOR 7
                    a = a + kendi
                    GOTO fff:
END IF
GOTO ff:

fff:
'Restart Cycle
cikli = cikli + 1
IF cikli = 20 THEN GOTO c:
kendi = kendi / 2
kov = 1
GOTO f:
SUB c (koef, a)
koef = koef
        END SUB

SUB InfDales (x0, y0, z0, v0, w0, xc, yc, tc, ac, vc,
xm, ym, a)
        g = (a - INT(a)) * .6 + INT(a)
        zg = y0 / COS(ATN(z0))
        z0 = (180 / 3.1415169542#) * ATN(z0)
        sh = a / .06
        LOCATE 6, 16: PRINT "RANGE                :"; x0;
            " meter"
        LOCATE 7, 16: PRINT "LAUNCHING ANGLE      :"; a;
            " Degree"
        LOCATE 8, 16: PRINT "TIME OF FLIGHT       :"; w0;
            " seconds"
        LOCATE 9, 16: PRINT "TERMINAL SPEED       :"; zg;
            "m/s"
        LOCATE 10, 16: PRINT "TERMINAL ANGLE      :"; z0;
            " Degree"
        LOCATE 12, 14: PRINT "Coordinates of Vertex
            (x,y) :"; "("; xm; ","; ym; ")"
```

```
      LOCATE 14, 18: PRINT "x-coordinate of a point[m]
         :"; xc
      LOCATE 15, 18: PRINT "Corresponding y [m]
         :"; yc
      LOCATE 16, 18: PRINT "Corresponding Time [sec]
         :"; tc
      LOCATE 17, 18: PRINT "Corresponding Speed [m/s]
         :"; vc
      LOCATE 18, 18: PRINT "Corresponding Angle [Deg]
         :"; ac
COLOR 7
LOCATE 22, 15: PRINT "_ Press [ P ] to Repeat and [
Esc ] to End _";
COLOR 10
LOCATE 22, 25: PRINT "P";
LOCATE 22, 45: PRINT "Esc";
COLOR 7
      cc$ = INPUT$(1)
      IF cc$ = CHR$(27) THEN SCREEN 9: CLS : END
END SUB
SUB InfHyres (xx, koef, Speed, xc1, n)
LOCATE 5, 13: INPUT "RANGE [meter] ="; xx
LOCATE 6, 13: INPUT "Initial Speed ="; Speed
LOCATE 7, 13: INPUT "Ballistics Coefficient = "; koef
LOCATE 8, 13: INPUT "INPUT n: n = 10"; n
LOCATE 12, 13: INPUT "x-coordinate of a point"; xc1
CLS
END SUB

SUB menu (cog, cof, xf, yf, xfu, yfu, t$)
COLOR cog, cof
LOCATE xf - 1, yf: PRINT t$
LOCATE xf, yf: PRINT "É" + STRING$(yfu - yf, 205) +
"»";
      FOR i = xf + 1 TO xfu
      LOCATE i, yf: PRINT "º" + SPACE$(yfu - yf) +
         "º";
      NEXT
      LOCATE xfu + 1, yf: PRINT "È" + STRING$(yfu -
         yf, 205) + "¼";
END SUB

SUB NPkoef (k, l, r, q, h, y1, z1, v1, w1)
      k = h * y1: l = h * z1
      r = h * v1: q = h * w1
END SUB

SUB NPxyzvw (nk, x, x0, y, y0, z, z0, v, v0, w, w0, h,
h0, k, l, r, q)
```

```
         IF nk = 1 THEN
               x = x0: y = y0: z = z0
               v = v0: w = w0: h = h0
         GOTO fund:
         END IF

         IF nk = 2 OR nk = 3 THEN
               x = x0 + (.5 * h): y = y0 + (.5 * k)
               z = z0 + (.5 * l): v = v0 + (.5 * r)
               w = w0 + (.5 * q)
         GOTO fund:
         END IF

         IF nk = 4 THEN
               x = x0 + h: y = y0 + k: z = z0 + l
               v = v0 + r: w = w0 + q
         END IF
fund:
END SUB

SUB y0z0 (y0, z0, Speed, a)
      y0 = Speed * (COS(a * 3.141516954# / 180))

      z0 = TAN(a * 3.141516954# / 180)
END SUB

SUB y1z1v1w1 (x, y, z, v, w, y1, z1, v1, w1, koef)

IF (y * SQR(1 + z ^ 2)) > 256! THEN

      y1 = -1 * koef * ((289.08 - .006328 * v) /
      289.08) ^ 4.4 * (1 / 3 - 80 / (y * SQR((1 + z
      ^ 2))))
ELSE
      y1 = -1 * koef * ((289.08 - .006328 * v) /
      289.08) ^ 4.4 * .0001212 * y * SQR((1 + z ^
      2))

END IF
      z1 = -9.80665 / y ^ 2
      v1 = z
      w1 = 1 / y
END SUB
```

APPENDIX C

QBasic PC Program
RANGEC.BAS

```
' FIND:     Range, Vertex of the Trajectory, etc.
'           Elements of Trajectory for a given abscisa x
' GIVEN:    Launching Angle, Ballistics Coefficient,
'           Projectile Speed

'------------------------------------------------------

' Control Data

' Input Data

' Input:    Initial x- Coordinate = 0, Initial y-
'           coordinate
' Input:    Launching Angle: 6.20, Initial Speed =
'           885,Ballistics Coefficient =0.2389
' Input:    Initial Time t0 = 0 , Integration Step h0 =1
' Input:    x-coordinate of a point on trajectory  xc =
'           4530
' Results
'           ' Range = 10000; Error in y-coordinate = -
'           0.02972; Terminal Speed = 419.45
'           ' Terminal Angle = -10.19; Coordinates of
'           vertex (5634, 346.81)
' Abscissa of a Point on the rajectory = 4530,
' Corresponding values of x = 4350 are: y-coordinate
' =330.51; Spped = 639.07; time = 6.047; angle =
' 1.649859

'------------------------------------------------------
```

```
'Functions, Subs

DECLARE SUB y1z1v1w1 (x, y, z, v, w, y1, z1, v1, w1,
   koef)
DECLARE SUB InfHyres (x0, y0, z0, v0, w0, a, koef,
   xc1, h0)
DECLARE SUB NPxyzvw (nk, x, x0, y, y0, z, z0, v, v0,
   w, w0, h, h0, k, L, r, q)
DECLARE SUB NPkoef (k, L, r, q, h, y1, z1, v1, w1)
DECLARE SUB menu (cog, cof, xf, yf, xfu, yfu, t$)
DECLARE SUB c (koef, a)

'Variables

DIM m(4, 4), v(4)
rendi = 4
cog = 7: cof = 0

'Solution
CLS

fillimi:

        menu cog, cof, 3, 10, 21, 70, "Initial Data"
        InfHyres x0, y0, z0, v0, w0, a, koef, xc1, h0
        c koef, a

f:
        FOR nk = 1 TO rendi
                NPxyzvw nk, x, x0, y, y0, z, z0, v, v0, w,
                   w0, h, h0, k, L, r, q
                y1z1v1w1 x, y, z, v, w, y1, z1, v1, w1,
                   koef
                NPkoef k, L, r, q, h, y1, z1, v1, w1
                m(nk, 1) = k: m(nk, 2) = L
                m(nk, 3) = r: m(nk, 4) = q
        NEXT nk

'Calculation

        FOR i = 1 TO rendi
                v(i) = 1 / 6 * (m(1, i) + 2 * m(2, i) + 2
                   * m(3, i) + m(4, i))
        NEXT i

'New Data

        x0 = x0 + h: y0 = y0 + v(1): z0 = z0 + v(2)
        v0 = v0 + v(3): w0 = w0 + v(4)
```

```
IF y0 >= 256 THEN
IF ABS(z0) < .00001 OR ABS(z0) <= .0001 THEN
ymax = v0
xmax = x0
END IF
END IF
IF y0 < 256 THEN
IF ABS(z0) < .0001 OR ABS(z0) < .001 THEN
ymax = v0
xmax = x0
END IF
END IF


IF ABS(x0 - xc1) <= .1 THEN
xc = x0
yc = v0
tc = w0
ac = (180 / 3.141592654#) * ATN(z0)
vc = y0 / COS(ATN(z0))
END IF



IF v0 <= .0001 THEN

'Display Results
menu cog, cof, 2, 20, 22, 76, "RESULTS:"

        LOCATE 4, 26: PRINT "Range  [m]                     = ";
          x0
        LOCATE 5, 26: PRINT "Error in y-coord [m]    = ";
          v0
        LOCATE 6, 26: PRINT "Terminal Speed [m/s]    = ";
          y0 * (1 + z0 ^ 2) ^ .5
        LOCATE 7, 26: PRINT "Terminal Angle [Deg.]  = ";
          ATN(z0) * 180 / 3.141593
        LOCATE 8, 26: PRINT "Time of Flight [s]      = ";
          w0
        LOCATE 11, 26: PRINT "Trajectory Vertex (xm, ym)
            = "; "("; xmax; ","; ymax; ")"
        LOCATE 14, 23: PRINT "Abscissa of a point on
          trajectory:   x = "; xc
        LOCATE 15, 23: PRINT "Corresponding ordinate of
          x:          y = "; yc
        LOCATE 16, 23: PRINT "Corresponding Speed:
                      v = "; vc
        LOCATE 17, 23: PRINT "Corresponding Time:
                      t = "; tc
```

```
      LOCATE 18, 23: PRINT "Corresponding Angle:
                    a = "; ac


      LOCATE 21, 24: PRINT "Ballistics Coefficient =
         "; koef


      ELSE
      GOTO f:

END IF
END

SUB c (koef, a)

koef = koef

END SUB

SUB InfHyres (x0, y0, z0, v0, w0, a, koef, xc1, h0)

LOCATE 5, 13: INPUT "Initial x-coordinate      = ";
x0
LOCATE 6, 13: INPUT "Initial y-coordinate      = ";
v0
LOCATE 7, 13: INPUT "Initial Time              = ";
w0
LOCATE 8, 13: INPUT "Launching Angle [Degree]  = ";
z0
LOCATE 9, 13: INPUT "Launching Speed [m/s]     = ";
y0
      a = z0
      y0 = y0 * COS(z0 * 3.141516954# / 180)
      z0 = TAN(z0 * 3.141516954# / 180)
LOCATE 10, 13: INPUT "Ballistics Coefficient   = ";
koef
LOCATE 12, 13: INPUT "x coordinate of a point on
Trajectory [m]  = "; xc1

LOCATE 14, 13: INPUT "Integration Step: h0=10 or h0 =
1 "; h0

CLS
END SUB

SUB menu (cog, cof, xf, yf, xfu, yfu, t$)

COLOR cog, cof
```

```
LOCATE xf - 1, yf: PRINT t$

LOCATE xf, yf: PRINT "É" + STRING$(yfu - yf, 205) +
"»";

     FOR i = xf + 1 TO xfu
          LOCATE i, yf: PRINT "º" + SPACE$(yfu - yf)
            + "º";
     NEXT
          LOCATE xfu + 1, yf: PRINT "È" +
            STRING$(yfu - yf, 205) + "¼";
END SUB

SUB NPkoef (k, L, r, q, h, y1, z1, v1, w1)

     k = h * y1: L = h * z1
     r = h * v1: q = h * w1
END SUB

SUB NPxyzvw (nk, x, x0, y, y0, z, z0, v, v0, w, w0, h,
h0, k, L, r, q)

     IF nk = 1 THEN
          x = x0: y = y0: z = z0
          v = v0: w = w0: h = h0
          GOTO fund:
     END IF

     IF nk = 2 OR nk = 3 THEN
          x = x0 + (.5 * h): y = y0 + (.5 * k)
          z = z0 + (.5 * L): v = v0 + (.5 * r)
          w = w0 + (.5 * q)
          GOTO fund:
     END IF

     IF nk = 4 THEN
          x = x0 + h: y = y0 + k: z = z0 + L
          v = v0 + r: w = w0 + q
     END IF

fund:
END SUB

SUB y1z1v1w1 (x, y, z, v, w, y1, z1, v1, w1, koef)

IF (y * SQR(1 + z ^ 2)) > 256! THEN
```

```
      y1 = -1 * koef * ((289.08 - .006328 * v) /
         289.08) ^ 4.4 * (1 / 3 - 80 / (y * SQR((1 + z
         ^ 2))))
ELSE
      y1 = -1 * koef * ((289.08 - .006328 * v) /
         289.08) ^ 4.4 * .0001212 * y * SQR((1 + z ^
         2))

END IF
      z1 = -9.80665 / y ^ 2
      v1 = z
      w1 = 1 / y

END SUB
```

APPENDIX D

```
'              QBasic PC Program
'                 INGAAN.BAS

' Using Ingalls' Function (English Units)
' FIND:    LAUNCHING Angle,Time of Flight, y-coordinate
'          of a point
' GIVEN:   RANGE, Ballistics Coefficient "C=m/id^2",
'          Launching Speed (Standard Atmosphere)
'-----------------------------------------------------
' CONTROL DATA
' INPUT:   Range = 30,000: Launching Speed = 2,600:
'          Ballistics Coefficient =13.44775,
'          x-coordinate of a point on Trajectory =
'          15,000
' RESULTS: Launching Angle = 5.0451 [Degree], Time of
'          Flight = 13.47 [s]
'          Terminal Angle = - 6.15744 [Degree],
'          Terminal speed = 1926.73 [ft./s]
'          Coordinates of Trajectory vertex (15760,
'          731)

'          For x =15000: y = 728.84 ft., Time = 6.231s,
'          Angle = 0.278 Degree, Speed = 2239 ft/s
'-----------------------------------------------------
' Note: Round the Input range to the nearest 10
'-----------------------------------------------------
'FUNCTIONS, SUBS.
```

```
DECLARE SUB y1z1v1w1 (x, y, z, v, w, y1, z1, v1, w1,
    koef)
DECLARE SUB InfHyres (xx, koef, Speed, xc1, n)
DECLARE SUB InfDales (x0, y0, z0, v0, w0, xc, yc, tc,
    ac, vc, xm, ym, a)
DECLARE SUB NPxyzvw (nk, x, x0, y, y0, z, z0, v, v0,
    w, w0, h, h0, k, l, r, q)
DECLARE SUB NPkoef (k, l, r, q, h, y1, z1, v1, w1)
DECLARE SUB menu (cog, cof, xf, yf, xfu, yfu, t$)
DECLARE SUB y0z0 (y0, z0, Speed, a)
DECLARE SUB c (koef, a)

'Variables

SCREEN 0
1 :
DIM m(4, 4), v(4)
rendi = 4
cog = 7: cof = 0
cikli = 0
a = 23              'Initial Guessed Angle 23
kendi = 22
kov = 1
gab = 1
tt = 1

'Solution
CLS
'Initial Data
        menu cog, cof, 3, 10, 8, 70, "INPUT"
        InfHyres xx, koef, Speed, xc1, n
        hap = 10

'Start
f:
        x0 = 0: v0 = 0: w0 = 0
        y0z0 y0, z0, Speed, a: h0 = hap
        c koef, a
ff:
        FOR nk = 1 TO rendi
                NPxyzvw nk, x, x0, y, y0, z, z0, v, v0, w,
                    w0, h, h0, k, l, r, q
                y1z1v1w1 x, y, z, v, w, y1, z1, v1, w1,
                    koef
                NPkoef k, l, r, q, h, y1, z1, v1, w1
                m(nk, 1) = k: m(nk, 2) = l
                m(nk, 3) = r: m(nk, 4) = q
        NEXT nk
```

```
'Calculation

        FOR i = 1 TO rendi
              v(i) = 1 / 6 * (m(1, i) + 2 * m(2, i) + 2
                 * m(3, i) + m(4, i))
        NEXT i


'New Point

        x0 = x0 + h: y0 = y0 + v(1): z0 = z0 + v(2)
        v0 = v0 + v(3): w0 = w0 + v(4)

IF ABS(x0 - xc1) <= 1.64 THEN
xc = x0
yc = v0
tc = w0
ac = (180 / 3.141592654#) * ATN(z0)
vc = y0 / COS(ATN(z0))
END IF


IF ABS(z0) <= .0001 THEN
xm = x0
ym = v0
END IF


'Test if the y-coordinate passes here

IF kov = 1 THEN kov = -1: GOTO ff:
IF v0 <= (gab * TAN(a * 3.1415954# / 180)) AND v0 >=
((-1 * gab) * TAN(a * 3.1415954# / 180)) AND ABS(x0 -
xx) < .01 THEN

c:
                'Display of results
                CLS
                PLAY "a8a16a32b8"
                menu 12, 0, 5, 10, 11, 70, "RESULTS:"
                COLOR 7
                InfDales x0, y0, z0, v0, w0, xc, yc, tc,
                    ac, vc, xm, ym, a
                CLS
                GOTO 1:
END IF

IF v0 > (gab * TAN(a * 3.1415954# / 180)) AND ABS(x0 -
xx) < .033 THEN
                t$ = "   * ? *"
                menu 18, 0, 10, 20, 14, 60, t$
```

```
                    COLOR 14
                    LOCATE 12, 30: PRINT "Wait a Moment Please
                        (+)";
                    LOCATE 12, 53: PRINT tt
                    tt = tt + 1
                    COLOR 7
                    a = a - kendi
                    GOTO fff:
END IF

IF v0 < ((-1 * gab) * TAN(a * 3.1415954# / 180)) THEN
                    t$ = "    * ? *"
                    menu 18, 0, 10, 20, 14, 60, t$
                    COLOR 14
                    LOCATE 12, 30: PRINT "Wait a Moment Please
                        (-)";
                    LOCATE 12, 53: PRINT tt
                    tt = tt + 1
                    COLOR 7
                    a = a + kendi
                    GOTO fff:
END IF
GOTO ff:

fff:
'Restart Cycle
cikli = cikli + 1
IF cikli = 40 THEN GOTO c:
kendi = kendi / 2
kov = 1
GOTO f:

SUB c (koef, a)

koef = koef

        END SUB

SUB InfDales (x0, y0, z0, v0, w0, xc, yc, tc, ac, vc,
xm, ym, a)
        g = (a - INT(a)) * .6 + INT(a)
        zg = y0 / COS(ATN(z0))
        z0 = (180 / 3.1415169542#) * ATN(z0)
        sh = a / .06
        LOCATE 6, 16: PRINT "RANGE                :"; x0;
            " feet"
        LOCATE 7, 16: PRINT "LAUNCHING ANGLE    :"; a;
            " Degree"
```

```
        LOCATE 8, 16: PRINT "TIME OF FLIGHT        :"; w0;
          " seconds"
        LOCATE 9, 16: PRINT "TERMINAL SPEED         :"; zg;
          "ft./s"
        LOCATE 10, 16: PRINT "TERMINAL ANGLE        :"; z0;
          " Degree"
        LOCATE 12, 14: PRINT "Coordinates of Vertex
          (x,y) :"; "("; xm; ","; ym; ")"
        LOCATE 14, 18: PRINT "x-coordinate of a point
          [ft.] :"; xc
        LOCATE 15, 18: PRINT "Corresponding y [ft.]
               :"; yc
        LOCATE 16, 18: PRINT "Corresponding Time  [s]
               :"; tc
        LOCATE 17, 18: PRINT "Corresponding Speed
          [ft./s]     :"; vc
        LOCATE 18, 18: PRINT "Corresponding Angle [Deg]
               :"; ac

COLOR 7
LOCATE 22, 15: PRINT "_ Press [ P ] to Repeat and [
Esc ] to End _";
COLOR 10
LOCATE 22, 25: PRINT "P";
LOCATE 22, 45: PRINT "Esc";
COLOR 7
           cc$ = INPUT$(1)
      IF cc$ = CHR$(27) THEN SCREEN 9: CLS : END
END SUB

SUB InfHyres (xx, koef, Speed, xc1, n)
LOCATE 5, 13: INPUT "RANGE [Feet]          ="; xx
LOCATE 6, 13: INPUT "Initial Speed [ft./s] ="; Speed
LOCATE 7, 13: INPUT "Ballistics Coefficient = "; koef
 koef = 1.422334331# / koef
LOCATE 8, 13: INPUT "INPUT n: n = 10"; n
LOCATE 12, 13: INPUT "x-coordinate of a point"; xc1
CLS
END SUB

SUB menu (cog, cof, xf, yf, xfu, yfu, t$)

COLOR cog, cof
LOCATE xf - 1, yf: PRINT t$

LOCATE xf, yf: PRINT "É" + STRING$(yfu - yf, 205) +
"»";

      FOR i = xf + 1 TO xfu
```

```
        LOCATE i, yf: PRINT "º" + SPACE$(yfu - yf) +
            "º";
        NEXT
        LOCATE xfu + 1, yf: PRINT "È" + STRING$(yfu -
            yf, 205) + "¼";
END SUB

SUB NPkoef (k, l, r, q, h, y1, z1, v1, w1)

        k = h * y1: l = h * z1
        r = h * v1: q = h * w1

END SUB

SUB NPxyzvw (nk, x, x0, y, y0, z, z0, v, v0, w, w0, h,
h0, k, l, r, q)

        IF nk = 1 THEN
                x = x0: y = y0: z = z0
                v = v0: w = w0: h = h0
                GOTO fund:
        END IF

        IF nk = 2 OR nk = 3 THEN
                x = x0 + (.5 * h): y = y0 + (.5 * k)
                z = z0 + (.5 * l): v = v0 + (.5 * r)
                w = w0 + (.5 * q)
                GOTO fund:
        END IF

        IF nk = 4 THEN
                x = x0 + h: y = y0 + k: z = z0 + l
                v = v0 + r: w = w0 + q
        END IF

fund:
END SUB

SUB y0z0 (y0, z0, Speed, a)
                y0 = Speed * (COS(a * 3.141516954# / 180))
                z0 = TAN(a * 3.141516954# / 180)
END SUB

SUB y1z1v1w1 (x, y, z, v, w, y1, z1, v1, w1, koef)

IF (y * SQR(1 + z ^ 2)) > 840! THEN

        y1 = -1 * koef * EXP(-.0000315914# * v) * (1 / 3
            - 262.1338667# / (y * SQR((1 + z ^ 2))))
```

```
ELSE
      y1 = -1 * koef * EXP(-.0000315914# * v) *
      .0001212 * y * SQR((1 + z ^ 2))

END IF
      z1 = -32.17405 / y ^ 2
      v1 = z
      w1 = 1 / y

END SUB
```

APPENDIX E

All 19 PC programs are published in the book "Exterior Ballistics of Small Arms—Companion to Exterior Ballistics with Applications", by G. Klimi, Xlibris, 2009.
List of PC Programs
Available upon Request on CD (or electronic mail)
Request a copy to the author at gklimi@pace.edu or iven24@aol.com.

I. **PC programs that calculate the launching angle and all the other elements of the trajectory when the "range," the "launching speed," and the "ballistics coefficient" are known**

1. ANGLE122.BAS (Standard Atmosphere, Standard Projectile - 122 mm cannon)
2. ANGLEC.BAS (Standard Atmosphere, Standard Projectile)
3. ANGLEM30.BAS (Standard Atmosphere, Standard Projectile - 0.30 M2 ball projectile)
4. ANGMET.BAS (Nonstandard Atmosphere, Nonstandard Projectile)
5. ANTA122.BAS (Antiair Shooting, Projectile 122mm)
6. ANTAIR.BAS (Antiair Shooting, 0.30 M2 ball projectile)
7. ANTIARC.BAS (Antiair Shooting)
8. DWNH1.BAS (Downhill Shooting)
9. INGAAN.BAS (Uses Ingalls's Function and English Units)

II. **PC programs that calculate the range and all the other elements of the trajectory when the "launching angle," the "launching speed," and the "ballistics coefficient" are known**

1. RANGE122.BAS (Standard Atmosphere, Standard Projectile - 122 mm cannon)
2. RANGEC.BAS (Standard Atmosphere, Standard Projectile)
3. RANBALL.BAS (Nonstandard Atmosphere, Nonstandard Projectile - 0.30 M2 ball projectile)
4. RANGEM30 (Standard Atmosphere, Projectile - 0.30 M2 ball)
5. RANGMET.BAS (Nonstandard Atmosphere, Nonstandard Projectile)
6. MORTAR.BAS (Nonstandard Atmosphere, Nonstandard Projectile, Small Projectile Speed)
7. INGRANG.BAS (Uses Ingalls's Function and English Units)
8. SIZERO.BAS (Standard Atmosphere, Sierra 0.257" caliber, 117 grain Spitzer Boat Tail bullet, Initial Speed 2,900 ft./s.)

III. **PC program to calculate the ballistics coefficient**

1. COEFF.BAS

IV. **Skydiving PC Programs**

1. PARACH.BAS   (Standard Atmosphere)
2. PARAMET.BAS (Nonstandard Atmosphere, Presence of Wind)

V. **Universal PC Programs**

Four Universal PC Programs are included in my book "Exterior Ballistics - A New Approach", Xlibris, 2010.

NOTE: To execute a QBasic program, you have to download the program QB.EXE and follow the following steps:

Double click on QB.EXE; Press **Esc**; Click **File**; Click **Open Program** Double Click on a program (Example, Anglec.bas); Read Control Data Click **Run**; Click **Start** to input the initial data.

APPENDIX F

```
'                    QBasic PC Program
'                       RPMELA.BAS
'                  Non Standard Atmosphere
'        Lapua GB528 Scenar 19.4g, Caliber 8.6mm Bullet
'                   Departure Speed 830m/s

'  FIND : Elements of Trajectory at any two points, as
well as the projectile
'         drop when departure angle is zero
'  GIVEN: Departure Angle, Initial speed, Ballistics
Coefficient BC
'
'  BC = 1000 d^2/m = 3.796   (Form Coefficient i = 1)

'---------------------------------------------------
'   DATA
'   Input: y-coordinate of FIREARM = 0, departure
Angle; 0, departure speed ,830
'          Temperature of Air, 15, temperature of
propellant, 21;
'          Pressure = 760, Pressure of Air vapor = 0,
Projectile mass, 0.0194;
'          Change in Projectile mass = 0, range wind,
0; cross wind, 0; BC = 3.796
'          x-coordinate of a Point= 1200; X- coordinate
of a Point = 1500,
'          Integration step = 10
```

```
'Results: x-coordinate  = 1200 m, y-coordinate (drop)
= -16.514m, time = 2.04s,
'          speed  = 423m/s, angle = -2.0254 degree.
'          X- coordinate = 1500, y-coordinate,(drop) =
- 29.904, time = 2.82s,
'          speed = 357m/s, angle = - 3.1491 degree.
'
'--------------------------------------------------------

'Functions & Subs.

DECLARE SUB y1z1v1w1 (x, y, z, v, w, y1, z1, v1, w1,
koef, pa1, wind, ys, yy, pa, ta1)
DECLARE SUB InfHyres (x0, y0, z0, v0, w0, a, h0, ta,
pa, ea, m, dm, tp, ta1, pa1, xx1, voo, vo1, wind,
koef, cw, vv, xax)
DECLARE SUB NPxyzvw (nk, x, x0, y, y0, z, z0, v, v0,
w, w0, h, h0, k, L, r, q)
DECLARE SUB NPkoef (k, L, r, q, h, y1, z1, v1, w1)
DECLARE SUB menu (cog, cof, xf, yf, xfu, yfu, t$)
DECLARE SUB c (koef, m, dm, BC)

'Variables

DIM m(4, 4), v(4)
rendi = 4
cog = 7: cof = 0

'Zgjidhja
CLS

fillimi:

        menu cog, cof, 3, 10, 21, 70, "INITIAL DATA"
        InfHyres x0, y0, z0, v0, w0, a, h0, ta, pa,
ea, m, dm, tp, ta1, pa1, xx1, voo, vo1, wind, koef,
cw, vv, xax
        c koef, m, dm, BC

f:
        FOR nk = 1 TO rendi
            NPxyzvw nk, x, x0, y, y0, z, z0, v, v0, w,
w0, h, h0, k, L, r, q
            y1z1v1w1 x, y, z, v, w, y1, z1, v1, w1,
koef, pa1, wind, ys, yy, pa, ta1
            NPkoef k, L, r, q, h, y1, z1, v1, w1
            m(nk, 1) = k: m(nk, 2) = L
            m(nk, 3) = r: m(nk, 4) = q
```

```
          NEXT nk

'Calculation

          FOR i = 1 TO rendi
            v(i) = 1 / 6 * (m(1, i) + 2 * m(2, i) + 2 *
m(3, i) + m(4, i))
          NEXT i

'New Data

          x0 = x0 + h: y0 = y0 + v(1): z0 = z0 + v(2)
          v0 = v0 + v(3): w0 = w0 + v(4)

IF ABS(z0) < .00001 THEN
ymax = v0
xmax = x0 + wind * w0
END IF

xxx = x0 + wind * w0
IF (xxx - xx1) <= .001 THEN
   xc = xxx
   yc = v0
   tc = w0
   ac = 180 * ATN(z0) / 3.141592654#
   vc = y0 * (1 + z0 ^ 2) ^ .5
   zc = cw * (w0 - xc / (voo * COS(a)))
END IF

xxT = x0 + wind * w0
IF (xxT - xax) <= .001 THEN
   tt = w0
   xt = xxT
   yt = v0
   at = 180 * ATN(z0) / 3.141592654#
   vt = y0 * (1 + z0 ^ 2) ^ .5
   zt = cw * (w0 - xt / (voo * COS(a)))
  END IF

ytt = yt
IF ABS(v0 - ytt) >= .1 THEN


'Display Results

menu cog, cof, 6, 20, 22, 72, "RESULTS:"

          LOCATE 7, 25: PRINT "x-coordinate of Point[m]
= "; INT((xc) + .5)
```

```
        LOCATE 8, 25: PRINT "Coresponding y-Coord [m]
= "; INT((yc) * 1000 + .5) / 1000
        LOCATE 9, 25: PRINT "Departure Angle [Deg.]
= "; a
        LOCATE 10, 25: PRINT "Time of Flight [s]
= "; INT((tc) * 100 + .5) / 100
        LOCATE 11, 25: PRINT "Terminal Speed [m/s]
= "; INT((vc) + .5)
        LOCATE 12, 25: PRINT "Terminal Angle [Deg.]
= "; INT((ac) * 10000 + .5) / 10000
        LOCATE 13, 25: PRINT "Cross-Wind Deflection
= "; INT((zc) * 100 + .5) / 100
        LOCATE 15, 25: PRINT "Second Point[m]
= "; "("; INT((xt) + .5); ","; INT((yt) * 1000 + .5) /
1000; ")"
        LOCATE 16, 25: PRINT "Time [s]
  = "; INT((tt) * 100 + .5) / 100
        LOCATE 17, 25: PRINT "Corresponding Speed [m/s]
  = "; INT((vt) + .5)
        LOCATE 18, 25: PRINT "Corresponding Angle [Deg]
  = "; INT((at) * 10000 + .5) / 10000
        LOCATE 19, 25: PRINT "Cross-Wind Deflection
= "; INT((zt) * 100 + .5) / 100
        LOCATE 20, 25: PRINT "Trajectory Vertex [m]
= "; "("; INT((xmax) + .5); ","; INT((ymax) * 100 +
.5) / 100; ")"
        LOCATE 22, 25: PRINT "Ballistics Coefficient
= "; BC
ELSE
GOTO f:
END IF
END

SUB c (koef, m, dm, BC)
BC = koef
koef = koef * (1 - dm / m)
END SUB

SUB InfHyres (x0, y0, z0, v0, w0, a, h0, ta, pa, ea,
m, dm, tp, ta1, pa1, xx1, voo, vo1, wind, koef, cw,
vv, xax)
```

```
LOCATE 5, 13: INPUT "y-coordinate of FIREARM        = ";
v0
LOCATE 6, 13: INPUT "Departure Speed [m/s]          = ";
y0
LOCATE 7, 13: INPUT "Departure Angle [Degree]       = ";
z0
LOCATE 8, 13: INPUT "Temperature of Air [C]         = ";
ta
LOCATE 9, 13: INPUT "Propellant Temperature[C]      = ";
tp
LOCATE 10, 13: INPUT "Atmospheric Pressure [mm]     = ";
pa
LOCATE 11, 13: INPUT "Pressure of Air Vapor [mm]    = ";
ea
LOCATE 12, 13: INPUT "Projectile Mass               = ";
m
LOCATE 13, 13: INPUT "Change in Projectile mass     = ";
dm
LOCATE 14, 13: INPUT "Range Wind                    = ";
wind
LOCATE 15, 13: INPUT "Cross Wind                    = ";
cw
LOCATE 16, 13: INPUT "x-coordinate of a point on
Trajectory = "; xx1
LOCATE 17, 13: INPUT "x-coordinate of a point on
Trajectory = "; xax
LOCATE 18, 13: INPUT "Ballistics Coefficient
 = "; koef
LOCATE 19, 13: INPUT "Integration Step,  10, 1, or 0.5
= "; h0
vv = v0: a = z0: voo = y0
ta = ta + 273.15
pa1 = ta / (1 - .3785 * ea / pa)
vo1 = (voo - .4 * voo * (dm / m) + .00125 * voo * (tp
- 21))
y0 = SQR(vo1 ^ 2 + wind ^ 2 - 2 * vo1 * wind * COS(a *
3.141592654# / 180))
y0 = y0 * COS(a * 3.141592654# / 180)
z0 = TAN(a * 3.141592654# / 180)
z0 = z0 / (1 - wind / (vo1 * COS(a * 3.141592654# /
180)))
CLS
END SUB

SUB menu (cog, cof, xf, yf, xfu, yfu, t$)

COLOR cog, cof
LOCATE xf - 1, yf: PRINT t$
```

```
LOCATE xf, yf: PRINT "É" + STRING$(yfu - yf, 205) +
"»";

    FOR i = xf + 1 TO xfu
        LOCATE i, yf: PRINT "º" + SPACE$(yfu - yf) +
"º";
    NEXT
        LOCATE xfu + 1, yf: PRINT "È" + STRING$(yfu -
yf, 205) + "¼";
END SUB

SUB NPkoef (k, L, r, q, h, y1, z1, v1, w1)

  k = h * y1: L = h * z1
  r = h * v1: q = h * w1
END SUB

SUB NPxyzvw (nk, x, x0, y, y0, z, z0, v, v0, w, w0, h,
h0, k, L, r, q)

  IF nk = 1 THEN
                x = x0: y = y0: z = z0
                v = v0: w = w0: h = h0
                GOTO fund:
  END IF

  IF nk = 2 OR nk = 3 THEN
                x = x0 + (.5 * h): y = y0 + (.5 * k)
                z = z0 + (.5 * L): v = v0 + (.5 * r)
                w = w0 + (.5 * q)
                GOTO fund:
  END IF

  IF nk = 4 THEN
                x = x0 + h: y = y0 + k: z = z0 + L
                v = v0 + r: w = w0 + q
  END IF

fund:
END SUB

SUB y1z1v1w1 (x, y, z, v, w, y1, z1, v1, w1, koef,
pa1, wind, ys, yy, pa, ta1)

ta1 = (288.15 / pa1) ^ .5
yy = y * SQR(1 + z ^ 2)
IF yy * ta1 > 280 THEN
y1 = -1 * koef * (pa / 760) * ((288.15 - .006328 * v)
/ 288.15) ^ 4.4 * (ta1 * yy - 237.808) / (6.6667 * yy)
```

```
END IF
IF yy * ta1 <= 280 AND ta1 > 136 THEN
y1 = -1 * koef * (pa / 760) * ((288.15 - .006328 * v)
/ 288.15) ^ 4.4 * (ta1 * yy - 53.956) / (42.3 * yy)
END IF
IF yy * ta1 <= 136 THEN
y1 = -1 * koef * (pa / 760) * ((288.15 - .006328 * v)
/ 288.15) ^ 4.4 * .0001924 * ta1 ^ 2 * yy ^ 2 / yy
END IF
z1 = -9.80665 / y ^ 2
v1 = z
w1 = 1 / y

END SUB
```

REFERENCES

Adair, Robert K. *The Physics of Baseball*. Harper and Row, Publishers, New York, 1990.

Alonso, Marcelo, and Finn, Edward. *Fundamental University Physics*, Vol 1, 2nd Ed.

Addison-Wesley Publishing Company, Inc.,1980.

Antipersonnel Weapons. Sipri, 1978.

Anton, Howard, Irl Bivens, and Davis, Stephen. *Calculus*, 7th Edition. John Wiley and Sons, Inc., 2002.

Arney, D. C., Melendez, B. S., and Schnelle, D. *Parachute Jumping, Falling, and Landing*.

http://www.dean.usma.edu/departments/math/pubs/mmm99/PDF/C5.PDF. [Web site]

Cline, Donna. *Trajectories.* Copyright © 2002.
    http://www.angelfire.com/poetry/u31240468/AeroBallistics.pdf. [Web site]

Cujko, V. S. *Vneshnaja Balistika*. Moskva, 1958.

De Mestre, Neville. *The Mathematics of Projectiles in Sport*. Cambridge University Press, 1990.

Granger, Robert A. *Fluid Mechanics*. Dover Publications, Inc., 1995.

Gubinim, S. G., and Gorovim, S. A. Ballistics, Handbook.

http://www.ssga.ru/AllMetodMaterial/metod_mat_for_ioot/metodichki/ballistica/index.htm. [Web site]

Halliday, David, Rescnik, Robert, and Walker, Jearl. *Fundamentals of Physics*, 6th Edition. John Wiley and Sons, Inc., 2001.

Hayden, Robert, Almgren, Ted, Thomas, Kevin, and McDonald, William T. *Sierra's Exterior Ballistics*.
http://www.exteriorballistics.com/ebexplained/index.cfm. [Web site]

Herrmann, Ernest E. Exterior Ballistics. U.S. Naval Institute, The College Press, 1935.

Hurley, James P., and Claude Garrod. *Principi Di Fisica*. Zanichelli, 1986.

Kallend, John. *Physics of Skydiving*.
http://www.iit.edu/~kallend/skydive/. [Web site]

McCoy, Robert L. *Modern Exterior Ballistics. The Launch and Flight Dynamics of Symmetric Projectiles*. Schiffer Publishing, Ltd., 1999.

McShane, Edward J., John L. Kelly, and Franklin Reno. *Exterior Ballistics*. The University of Denver Press, 1953.

Meade, D. B., Struthers, A. A. *Differential Equations in the New Millennium: The Parachute Problem*. Tempus Publication, 1999.
http://www.math.sc.edu/~meade/papers/ijee-parachute99.pdf. [Web site]

Mori, Edoardo. *Balistica Teorica e Pratica*.
http://www.earmi.it/balistica. [Web site]

Mucinov, S.S., and Shevcenko, N.A. Zadacnik po Osnovami Strelbi is Strelkovogo Oruzie. 1964.

Okunev, B. H. *Fundamentals of Ballistics*, Vol.1, Book 2. Moskva, 1943.

Pejsa, Arthur J. *Modern Practical Ballistics*, 2nd Edition. Kenwood Publishing, 1990.

Rinker, Robert A. *Understanding Firearm Ballistics*, 6th Edition. Mulberry House Publishing, 2005.

Roller, Duane E., and Ronald Blum. *Fisica. Meccanica, Onde, Termodinamica*, Vol.1. Zanichelli, 1984.

Sellier, Karl G., and Beat P. Kneubuehl. *Wound Ballistics And The Scientific Background*. Elsevier Science B.V., 1994.

Shapiro, J. M. *Vneshnaja Balistika*. Oborongiz, 50'.

Thomas, George B. Jr., Maurice D. Weir, Joel Hass, and Frank R. Giordano. *Thomas' Calculus, Early Transcendentals*, 11th Edition. Pearson Education, Inc., 2006.

Ugolini, Antonio. *L'Esperto Balistico*, Vol. I. Editoriale Olimpia, 1983.

Whelan, P.M., and Hodgson, M.J. *Essential Principles of Physics*. John Murray, 1979.

Zill, Dennis G., and Michael R. Cullen. *Differential Equations with Boundary-Value*, 5th Edition. Books/Cole, 2001.

INDEX